Plants, Lichens, Fungi and Algae Ingredients for Nutrition and Health

Plants, Lichens, Fungi and Algae Ingredients for Nutrition and Health

Guest Editors

Silvia Mironeasa
Mădălina Ungureanu-Iuga

MDPI

Basel • Beijing • Wuhan • Barcelona • Belgrade • Novi Sad • Cluj • Manchester

Guest Editors
Silvia Mironeasa
Faculty of Food Engineering
"Ștefan cel Mare" University
of Suceava
Suceava
Romania

Mădălina Ungureanu-Iuga
Integrated Center for Research,
Development and Innovation in
Advanced Materials,
Nanotechnologies, and Distributed
Systems for Fabrication and
Control (MANSiD)
"Ștefan cel Mare" University
of Suceava
Suceava
Romania

Editorial Office
MDPI AG
Grosspeteranlage 5
4052 Basel, Switzerland

This is a reprint of the Special Issue, published open access by the journal *Applied Sciences* (ISSN 2076-3417), freely accessible at: https://www.mdpi.com/journal/applsci/special_issues/plants_lichens_fungi_ingredients.

For citation purposes, cite each article independently as indicated on the article page online and as indicated below:

Lastname, A.A.; Lastname, B.B. Article Title. *Journal Name* **Year**, *Volume Number*, Page Range.

ISBN 978-3-7258-4145-5 (Hbk)
ISBN 978-3-7258-4146-2 (PDF)
https://doi.org/10.3390/books978-3-7258-4146-2

Contents

About the Editors

Silvia Mironeasa

Silvia Mironeasa (Professor Habilitate, PhD. Eng. at the "Ştefan cel Mare" University of Suceava) has expertise in food engineering, food processing, food quality analysis, and in the design of experiments and data analysis. Her research activities focus on enhancing the characteristics of raw materials by applying physical treatments and the valorization of vegetable by-products to improve the nutritional profile of foods. She has been involved in 33 interdisciplinary research projects, has published more than 165 scientific papers, and is the author of 21 patents under evaluation and 14 published patents.

Mădălina Ungureanu-Iuga

Mădălina Ungureanu-Iuga (Researcher, PhD. Eng. at the "Ştefan cel Mare" University of Suceava and at the Romanian Academy (CE-MONT, INCE)) has expertise in the physico-chemical characterization of food and its microstructure, texture, rheology, thermal processing, and by-product valorization for value-added product development. Her research experience comprises 50 articles published in WoS journals, 20 articles in journals indexed in other international data bases, 10 patent proposals and 3 patents. She has successfully participated in 4 research projects as a team member, 1 researcher mobility project as a director, and she was a Management Committee member for Ocean4biotech COST action (CA18238).

applied sciences

MDPI

Editorial

Plants, Lichens, Fungi and Algae Ingredients for Nutrition and Health

Silvia Mironeasa [1] and Mădălina Ungureanu-Iuga [2,3,*]

[1] Faculty of Food Engineering, Ștefan cel Mare University of Suceava, 13 Universitatii Street, 720229 Suceava, Romania; silviam@fia.usv.ro
[2] Integrated Center for Research, Development and Innovation in Advanced Materials, Nanotechnologies, and Distributed Systems for Fabrication and Control (MANSiD), "Ştefan cel Mare" University of Suceava, 13th University Street, 720229 Suceava, Romania
[3] Mountain Economy Center (CE-MONT), "Costin C. Kiriţescu" National Institute of Economic Researches (INCE), Romanian Academy, 49th, Petreni Street, 725700 Vatra Dornei, Romania
* Correspondence: madalina.iuga@usm.ro

There is a high awareness in the industry of the need to develop food products enriched with health-promoting ingredients and to avoid nutrition-related disorders. An increasing amount of research has been conducted aiming to further understanding of these issues, as well as research focusing on the opportunity to use various plants, lichens, and algae products and by-products in the food industry and in animal feed. These studies deal with the valorization of unconventional sources of bioactive compounds, aiming to increase the nutritional value of staple food or to improve meat quality in a sustainable way.

The use of ingredients rich in proteins and high-quality fatty acids in bird feed impacts the bird's welfare, egg quality, and overall production performance. The exploration of alternative sources for poultry nutrition may lead to difficulties regarding food security. In this context, microalgae represent a feasible and eco-friendly solution in animal and poultry diet formulation due to their high protein content and high levels of omega-3 fatty acids. Microalgae are also rich in polysaccharides, vitamins, pigments, enzymes, etc., while the biologically active compounds present in microalgae are involved in disease prevention and helping the immune system due to their antioxidant, antimicrobial, and immunomodulatory properties. Additionally, microalgae are regarded as eco-friendly because they have a low influence on land and water resources. Another plant ingredient that is used as a protein supplement due to its high-quality protein is soybean meal. Soybean meal is recognized as a balanced amino acid source with a very high protein digestibility, which is essential for the overall growth of birds and egg production. An alternative, used to supersede part of the soybean meal included in poultry diets, is represented by Chlorella (*Chlorella vulgaris*), a naturally single-celled green microalga. This food source, appreciated due to its particular and variate composition of beneficial macro- and micro-nutrients, includes proteins, omega-3 polyunsaturated fatty acids, polysaccharides, vitamins, and minerals. Panaite et al. [1] illustrate the possible influences of chlorella and spirulina, which can partially substitute soybean meal at equivalent inclusion levels in the diet of laying hens, on poultry productivity and egg quality. A significant improvement in egg quality and nutritional profile in terms of egg weight and size, intensity of yolk color, beta carotene content, and antioxidant capacity was observed when chlorella and spirulina at a concentration of 2% each were used to supplement laying hens' diet. Moreover, the addition of chlorella led to a remarkable rise in omega-3 polyunsaturated fatty acids, with crucial implications beyond poultry production.

Wheat, one of the best-known cereal crops in the world, represents the main food source of most communities. The varieties of wheat present different grain characteristics related to their genetic diversity, with an impact on the milling and bakery sectors. Romania is recognized as one of the most important producers of wheat in Europe and the world.

Golea et al. [2] investigated the genetic varieties by employing Inter-Simple Sequence Repeats (ISSR) markers of 31 wheat samples cultivated in Romania, from various sources, and maintained in the active collection of the "Mihai Cristea" Suceava Plant Genetic Resources Bank. The physical, chemical, and genetic diversity properties of the various wheat varieties grown in Romania exhibited significant variations between ancient and modern wheat species in terms of ash, protein, wet gluten, lipid, and starch contents, as well as their falling number and damaged starch values. The ISSR-type molecular markers are useful and relevant to genetic diversity research. The findings of this study suggest that knowledge of the genetic diversity of germplasm collection and investigations into the magnitude and nature of genetic variations in wheat are crucial for breeding programs and the preservation of genetic resources. Additionally, when taking into account the wheat samples that were investigated, the ancient ones presented the best quality for bread-making, as they have the greatest quantity of fat and minerals.

White wheat flour is usually the basic ingredient in bread, but it has a lower nutritional quality due to the considerable loss of dietary fiber, vitamins, minerals, and phytochemicals during the refining process. Researchers are concerned with the development new functional food and creation of sustainable food systems, taking into account the alternative uses of vegetable by-products. Many studies have focused on enhancing wheat bread with valuable food industry waste products with high levels of phenolic antioxidants. The impact of the incorporation of fresh or freeze-dried pistachio hulls and grape seeds on the total phenolic content, antioxidant activity, moisture content, crust and crumb color, specific volume, baking loss, and textural and sensory profiles of white wheat bread was specifically investigated by the Koç and Atar Kayabaşi [3]. The obtained results show that the majority of bread properties are influenced by the amount of the ingredient used; thus, it is crucial to control the quantity of these by-products in bread. The results reveal that pistachio hull powder is rich in antioxidants, polyphenols, and fibers, and thus can be considered a valuable functional ingredient in breadmaking. An advantage of the use of pistachio hull powder in bread is that it led to a significant increase in phenolic content and antioxidant activity; however, the sensory profile of the final product was affected and lower scores were obtained compared to the control.

Various types of breakfast baked goods, mostly manufactured from flour, sugar, fats, and eggs, present a lipid content of up to 24–40% relative to the flour content, whether of animal or vegetable origin. Researchers face a notable challenge in their attempts find a feasible solution to reducing the lipid content in bakery products, since it is known that fat influences the sensory and texture profile of the final product. Many studies are centered on diminishing the amount of lipids in muffin cakes by replacing them with various lipids mimetics, such as starch, chia seed mucilage, inulin, cocoa fibers, gums, polydextrose, and legume/fruit purees. One alternative to fat in muffins is represented by oleogels, which also benefit human health compared to shortenings and saturated fats. In this context, Banu et al. [4] studied the influence of soy-, lupin-, and yeast protein-based emulsions as oil substituents on the rheological properties of muffin batter and the characteristics of the final product. The results revealed different contributions of the emulsions to the rheological properties of dough, depending on the ingredients that were used. Changes in starch gelatinization and Mixolab torque values were observed when sunflower oil was replaced with various emulsions. The final products exhibited higher moisture, increased firmness, good CO_2 retention, and acceptable sensory characteristics compared to the control. This research confirmed the suitability of replacing fat with some protein emulsions in muffins.

Some researchers focused on the development of innovative packaging or bioactive packaging that liberates the compounds. A novel approach to the manufacturing of bioactive films is the addition of various types of fruit juices into the film-forming solution (FFS) to obtain an intake of antioxidants. As considered by Avramia and Amariei [5], yeast cell walls or the β-glucans extracted by purification from the cell walls can be the proper matrices of FFS, with immunostimulatory benefits. A study was conducted to analyze the bioactive films manufactured from β-glucan and the three kinds of fruit juices regarding the uniformity of film thickness, moisture vapor transmission rate, water vapor permeability, and dissolution time. As stated by these authors, the quantity of the juice and gelling agents plays an essential role in changing the physicochemical characteristics of the bioactive films.

At present, non-traditional flours made of different raw products are attracting the attention of researchers and the food industry. Plant-based products are becoming increasingly popular on the market and there is a growing diversity of flours that can be used in food preparation. The differences between these flours consist mainly in their ingredients, origin, and processing process. These flours can be used to improve the rheological profile of doughs through their inclusion in wheat flour used for baked goods, to improve the texture characteristics of foods, and/or to enhance their nutritional profile. Vivar-Quintana et al. [6] evaluated the nutritional properties of ten commercial flours containing rice, pea, chickpea, soybean, and hemp concerning current nutritional guidelines and dietary requirements. Even if plant-based flours are included with a generic expression, their nutritional value differs according to their distinct composition. The flours considered in the research showed carbohydrates as a main component, and exhibited similar energy values. The results revealed that pea and chickpea are a great source of proteins and fibers, while soybean and hemp are rich in fat. The main macro- and microelements found in all flours are potassium, phosphorus, and iron.

Consumers are currently aware of the high-quality chicken meat usually obtained in organic systems and the valorization of organic food industry by-products. The literature highlights the possibility of using microbial pigment sources in poultry nutrition with many advantages related to the time needed, costs, and environmental protection. A good source of carotenoids, lipids, and enzymes is represented by yeast. In this regard, Grigore et al. [7] evaluated the singular and the interaction effects of the dietary yeasts' lyophilized lysate supplements when included in broilers' diet on the carcass yield and breast and thigh meat properties in terms of pH, color, nutritional profile, texture profile analysis, and sensorial characteristics. As observed in their work, the inactive yeasts *Saccharomyces cerevisiae* and *Rhodotorula mucilaginosa* influenced the physical and mechanical properties of meat. Bivalent yeast supplementation resulted in an enhancement of meat quality via an increase in moisture, lightness, and redness, and a reduction in the browning index. This sensory evaluation of meat confirmed the possibility of using yeast supplements as a feasible alternatives to feed additives.

Carrot (*Daucus carota* subsp. *Sativus*) is a popular vegetable used in the human diet that has many nutritional benefits. It is a valuable source of vitamin A (beta-carotene) and dietary fiber, but it also contains significant amounts of potassium and antioxidants like luteolin, which benefits the cardiovascular system. Chomanov et al. [8] evaluated the effect of carrots on the nutritional and biological value of canned goat meat. According to them, the replacement of a part of the goat meat with carrots had a considerable impact on the chemical, amino acid, vitamin, and mineral profile of canned food. A variation in the characteristics according to the amount of carrot added was observed, whereas the sensorial characteristics presented differences in color, taste, consistency, and odor. According to these authors, the inclusion of carrots can lead to an increase in the final product's biological value and an enhancement of the canned goat meat's palatability.

There is a growing interest in healthier food, promoting the reformulation of some products through reducing or replacing some components, like carbohydrates or lipids, or using natural additives while maintaining the sensory and safety of the product. Metri-Ojeda et al. [9] investigated mayonnaise made of egg yolk, *A. platensis* protein, and sodium alginate, motivated by Mexican consumers who consider proteins to be a natural ingredient with acceptable characteristics. According to the results obtained, the emulsion-filled gels are a good alternative for low-fat mayonnaise because they present acceptable stability, a rheological behavior close to that of commercial mayonnaise, and a rise in the bio-availability of essential amino acid content, after buccal and gastric digestion. Furthermore, there is the option of creating an acceptable product from vegetable protein with a small amount of oil. Considering the purchase intention results, the best mayonnaise formulations would be made of soy protein isolate with 22.5–30% oil or *A. platensis* protein concentrate with 30% oil.

Usnea barbata (L.) Weber ex F.H. Wigg (*U. barbata*) is a medicinal belonging to the lichens category from the *Usnea* genus (*Parmeliaceae*, lichenized *Ascomycetes*), which is a great source of bioactive secondary metabolites. Popovici et al. [10] investigated the physico-chemical characteristics of two distinct parts of the thallus layers, namely the medulla–cortex and central cord, and the whole dried *U. barbata* thallus. The results revealed that the central cord fraction has the smallest mineral content, color intensity, lightness, and total polyphenols content, followed by the whole lichen and medulla cortex fraction. The greatest polyphenols content, correlated with the antioxidant activity, was observed in the medulla cortex fraction, with the results depending on the solvent used. The extraction of the bioactive metabolites from the medulla cortex fraction could increase the yield and selectivity.

The current trends regarding the addition of functional ingredients to monogastric animal feed were observed, aiming to enhance the performance and increase the sustainability under different raising conditions. The nutritional value and the potential benefits of Jerusalem artichoke (*Helianthus tuberosus* L.) made it a great candidate to improve monogastric diets. In this sense, Cornescu et al. [11] reviewed the existing literature regarding the opportunity to use Jerusalem artichoke in monogastric animal feed and its effects on the production performances, along with the potential prebiotic activity of this ingredient. Based on the existing data, the authors concluded that Jerusalem artichoke has prebiotic properties and contributes to the enhancement of the gastrointestinal microbiota and production parameters.

The papers published in this Special Issue confirm the actual trends regarding the use of plant ingredients to enhance the nutritional, physical, and sensory characteristics of various food products. Furthermore, the positive effects of plant ingredients, when used in animal feed, on the quality of meat are also noted.

Author Contributions: S.M. and M.U.-I. contribute equally to this paper. All authors have read and agreed to the published version of the manuscript.

Funding: This research received no external funding.

Institutional Review Board Statement: Not applicable.

Informed Consent Statement: Not applicable.

Acknowledgments: I want to thank all the authors and peer reviewers for their valuable contributions to this Special Issue. In addition, thanks to the MDPI management of *Applied Sciences* and their editorial support.

Conflicts of Interest: The authors declare no conflicts of interest.

References

1. Panaite, T.D.; Cornescu, G.M.; Predescu, N.C.; Cismileanu, A.; Turcu, R.P.; Saracila, M.; Soica, C. Microalgae (*Chlorella vulgaris* and *Spirulina platensis*) as a Protein Alternative and Their Effects on Productive Performances, Blood Parameters, Protein Digestibility, and Nutritional Value of Laying Hens' Egg. *Appl. Sci.* **2023**, *13*, 10451. [CrossRef]
2. Golea, C.M.; Galan, P.-M.; Leti, L.-I.; Codină, G.G. Genetic Diversity and Physicochemical Characteristics of Different Wheat Species (*Triticum aestivum* L., *Triticum monococcum* L., *Triticum spelta* L.) Cultivated in Romania. *Appl. Sci.* **2023**, *13*, 4992. [CrossRef]
3. Koç, B.; Atar Kayabaşi, G. Enrichment of White Wheat Bread with Pistachio Hulls and Grape Seeds: Effect on Bread Quality Characteristics. *Appl. Sci.* **2023**, *13*, 3431. [CrossRef]
4. Banu, I.; Patrașcu, L.; Vasilean, I.; Dumitrașcu, L.; Aprodu, I. Influence of the Protein-Based Emulsions on the Rheological, Thermo-Mechanical and Baking Performance of Muffin Formulations. *Appl. Sci.* **2023**, *13*, 3316. [CrossRef]
5. Avramia, I.; Amariei, S. A Comparative Study on the Development of Bioactive Films Based on β-glucan from Spent Brewer's Yeast and Pomegranate, Bilberry, or Cranberry Juices. *Appl. Sci.* **2023**, *13*, 2807. [CrossRef]
6. Vivar-Quintana, A.M.; Absi, Y.; Hernández-Jiménez, M.; Revilla, I. Nutritional Value, Mineral Composition, Fatty Acid Profile and Bioactive Compounds of Commercial Plant-Based Gluten-Free Flours. *Appl. Sci.* **2023**, *13*, 2309. [CrossRef]

7. Grigore, D.-M.; Mironeasa, S.; Ciurescu, G.; Ungureanu-Iuga, M.; Batariuc, A.; Babeanu, N.E. Carcass Yield and Meat Quality of Broiler Chicks Supplemented with Yeasts Bioproducts. *Appl. Sci.* **2023**, *13*, 1607. [CrossRef]
8. Chomanov, U.; Kenenbay, G.; Tursynov, A.; Zhumalieva, T.; Tultabayev, N.; Suychinov, A. Nutritive Profile of Canned Goat Meat Food with Added Carrot. *Appl. Sci.* **2022**, *12*, 9911. [CrossRef]
9. Metri-Ojeda, J.; Ramírez-Rodrigues, M.; Rosas-Ordoñez, L.; Baigts-Allende, D. Development and Characterization of a Low-Fat Mayonnaise Salad Dressing Based on *Arthrospira platensis* Protein Concentrate and Sodium Alginate. *Appl. Sci.* **2022**, *12*, 7456. [CrossRef]
10. Popovici, V.; Bucur, L.; Gîrd, C.E.; Calcan, S.I.; Cucolea, E.I.; Costache, T.; Rambu, D.; Ungureanu-Iuga, M.; Oroian, M.; Mironeasa, S.; et al. Advances in the Characterization of *Usnea barbata* (L.) Weber ex F.H. Wigg from Călimani Mountains, Romania. *Appl. Sci.* **2022**, *12*, 4234. [CrossRef]
11. Cornescu, G.M.; Panaite, T.D.; Soica, C.; Cismileanu, A.; Matache, C.C. Jerusalem Artichoke (*Helianthus tuberosus* L.) as a Promising Dietary Feed Ingredient for Monogastric Farm Animals. *Appl. Sci.* **2023**, *13*, 12748. [CrossRef]

applied sciences

MDPI

Article

Microalgae (*Chlorella vulgaris* and *Spirulina platensis*) as a Protein Alternative and Their Effects on Productive Performances, Blood Parameters, Protein Digestibility, and Nutritional Value of Laying Hens' Egg

Tatiana Dumitra Panaite [1,*], Gabriela Maria Cornescu [1], Nicoleta Corina Predescu [2], Ana Cismileanu [1], Raluca Paula Turcu [3], Mihaela Saracila [3] and Cristina Soica [1]

1 Nutrition Physiology Laboratory, National Research and Development Institute for Biology and Animal Nutrition, 077015 Balotesti, Ilfov, Romania; gabriela_cornescu@yahoo.com (G.M.C.); ana_cismileanu@yahoo.com (A.C.); cristina05101988@yahoo.com (C.S.)

2 Faculty of Veterinary Medicine of Bucharest, 105 Splaiul Independentei, District 5, 050097 Bucharest, Romania; corina.predescu@fmvb.usamv.ro

3 Food and Feed Quality Laboratory, National Research and Development Institute for Biology and Animal Nutrition, 077015 Balotesti, Ilfov, Romania; raluca.turcu@ibna.ro (R.P.T.); mihaela.saracila@ibna.ro (M.S.)

* Correspondence: tatiana.panaite@ibna.ro; Tel.: +40-21-351-2082

Abstract: Protein is an essential nutrient for laying hens, playing a crucial role in egg production and supporting their overall health. An 8-week feeding trial was conducted on 120 Lohmann Brown laying hens (aged 38 weeks). The layers were assigned randomly to three groups and housed in cages (twenty replicates × two birds/cage). All groups were fed a corn–soybean meal basal diet (2750 kcal/kg metabolizable energy (ME) with 17.8% crude protein (CP)). In contrast to conventional diet (CON), the experimental groups were supplemented primarily at the expense of soybean meal with 2.0% *Chlorella vulgaris* (CV2%) and 2.0% *Spirulina platensis* (SP2%). Their high concentrations of chlorophyll *a* (5.56; 9.06 mg/g), chlorophyll *b* (0.88; 1.34 mg/g), and antioxidant activity expressed as 2,2-diphenyl-1-picrylhydrazyl (73.29; 81.27 DPPH% inhibition) improved egg yolk quality. At the end of the trial, eighteen eggs/group (six yolk samples/group, three eggs/sample) were collected to determine the egg quality and nutritional parameters (fatty acids profile, cholesterol, β-carotene, yolk color, and antioxidant capacity). To determine the activity of antioxidant enzymes, including superoxide dismutase (SOD), catalase (CAT), glutathione peroxidase (GSH), and total antioxidant capacity (TAC), blood samples were collected at the end of the period. Microalgae inclusion increased ($p < 0.05$) the fatty acid content, ß-carotene concentration ($p < 0.001$), antioxidant capacity ($p < 0.0001$), and yolk color intensity ($p < 0.001$) significantly, especially the yolk redness a* color parameter, but without any significant results concerning cholesterol concentration. Boiling the eggs for 10 min significantly ($p < 0.001$) increased the b* color parameter on microalgae treatments. The supplementation of laying hens' diet with microalgae positively influenced egg quality and nutritional properties.

Keywords: antioxidant capacity; blood antioxidant enzymes; cholesterol; fatty acids; laying hens; microalgae; yolk color

1. Introduction

The protein source in a hen's diet is of utmost importance as it directly influences the bird's growth and health, egg quality, and overall production performances [1].

Soybean meal is considered a high-quality protein source, widely used in poultry diet formulation [2], which contains a balanced amino acid profile, crucial for egg production and overall growth. These amino acids contribute to the synthesis of proteins, enzymes, and hormones, supporting various physiological functions in hens [3]. Soybean protein can be compared to proteins found in meat, milk, and eggs. Among plant-based protein sources, soybean protein is widely regarded as having the highest biological value [4]. Alshelmani et al. [5] consider that the increasing competitiveness of feedstuffs for poultry nutrition presents a challenge to food security; therefore, ongoing efforts are made to explore alternative protein sources that can partially replace soybean meal in poultry diets.

Microalgae are being increasingly explored as a valuable and sustainable alternative in animal and poultry nutrition due to high protein content [6], and are primary sources of omega-3 fatty acids, such as eicosapentaenoic acid (EPA) and docosahexaenoic acid (DHA) [7]. Moreover, they are considered environmentally friendly due to their minimal impact on land and water resources [8]. Furthermore, microalgae contain bioactive compounds that exhibit antioxidant [9], antimicrobial, and immunomodulatory properties [10], thereby contributing to disease prevention and supporting the immune system.

Chlorella (*Chlorella vulgaris*) is a naturally single-celled green microalga considered as an alternative for partially replacing soybean meal included in poultry diets [11]. Previous studies found that *Chlorella* has a positive impact on egg production and quality, enhancing the intensity of the yolk color due to carotenoid transfer (canthaxanthin and β-carotene) [12]. Furthermore, it promotes the growth of lactic-acid-producing bacteria in the intestines and lowers the total cholesterol and triglyceride level concentration in serum and liver [13].

Spirulina (*Sp. platensis*), a blue-green spiral filamentous alga [14], is a natural product with high nutritional value and increased antioxidant potential. Its utilization improves production efficiency, egg production [15], and yolk redness, while also exhibiting favorable amino acid profiles and high digestibility [16]. Spirulina is recognized as a sustainable protein source with a low-impact environmental footprint that can vary significantly depending on factors such as the production system and regional climate [17].

Despite previous research on this topic, the results of previous experiments involving the inclusion of microalgae in poultry diets have generated inconsistent findings with regard to both poultry productivity and egg quality. As a result, our study aims to investigate the partial substitution of soybean meal in the diet of laying hens, and to examine the potential impacts and effects of chlorella and spirulina, both at equivalent inclusion levels, on these specific parameters.

2. Materials and Methods

2.1. Ethical Statement

The study was carried out at the Laboratory of Animal Physiology, National Research-Development Institute for Animal Biology and Nutrition (IBNA), Balotesti, Romania. The feeding, handling, and slaughtering procedures of the study were performed in accordance with Directive 2010/63/EU on the protection of animals used for scientific purposes, and the experimental procedures, according to an experimental protocol (No. 6252/27.10.2021), were approved by the Research Ethics Committee for Animal Production studies of IBNA.

2.2. Microalgae Purchase and Chemical Analyses

Microalgae chlorella and spirulina powder were purchased from the agri-food market. Triplicate analyses were conducted on samples of chlorella and spirulina powder to determine the following: dry matter (DM), ash, organic matter (OM), crude protein (CP), ether extract (EE), crude fiber (CF), and non-fermentable extractive substance (NFE); in vitro nutrient digestibility of protein, dry matter, and organic matter (DCP, DDM, DOM); and antioxidant activity and fatty acid profile.

Metabolizable energy (ME) of the microalgae was calculated using formula (1), according to [18] and cited by [19]:

$$\text{ME (kcal/kg)} = (35.3 \times \text{CP \%}) + (79.5 \times \text{EE \%}) + (40.6 \times \text{NFE \%}) + 199.0 \quad (1)$$

2.3. Animals, Housing, and Experimental Diets

An eight-week feeding trial was conducted on 120 Lohmann Brown layers (38 weeks), individually weighed and assigned in 3 treatments (CON, CV2%, and SP2%). The layers were randomly placed in twenty replicates with 2 birds per treatment, housed in metabolic cages (50 cm width × 40 cm height × 50 cm length) under controlled environmental conditions monitored by a ViperTouch computer (16 h light/24 h; T = 23.08 ± 0.98 °C; H = 66.35 ± 5.68%). Each replicate was considered an experimental unit and performance parameters were evaluated per pen. The feed was administrated daily at 08:30 a.m. and water was available at all times. Throughout the experimental period, no vaccination treatment was applied to the birds.

The isocaloric and isonitrogenous three experimental treatments (in mash form) were formulated by a nutritional optimization program (HYBRIMIN® Futter5) to meet the nutrient requirements for laying hens as given by [20]. All groups were fed a corn–soybean meal basal diet (17% crude protein and 2750 kcal ME/ kg feed) as follows: CON—a commercial diet without microalgae (chlorella or spirulina); CV2%—a control diet containing 2.0% chlorella powder; and SP2%—a control diet containing 2.0% spirulina powder, as shown in Table 1. A quantity of 500 g feed samples from each group were taken and analyzed by chemical composition as described previously for the microalgae samples. Following the manufacturing of the diets, the feed was packaged, appropriately labeled, and stored under optimal conditions, specifically in a cool environment, in preparation for the experimental procedures.

Table 1. Ingredients and chemical composition of the diets (% as fed).

Ingredients, %	**Experimental Diets**		
	Control (CON)	**Chlorella Powder (CV2%)**	**Spirulina Powder (SP2%)**
Corn	40.00	40.00	40.00
Wheat bran	22.49	23.72	23.60
Chlorella vulgaris	-	2.00	-
Spirulina platensis	-	-	2.00
Soybean meal	24.36	21.58	21.63
Vegetable oil	1.48	0.97	1.04
L-lysine HCl	-	0.01	0.01
DL–methionine	0.16	0.17	0.17
Calcium carbonate	8.83	8.84	8.84
Monocalcium phosphate	1.32	1.33	1.33
Salt	0.33	0.33	0.33
Choline premix	0.04	0.04	0.04
Vitamin–mineral premix *	1.00	1.00	1.00
Total	100.00	100.00	100.00
	Calculated analysis (%) **		
Metabolizable energy (Kcal/kg)	2.750.00	2.750.00	2.750.00
Crude protein	17.00	17.09	17.00
Lysine	0.87	0.80	0.80
Methionine+Cystine	0.73	0.71	0.71
Threonine	0.64	0.59	0.59
Calcium	3.90	3.90	3.90
Phosphorus	0.63	0.62	0.62

Where: CON, conventional diet; CV2%, conventional diet supplemented with 2% chlorella powder; SP2%, conventional diet supplemented with 2% spirulina powder; * 1 kg diet contains: = 11,000 IU/kg vit. A; 2000 IU/kg vit. D3; 27 IU/kg vit. E; 3 mg/kg vit. K; 2 mg/kg Vit. B1; 4 mg/kg vit. B2; 14.85 mg/kg pantothenic acid; 27 mg/kg nicotinic acid; 3 mg/kg vit. B6; 0.04 mg/kg Vit. B7; 1 mg/kg vit. B9; 0.018 mg/kg vit. B12; 20 mg/kg vit. C; 80 mg/kg manganese; 80 mg/kg iron; 5 mg/kg copper; 60 mg/kg zinc; 0.37 mg/kg cobalt; 1.52 mg/kg iodine; 0.18 mg/kg selenium. ** Calculated according to NRC [20].

2.4. Laying Hens Performance

During the 8-week feeding trial, the performance parameters of the laying hens (daily feed intake (DFI; g/day/layer), feed conversion ratio (FCR; g feed/g egg), hen day egg production (HDEP; %), egg weight (EW; g), and egg size classification (%)) were monitored. At the initial and the final period, body weight (g/hen) was measured, and eggs were collected and weighed every day. Hen day egg production was calculated using the following formula [(100 × number of eggs laid)/(number of hens × days)] and classified according to the European Council Directive (2006). Data on feed intake and egg mass were used to calculate the feed conversion ratio (feed intake/egg mass; g/g). All performance parameters were determined for each replicate of treatment groups.

2.5. Nutrient Digestibility Trial

During the last week of the feeding trial (the 8th wk), 6 cages per group (2 birds per cage) were randomly selected from the digestibility trial to measure the apparent nutrient digestibility. For 5 days, both feed leftovers and excreta were collected and weighed daily to determine nutrient intake. During the balance period, fecal samples were stored in a refrigerator at a constant temperature of 4 °C. Finally, each sample was homogenized, and approximately 200 g samples were extracted and dried for 48 h at a constant temperature of 65 °C in an oven (ECOCELL Blueline Comfort, Nuremberg, Germany). After drying, the samples were ground (using a Grindomix GM 200 knife mill, Retsch, Germany) and analyzed for chemical composition. The values obtained from the laboratory chemical analysis were used to calculate the apparent digestibility of nutrients (DDM, DOM, DCP, DEE, and DNFE) as described earlier by [21] using the following formula:

$$\text{Apparent nutrient digestibility } (\%) = \frac{(\text{nutrient intake} - \text{nutrient excreta})}{\text{nutrient intake}} \times 100 \quad (2)$$

2.6. Blood Collection and Analysis

On the final day of the experiment, approximately 3 mL of venous blood samples per birds were aseptically collected from 18 laying hens from the sub-axial region into 9 mL anticoagulant-free Vacutainers containing 14.3 U/mL of lithium heparin (Vacutest®, Arzergrande, Italy). Further, these samples were used to determine the activity of blood antioxidant enzymes including superoxide dismutase (SOD), catalase (CAT), glutathione peroxidase (GSH), and total antioxidant capacity (TAC). Blood samples were separated by centrifugation at 3000× *g* in a refrigerated centrifuge (Eppendorf Centrifuge 5430R; Eppendorf, Hamburg, Germany) for 25 min at 4 °C. Afterwards, the supernatant obtained from serum samples were carefully transferred to plastic vials and stored at −20 °C until the analysis.

2.7. Egg Quality Measurement

A total of 306 eggs were collected during the experiment. The collected eggs (18 eggs/group: 3 eggs/cage, 6 cages/lot; each cage representing a sample) were analyzed at the end of the experiment (2 months) to evaluate the impact of microalgae-based diets, specifically those containing chlorella and spirulina, on the fatty acid composition. The antioxidant profile of the yolk, as well as the internal and external quality parameters of the eggs, were determined at the end of the experiment (8 weeks) using a Digital Egg Tester DET-6500 (NABEL Co., Ltd., Kyoto, Japan). First, the eggs were weighed whole and then cracked, and the yolks were separated from the albumen and shell; every yolk was rolled onto a paper towel to remove any adherent albumen or chalazae membrane as described by [22]. Each egg component was weighed with a Kern scale (precision 0.001). The yolk color intensity was measured using the portable colorimeter 3 nh YS3020 (Shenzhen ThreeNH Technology Co., Ltd., Beijing, China), and the temperature and pH of yolk and albumen were measured using a portable pH meter (Five Go F2-Food, Greifensee, Switzerland) and Haugh unit. After measuring the internal and external quality parameters of the eggs, the yolk samples were dried for 48 h at a constant temperature of 65 °C in an oven (ECOCELL Blueline Comfort, Nuremberg, Germany) for further chemical analysis, such as the concentration of β-carotene (μg/g), total polyphenols (mg/g GAE), antioxidant activity (expressed as DPPH % inhibition and μM Trolox), fatty acid profile (g acid/100 g total FAME), and cholesterol concentration (g/egg).

To assess the yolk color stability after boiling, at the end of the experiment, 90 eggs were collected (10 eggs/group, 30 eggs/period) and boiled for 10, 15, and 20 min, respectively.

2.8. Chemical Analysis of Samples

2.8.1. Determination of In Vitro Digestibility of Nutrients

The in vitro digestibility of nutrients was determined following the method proposed by [23] and adapted for poultry as described by [21] using a Daisy Incubator (ANKOM Technology, Macedon, NY, USA) in a 2-step procedure: two successive incubations with pepsin and pancreatin. The samples were introduced into F57 bags (Ankom) and incubated in Daisy Incubator jars with 0.1 M phosphate pH 2.0 buffer with 0.3 g of pepsin (porcine, 2000 FIP U/g) per liter for 6 h at 39 °C. After draining the buffer and washing bags with slightly warm tap water, the next 0.04 M phosphate buffer pH 6.8 with 1 g of pancreatin (porcine, grade IV, reference Sigma P-1750) per liter was added to the jars. Incubation lasted for 18 h at 39 °C and finally the bags were dried in a forced draught oven at 65 °C for 48 h. The final weight after digestion of each bag was recorded for in vitro digestibility of dry matter calculation. Some of the bags was retained for nitrogen analysis and consequently for calculation of the in vitro digestibility of nitrogen. The remaining bags were subsequently subjected to incineration in a muffle furnace at a temperature of 550 °C for a duration of 5 hours. The resulting ash was utilized for the purpose of residue digestion and in vitro calculation of organic matter digestibility. The results were expressed as mean ± standard deviation of five replicate analyses.

2.8.2. Pigment Extraction from *Spirulina platensis* and *Chlorella*

To extract pigments from feed and dried *Spirulina platensis* and *Chlorella*, we used a combined method of sonication–solvent extraction followed by stirring on a magnetic stirrer [24]. Acetone solvent ratio was 1:100, w:v; sonication was performed for 30 min, and magnetic stirring was applied for 60 min. The extract was obtained by centrifugation (SIGMA 2-16KL refrigerated centrifuge) at 2599× *g* for 10 min. The resulting precipitate was extracted until no color was observed. Pigment extracts were then analyzed using a spectrum UV-Vis with wavelengths between 400 and 700 nm and absorbance at 470, 645, and 663 nm (JASCO V-670 spectrophotometer), in triplicate. The pigment levels, including chlorophyll *a* (C*a*), chlorophyll *b* (C*b*), and total carotenoids (Cc), were estimated with Equations (3)–(5). The results were reported, taking into account the dilution factor (DF) as mg/g DW (dry weight) for *Spirulina platensis* and *Chlorella vulgaris* powder and μg/g feed.

$$Ca = 11.24 \times A_{663} - 2.04 \times A_{645} \times \mathrm{DF} \quad (3)$$

$$Cb = 20.13 \times A_{645} - 4.19 \times A_{663} \times \mathrm{DF} \quad (4)$$

$$\mathrm{Cc} = \frac{(1000 \times A_{470} - 1.90 \times Ca - 63.14 \times Cb)}{214} \times \mathrm{DF} \quad (5)$$

2.8.3. Measurement of Some Antioxidant Enzyme Activity and GSH in Blood Serum

The activity of superoxide dismutase (SOD) was determined following the method described by [25]. Blood serum was added to the assay mixture containing 66 mM phosphate buffer with a pH of 7.8, 0.1 mM EDTA, 5.7 M nitro blue tetrazolium (NBT), 9.9 mM L-methionine, and 2.5% (*w*/*v*) Triton X100 and riboflavin (0.01 mL of 4.4%, *w*/*v*) was finally added to initiate the reaction. NBT reduction was measured at 560 nm in a Jasco V-670. The activity of SOD was calculated in units of enzyme/mL.

The activity of catalase (CAT) was determined by the classical method developed by [26]. CAT decomposes H_2O_2 (the substrate) and can be directly measured by decreased absorbance at 240 nm. Freshly prepared reagents prior to assays were phosphate buffer

(66 mM, pH 7.0) and 30 mM H_2O_2 in a phosphate buffer. The final volume was 1 mL and the reaction was started by the addition of H_2O_2. To correct for any non-enzymatic reaction, a blank assay containing buffer instead of substrate was used. CAT activity is defined in specific units/mL.

The level of reduced glutathione (GSH) was measured according to the method described by [27] and was determined based on the reaction of GSH with 5,5′-dithiobis (2-nitrobenzoic acid). The resulting chromophore, TNB (5-thio-2-nitrobenzoic acid), has a maximum absorbance of 412 nm. The TNB formation rate is proportional to the sample GSH level. Blood serum was treated with 0.6% sulfosalicylic acid and centrifugated. The supernatant was added to the assay mixture containing 100 mM phosphate buffer with a pH of 7.5. A 3 mM stock solution of the DTNB reagent was prepared in phosphate buffer with a pH 7.5, and diluted to a final concentration of 10 μM. The reaction between GSH and DTNB was monitored at a wavelength of 412 nm using a Jasco UV/Vis V-670 spectrophotometer. The concentration of GSH in blood serum was calculated with the linear equation generated from a GSH standard curve.

Total antioxidant capacity (TAC) was analyzed by scavenging of DPPH (2,2-diphenyl-1-picrylhydrazyl) radical activity [28]. Blood serum proteins were removed with one volume of acetonitrile, incubated for 5 min and centrifugated for 10 min at 9000× *g*. Supernatant (25 μL) was added to the assay mixture containing 970 μL of methanol and 5 μL of 10 mM of DPPH radical methanolic solution. After 30 min, the absorbance was read at 517 nm by a Jasco UV/Vis V-670 spectrophotometer. In parallel, a negative control with 25 μL acetonitrile, instead of deproteinated blood serum was prepared. All determinations were performed in triplicate and the serum scavenging effect (Sc%) was calculated according to Equation (6).

$$\text{Scavenging \%} = \frac{1 - A_{517\ sample}}{A_{517\ negativ\ control}} \times 100 \tag{6}$$

2.8.4. Egg Yolk β-Carotene and Antioxidant Activity Determination

The β-carotene concentration of egg yolk was determined using spectroscopy method [29]. A quantity of 0.5 g of well-mixed egg yolk from each fresh or lyophilized form was taken in a 50 mL conical flask. First, 25 mL of acetone was added and the vortex was used to make a smooth paste. The solution was mixed well for 10 min and filtered (Whatman No. 1, Merck KGaA, Darmstadt, Germany). The remaining solid was re-extracted with another 20 mL of acetone using the vortex. The two filtrates were combined and the acetone extract was diluted to 50 mL. The egg yolk pigments expressed as μg β-carotene/g were measured at 450 nm wavelength (E1% 2500) using a JASCO V-670 spectrophotometer.

The total phenolic content of egg yolk samples was determined by the Folin–Ciocalteu colorimetric method [30]. The absorbance was recorded at 732 nm using a spectrophotometer (Jasco V-530, Japan Servo Co., Ltd., Tokyo, Japan). Gallic acid was used as standard solution. The total phenolic content is expressed as mg gallic acid equivalents (GAE)/ g of the sample on the basis of a standard curve of gallic acid.

The antioxidant capacity of egg yolk samples was measured using the DPPH (2,2-diphenyl-1-picrylhydrazyl) radical-scavenging activity method described by [31]. The absorbance of the solution was measured at 517 nm with the help of a spectrophotometer (Jasco V-530, Japan Servo Co., Ltd., Japan). Trolox solution was used as standard. The results were expressed as mM Trolox equivalents (TE).

2.8.5. Egg Yolk Cholesterol Content and Fatty Acids Profile

The cholesterol content of dried yolk was determined using the gas chromatography (GC) method (AOAC, 1996) as described by [32]. The sample was saponified in a methanol–potassium hydroxide solution, extracted with petrol ether, concentrated using a rotavapor, and subjected to chloroform addition before being analyzed using a GC (Perkin Elmer Clarus-500, with a flame ionization detector). Separation was achieved using an HP-5 capillary column (30 m length, 0.32 mm internal diameter, 0.1 um film thickness), and the results were expressed as grams of cholesterol per whole egg. Fatty acid profile of dried yolk was determined as described by [21], using GC (Perkin ElmerClarus 500, Mass Spectrometer System) of fatty acid methyl esters (FAME) equipped with a flame ionization detector (FID) and a BPX70 capillary column (60 m × 0.25 mm ID, 0.25 μm film thickness). The column temperature was set at 5 °C/min^{-1} ramped from 180 °C to 220 °C. The carrier gas was hydrogen (linear velocity 35 cm/s at 180 °C), and the split ratio was 1:100. The injector and detector temperatures were 250 °C and 260 °C, respectively. The results were expressed as g fatty acid per 100 g total fatty acids. The average amount of each fatty acid was used to calculate the sum of the total saturated (SFAs), total monounsaturated (MUFAs), and total polyunsaturated (PUFAs) fatty acids.

2.8.6. Color Measurement of Fresh and Boiled Eggs

Yolk color intensity was measured using a portable colorimeter as previously described by [32]. The yolk was separated from the albumen and subsequently positioned on a Petri dish (∅ = 50 mm) prior to measurement. The color parameters of L* (lightness), a* (red-green intensity), and b* (yellow-blue intensity) of the CIE-Lab system (Commission Internationale de l'Eclaraige) were determined by reflectance CIE—L* a* b* color coordinates. The instrument was calibrated with a white calibration before the measurements. All measurements were performed in triplicate.

2.9. Statistical Analysis

The results obtained from feed nutritional composition, apparent nutrient digestibility, laying hens' performances, antioxidant enzyme activity, egg quality parameters, fatty acids, and yolk cholesterol content were analyzed using a randomized complete block design and the general linear model (GLM) procedures of SAS (Statistical Analysis System, Minitab version 17, SAS Institute Inc., Cary, NC, USA) considering a cage as an experimental unit, according to the following linear model:

$$Y_{ij} = \mu + A_j + e_{ij}, \quad (7)$$

where Y_{ij} means value of trait (the dependent variable); μ, overall mean; A_j, the treatment effect; and e_{ij}, random observation error.

The effects of boiling time on fresh vs. boiled yolk color were analyzed to determine whether the factors studied (treatment and boiling time) affected the fatty acid concentration and yolk color of eggs for different time periods. The data obtained were analyzed by two-way ANOVA using the Tukey test, following the statistical model:

$$Y_{ijk} = \mu + \alpha_i + \beta_j + \alpha_i\beta_j + e_{ijk} \quad (8)$$

where Y_{ijk} = variable measured for the kth observation of the ith treatment and jth feeding or boiling time; μ is the sample mean; α_i is the effect of the ith treatment; βj is the effect of the jth feeding or boiling time; αiβj is the interaction of the ith treatment and jth feeding or boiling time, and ε_{ijk} is the effect of error. The differences were highly significant when $p < 0.001$, significant if $p < 0.05$, and a tendency of influence was considered when $p < 0.10$.

The graphs for antioxidant enzyme activities were created using GraphPad Prism 9.1.2 software (GraphPad Software, La Jolla, CA, USA). Differences were considered significant when $p < 0.05$.

3. Results

3.1. Nutritional Value of Chlorella and Spirulina Powder

The nutritional values of the microalgae are presented within Table 2. Both microalgae present a high nutrient content, with a high concentration of easily digestible proteins, metabolizable energy, and a low level of cellulose. Compared to chlorella, spirulina is characterized by a higher antioxidant capacity, with a significantly higher content of chlorophylls *a* and *b*. The concentration of carotenoids is high in both chlorella and spirulina.

Chlorophylls are green pigments found in plants and algae. Chlorella has a chlorophyll *a* content of 5.56 mg/g and chlorophyll *b* content of 0.88 mg/g. Spirulina platensis has a higher chlorophyll content compared to chlorella, with chlorophyll *a* at 9.06 mg/g and chlorophyll *b* at 1.34 mg/g. It is known that carotenes are orange or red pigments that serve as powerful antioxidants. Chlorella contains 1.52 mg/g of carotenes and spirulina has a slightly higher carotene content at 1.68 mg/g. Polyphenols are compounds found in plants and algae known for their antioxidant properties and potential health benefits. The total polyphenol content of chlorella was assessed to be 1.16 mg/g of total polyphenols, while spirulina has a slightly higher content of 1.35 mg/g of total polyphenols. DPPH is a common method used to evaluate the antioxidant activity of substances. Trolox is a synthetic antioxidant used as a standard for comparison. Chlorella exhibited an antioxidant capacity of 73.29% inhibition of DPPH (2,2-diphenyl-1-picrylhydrazyl) and an antioxidant capacity of 15.49 μM Trolox. On the other hand, spirulina demonstrated a higher antioxidant capacity compared to chlorella, with 81.27% inhibition of DPPH and an antioxidant capacity of 16.78 μM Trolox. Both chlorella and spirulina exhibited antioxidant activity, with spirulina generally showing higher values in terms of chlorophyll content, carotenes, total polyphenols, and antioxidant capacity. These findings suggest that spirulina demonstrated stronger antioxidant properties compared to chlorella.

Concerning the fatty acid concentrations, spirulina had a higher content of total SFAs at 51.03 g FAME/100 g Total FAME compared to chlorella, which had a total SFA content of 33.77 g FAME/100 g Total FAME. The monounsaturated fatty acids (MUFAs) of chlorella contained 9.05 g FAME/100 g Total FAME, while spirulina contained a higher content of MUFA at 27.12 g FAME/100 g Total FAME. MUFAs are considered to be healthier fats compared to SFAs. Polyunsaturated fatty acids (PUFAs) include both omega-3 and omega-6 fatty acids. Chlorella has a higher total PUFA content of 56.56 g FAME/100 g Total FAME compared to spirulina (21.85 g FAME/100 g Total FAME). The data presented suggest that chlorella has a higher content of total PUFAs, particularly PUFA n-3, compared to spirulina. On the other hand, spirulina has higher contents of SFAs, MUFAs, and PUFA n-6. The nutritional value and potential health benefits of these microalgae may be influenced by their fatty acid composition. Overall, the fatty acid profile indicates that chlorella contains a significantly higher amount of polyunsaturated fats (PUFAs), with over 67% being represented by omega-3 polyunsaturated fats, particularly α-linolenic acid (ALA).

3.2. Nutritional Value of the Experimental Diets

The nutritional content of the diets is presented in Table 3. The antioxidant activity characterized by chlorophyll a is significantly higher ($p < 0.0001$) in the SP2% group compared to the CV2% and CON groups, but the CV2% group is also significantly higher ($p < 0.0001$) compared to the CON group. Chlorophyll *b* was found in significantly higher ($p < 0.0001$) concentrations in both the CV2% and SP2% groups compared to the CON group. The antioxidant capacity was significantly higher ($p < 0.0001$) in the SP2% group compared to the CV2% and CON groups, but the CV2% group also registered a significantly higher ($p < 0.0001$) antioxidant capacity compared to the CON group. The ΣSFA was significantly higher ($p < 0.001$) in the CON group compared to the SP2% and CV2% groups. Additionally, ΣMUFA concentration was significantly higher ($p < 0.03$) in the CON group compared to the CV2% group. The $\sum$ PUFA concentration was observed to be significantly higher ($p < 0.003$) in the SP2% and CV2% groups compared to the CON group. The highest value concentration of Σ n-3 PUFA ($p < 0.007$) was noticed in the CV group compared to the SP2% and CON groups. Concerning the $\sum$ n-6 PUFA, a statistically higher concentration was determined for the CV2% and SP2% groups compared to the CON group. The $\sum$ n-6/$\sum$ n-3 ratio was highly significant for the SP2% and CV2% groups compared to the CON group.

Table 2. Proximate composition, antioxidant activity, and fatty acid profile of chlorella and spirulina.

Parameters	Chlorella Powder	Spirulina Powder
	$\overline{X} \pm s_{\overline{X}}$	
Proximate composition *		
Calculated metabolizable energy (ME), MJ/kg	9.77 ± 0.41	9.68 ± 0.25
Crude protein (CP), %	51.06 ± 0.35	67.02 ± 0.04
Dry matter (DM), %	94.68 ± 0.80	92.70 ± 0.62
Organic matter (OM), %	88.33 ± 0.75	87.75 ± 0.50
Ether extract (EE), %	3.56 ± 0.50	0.48 ± 0.03
Crude fiber (CF), %	0.49 ± 0.09	0.19 ± 0.11
Non-fermentable extractive substance (NFE), %	33.22 ± 0.39	20.06 ± 0.07
Ash, %	6.35 ± 0.25	4.95 ± 0.18
In vitro nutrient digestibility **		
Digestible crude protein (DCP), %	96.59	96.71
Digestible dry matter (DDM), %	99.56	99.05
Digestible organic matter (DOM), %	99.59	99.12
Antioxidant activity *		
Chlorophyll *a*, mg/g	5.56 ± 1.08	9.06 ± 0.79
Chlorophyll *b*, mg/g	0.88 ± 0.20	1.34 ± 0.34
Carotenes, mg/g	1.52 ± 0.19	1.68 ± 0.31
Total polyphenols, mg/g GAE	1.16 ± 0.16	1.35 ± 0.05
Antioxidant capacity (DPPH % inhibition)	73.29 ± 2.93	81.27 ± 1.60
Antioxidant capacity (μM Trolox)	15.49 ± 3.87	16.78 ± 2.47
Fatty acids (g FAME/100 g Total FAME) *		
Caproic (C 6:0)	0.65 ± 0.05	0.16 ± 0.02
Caprilic (C 8:0)	0.25 ± 0.02	7.73 ± 0.65
Capric (C 10:0)	1.20 ± 0.10	-
Lauric (C 12:0)	0.07 ± 0.001	0.68 ± 0.06
Miristic (C 14:0)	0.96 ± 0.08	0.88 ± 0.07
Pentadecanoic (C 15:0)	0.11 ± 0.001	-
Palmitic (C 16:0)	27.25 ± 2.32	34.71 ± 3.12
Stearic (C 18:0)	2.51 ± 0.21	6.87 ± 0.55
Heneicosanoic (C 21:0)	0.17 ± 0.01	-
Behenic (C 22:0)	0.60 ± 0.05	-
∑ SFA	33.77 ± 3.00	51.03 ± 4.34
Pentadecenoic (C 15:1)	0.05 ± 0.004	-
Palmitoleic (C 16:1)	0.19 ± 0.02	4.87 ± 0.46
Heptadecenoic (C 17:1)	0.08 ± 0.007	-
Oleic cis (C 18:1)	8.73 ± 0.72	22.25 ± 2.14
∑ MUFA	9.05 ± 0.19	27.12 ± 1.3
Linoleic cis (C 18:2n6)	14.13 ± 1.20	16.84 ± 1.43
Linolenic γ (C 18:3n6)	0.16 ± 0.01	-
Linolenic α (C 18:3n3)	37.37 ± 3.18	4.41 ± 0.39
Octadecatetraenoic (C18:4n3)	0.73 ± 0.06	-
Eicosadienoic (C20:2n6)	0.33 ± 0.03	-
Eicosatrienoic (C20:3n6)	3.61 ± 0.31	-
Arachidonic (C20:4n6)	0.21 ± 0.01	0.60 ± 0.05
∑ PUFA	56.56 ± 4.80	21.85 ± 1.94
∑ PUFA n-3	38.10 ± 3.65	4.41 ± 0.41
∑ PUFA n-6	18.45 ± 1.77	17.44 ± 1.48
∑ PUFA n-6/∑ PUFA n-3	0.48 ± 0.05	3.95 ± 0.35
Other fatty acids	0.62 ± 0.05	-

Where: ∑SFA, sum of saturated fatty acid; ∑MUFA, sum of monounsaturated fatty acid; ∑P sUFA, sum of polyunsaturated fatty acid; * Mean ± standard deviation of three replicate analyses; ** Mean ± standard deviation of five replicate analyses.

Table 3. Nutritional compounds of the feed (% as fed).

Parameters	Dietary Treatments			SEM	p-Value
	CON	CV2%	SP2%		
	Antioxidant activity				
Chlorophyll a, µg/g	10.31 [c]	344.76 [b]	383.22 [a]	5.41	<0.0001
Chlorophyll b, µg/g	2.77 [b]	40.93 [a]	38.92 [a]	1.44	<0.0001
Carotenes, µg/g	31.45 [b]	84.92 [a]	81.70 [a]	1.08	<0.0001
Total polyphenols, mg/g GAE	1.49	1.67	1.64	0.177	0.933
Antioxidant capacity (DPPH % inhibition)	32.00 [c]	57.57 [b]	67.23 [a]	1.23	<0.0001
Antioxidant capacity (µM Trolox)	5.19	6.25	5.79	0.403	0.620
	Fatty acid composition (% of total fat)				
ΣSFA	29.03 [a]	19.72 [b]	22.44 [b]	0.893	0.001
ΣMUFA	47.42 [a]	37.06 [b]	40.12 [ab]	2.06	0.030
∑ PUFA, from which:	23.16 [b]	42.95 [a]	37.11 [a]	2.36	0.003
Σ n-3 PUFA	1.03 [b]	1.43 [a]	1.21 [ab]	0.06	0.007
∑ n-6 PUFA	22.14 [b]	41.52 [a]	35.90 [a]	2.32	0.003
∑ n-6/∑ n-3	21.58 [b]	29.18 [a]	29.47 [a]	1.22	0.006

Where: CON, conventional diet; CV2%, conventional diet supplemented with 2% chlorella powder; SP2%, conventional diet supplemented with 2% spirulina powder; $n = 5$ samples per group; SEM, standard error of the mean; [abc] Mean values within a row with different letters are significantly different at $p \leq 0.05$. Abbreviations: ΣSFA, sum of saturated fatty acids; ∑MUFA, sum of monounsaturated fatty acid; ΣPUFA, sum of polyunsaturated fatty acids; Σ n-3 PUFA = C18:3n-3 + C18:4n-3; Σ n-6 PUFA = C18:2n-6 + C20:2n-6 + C20:3n-6 + C20:4n-6.

3.3. Production Performances

The production performance values are shown within Table 4. There were no significant differences concerning the initial or final body weight ($p = 0.580$; $p = 0.688$) of hens in the CON, CV2%, and SP2% groups. The feed conversion ratio (g feed/g egg) registered the optimum value in the SP2% group compared to CON and CV2% ($p < 0.0001$). There was no significant difference in the daily feed intake among the dietary treatments ($p = 0.608$). The mean daily feed intake was 112.78 g/day/layer for CON, 112.40 g/day/layer for CV2%, and 113.46 g/day/layer for SP2%. A highly significant difference ($p < 0.0001$) was noticed in the feed conversion ratio between all groups. There was a highly significant difference ($p < 0.0001$) concerning egg weight among the dietary treatments. The mean egg weight was 59.78 g for CON, 62.42 g for CV2%, and 62.44 g for SP2%. There was a significant difference ($p = 0.046$) in the hen day egg production among the dietary treatments; the highest percentage was registered for the SP2% group (95.56%) compared to CON and CV2%. There were significant differences in the percentages of eggs classified into different sizes (XL, L, M, and S) categories among the dietary treatments ($p < 0.0001$). The data show that the proportions of eggs in each size category varied among the treatment groups, with significant differences observed for all categories ($p < 0.0001$). The CV2% group presented a higher percentage of eggs classified as "M" (medium) compared to the SP2% and CON groups. This suggests that chlorella supplementation might have influenced egg size distribution, potentially leading to a significantly higher proportion of medium-sized eggs ($p < 0.0001$). The SP2% group showed a significantly higher ($p < 0.0001$) percentage of eggs classified as "L" (large) compared to the CV2% and CON groups.

3.4. Serum Antioxidant Status

The effect of experimental diets CV2% and SP2% on the blood antioxidant enzymes SOD, CAT, GSH, and TAC activity is presented within Figure 1. A significant increase ($p < 0.001$; $p < 0.0001$) of SOD concentration in serum of laying hens fed with microalgae was noticed compared to the CON group. Also, a highly significant difference ($p < 0.001$) was registered between experimental groups, where SP2% was characterized by an increased antioxidant activity which assumes a high antioxidant status of this group. The chlorella and spirulina powder diet supplementations improved the enzymatic activity of CAT, increasing ($p < 0.01$) its serum concentration value significantly compared to CON. There were no differences ($p \geq 0.05$) noticed between the two experimental groups.

Table 4. Performances of laying hens fed with *Ch. vulgaris* and *Sp. platensis*.

Parameters	Dietary Treatments			SEM	p-Value
	CON	CV (2%)	SP (2%)		
Initial body weight (g/layer)	1561.30	1599.17	1556.88	23.23	0.580
Final body weight (g/layer)	1670.68	1681.52	1643.96	23.63	0.688
Daily feed intake (g/day/layer)	112.78	112.40	113.46	0.441	0.608
Feed conversion ratio (g feed/g egg)	2.04 ^{b}	2.17 ^{a}	1.92 ^{c}	0.024	<0.0001
Egg weight (g)	59.78 ^{b}	62.42 ^{a}	62.44 ^{a}	0.110	<0.0001
Hen day egg production (%)	94.61 ^{ab}	93.32 ^{b}	95.56 ^{a}	0.412	0.046
Egg classification *, %					
"XL" (>73 g), extra large	0.15 ^{b}	2.99 ^{a}	2.08 ^{a}	0.350	<0.0001
"L" (63–73 g), large	22.87 ^{b}	38.50 ^{a}	39.65 ^{a}	1.01	<0.0001
"M" (53–63 g), medium	71.22 ^{a}	57.321 ^{b}	56.50 ^{b}	1.03	<0.0001
"S" (<53 g), small	5.76 ^{a}	1.19 ^{b}	1.77 ^{b}	0.346	<0.0001

Where: CON, conventional diet; CV2%, conventional diet supplemented with 2% chlorella powder; SP 2%, conventional diet supplemented with 2% spirulina powder; * C.E Regulation no. 852/2004 on the general rules of food hygiene with subsequent amendments and completions and Directive 2000/13 / C.E.; SEM, standard error of the mean. ^{abc} Mean values within a row with different letters are significantly different at $p \leq 0.05$.

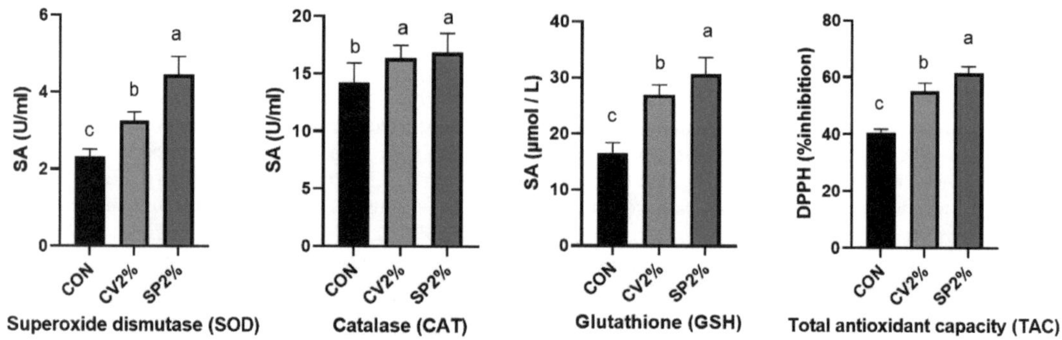

Figure 1. Effect of chlorella (CV2%) and spirulina powder (SP2%) supplementation on the enzymatic specific activity (SA) of superoxide dismutase (SOD), catalase (CAT), glutathione (GSH), and total antioxidant capacity (TAC), expressed as U /l or % inhibition of sanguine serum; (n = 6). ^{abc} Mean values within a column having different letters are significantly different at $p \leq 0.05$.

GSH serum concentration significantly increased ($p < 0.0001$) in the SP2% and CV2% groups compared to CON. Also, significant statistically differences ($p < 0.05$) were observed between the experimental groups. The same trend was noticed for total antioxidant capacity value from serum, but with highly statistical differences ($p < 0.001$) between the two experimental groups.

3.5. Digestibility Trial

The coefficients of apparent nutrient digestibility of laying hens fed with chlorella and spirulina powder diets are presented within Table 5. The supplementation of the conventional diet with chlorella or spirulina powder had no effect on the digestibility of dry matter, organic matter, crude protein, or digestible NFE because there were no significant differences ($p > 0.05$) between the CON diet and algae-supplemented diets. We registered only a tendency for crude fat digestibility to decrease just for chlorella powder.

Table 5. Apparent nutrient digestibility (%) of laying hens fed with *Ch. vulgaris* and *Sp. platensis* powder diets.

Parameters	Dietary Treatments			SEM	*p*-Value
	CON	CV2%	SP2%		
Digestible dry matter (DDM),%	73.17	71.29	72.61	0.835	0.288
Digestible organic matter (DOM), %	73.05	70.52	71.62	0.806	0.112
Digestible crude protein (DCP), %	85.81	85.20	84.85	0.607	0.539
Digestible crude fat (DCF), %	90.58 ^{a}	87.82 ^{b}	89.07 ^{ab}	0.740	0.052
Digestible non-fermentable extractive substance (DNFE), %	68.60	66.27	67.59	0.768	0.128

Where: CON, conventional diet; CV2%, conventional diet supplemented with 2% chlorella powder; SP2%, conventional diet supplemented with 2% spirulina powder; $n = 6$ samples per group; SEM, standard error of the mean. ^{ab} Mean values within a row having different letters are significantly different at $p \leq 0.05$.

3.6. Nutritional Egg Quality Parameters

Table 6 presents the effect of dietary chlorella and spirulina powder on laying hens' egg nutrients and external/internal quality parameters. The microalgae dietary supplementation influenced the nutritional egg quality positively. The β-carotene concentration registered a highly significant increase ($p < 0.0001$) in experimental diets, the results being positively correlated with the concentration of chlorophylls *a* and *b* and carotene of both dietary microalgae. The same observation could be noted for both antioxidant capacity expressed as DPPH% inhibition ($p < 0.0001$) and μM Trolox ($p < 0.033$).

Table 6. Effect of dietary *Ch. vulgaris* and *Sp. platensis* powder in laying hens' diets on egg yolk nutrients and external and internal egg quality parameters.

Parameters	Dietary Treatments			SEM	*p*-Value
	CON	CV2%	SP2%		
Nutrition quality of egg yolk					
β-carotene, (μg/g)	30.77 ^{b}	38.82 ^{a}	38.25 ^{a}	0.432	<0.0001
Total polyphenols (mg/g GAE)	0.534	0.598	0.574	0.027	0.271
Antioxidant capacity (DPPH% inhibition)	16.84 ^{b}	25.14 ^{a}	28.15 ^{a}	0.869	<0.0001
Antioxidant capacity (μM Trolox)	0.74 ^{b}	0.79 ^{a}	0.82 ^{a}	0.019	0.033
External and internal egg quality parameters					
Egg weight (g), of which:	61.28	61.53	61.82	0.389	0.617
albumen white (g)	37.00	37.76	37.38	0.352	0.320
egg yolk (g)	16.37	15.65	16.26	0.242	0.087
eggshell (g)	7.91	8.12	8.17	0.137	0.358
Albumen pH (value)	8.62 ^{ab}	8.44 ^{b}	8.80 ^{a}	0.069	0.002
Yolk pH (value)	6.51	6.38	6.46	0.093	0.625
t° albumen (°C)	19.10 ^{a}	18.57 ^{ab}	17.79 ^{b}	0.258	0.003
t° yolk (°C)	19.99 ^{a}	19.12 ^{b}	19.92 ^{a}	0.134	0.011
White height, mm	10.96	11.74	11.48	0.292	0.168
Haugh units (value)	102.92	106.04	104.82	1.150	0.163
Yolk color, (value)	4.06 ^{c}	8.11 ^{b}	11.06 ^{a}	0.113	<0.0001

Where: CON, conventional diet; CV2%, conventional diet supplemented with 2% chlorella powder; SP2%, conventional diet supplemented with 2% spirulina powder; $n = 6$ samples per group (3 eggs/ sample); SEM, standard error of the mean. ^{abc} Mean values within a row having different letters are significantly different at $p \leq 0.05$.

Concerning the egg quality parameters, no effects ($p > 0.05$) of chlorella or spirulina powder were observed on egg weight or its components (albumen, yolk, and shell). The albumen pH was influenced by the presence of algae compared to the CON diet: lower value for chlorella diet, and higher value for spirulina diet. The yolk pH and the albumen height were not influenced by the dietary treatment. The Haugh unit, the most widely accepted indicator of internal egg quality, had an increasing tendency in eggs obtained from algae supplementation diets, especially for the chlorella diet. The egg yolk color in groups fed diets with chlorella or spirulina were significantly higher ($p < 0.0001$) compared to the CON diet.

3.7. Yolk Cholesterol Content and Fatty Acid Profile

Table 7 presents the effects of chlorella and spirulina powder dietary inclusion on egg yolk cholesterol content and fatty acid profile. There were no statistical differences ($p = 0.061$) concerning cholesterol content groups. When examining the fatty acid composition, there were variations observed among the different experimental groups depending on how the results were reported. For the health of consumers, it is important that results are be expressed as mg fatty acids/egg. Concerning the total SFAs ($p = 0.825$) and the total MUFAs ($p = 0.280$), there were no differences registered. Regarding polyunsaturated fatty acids (PUFAs), expressed as mg fatty acids/egg, there were no differences ($p = 0.102$) between groups, even though when expressed as g FAME/ 100 g Total FAME, both CV2% and SP2% groups led to a decrease in the overall amount of PUFAs compared to the control group. The highest ($p < 0.0001$) concentration of omega-3 was observed in the chlorella group (1.72 g FAME/100 g Total FAME; 25.001 mg fatty acids/egg) compared to spirulina (1.21 g FAME/100 g Total FAME; 18.364 mg fatty acids/egg) and control groups (1.23 g FAME/100 g Total FAME; 18.887 mg fatty acids/egg), which influenced the Ω6/Ω3 ratio; highly significantly ($p < 0.0001$) lower for the chlorella group.

3.8. The Effect of Dietary Chlorella and Spirulina Powder on Egg Yolk Color in Fresh and Boiled Eggs

Table 8 presents the effects of chlorella and spirulina powder dietary inclusion on egg yolk color in fresh and boiled eggs. The treatment factor (CON, CV2%, SP2%) exhibited a significant effect on the L*, a*, and b* parameters. The p-values registered for all three parameters indicated a highly significant difference ($p < 0.0001$) between the treatments. The boiling time factor (fresh yolk, 10, 20, and 30 min.) also had a highly significant effect ($p < 0.0001$) on the L*, a*, and b* parameters, which indicated that the boiling time had a significant impact on the egg yolk color. The interaction between treatment and boiling time showed highly significant ($p < 0.0001$) effects on the a* and b* parameters. However, the interaction effect was not statistically significant for the L* parameter ($p = 0.293$). The L* parameter represents the lightness of the color, and therefore higher values indicate lighter or brighter colors, while lower values indicate darker colors. Comparing the values across different treatments and boiling times, it can be observed that as the boiling time increased, the L* values showed a tendency to increase as well. This suggests that the egg yolks become lighter in color as they are boiled for a longer duration. The darkest color values were noticed for SP2%, followed by CV2% ($p < 0.0001$). The a* parameter values (red-green color axis, positive values indicate redness, negative values indicate greenness) varied across different treatments and boiling times. The highest value for a* parameters was noticed on SP2%, followed by CV2% ($p < 0.0001$), compared to CON. However, there was no consistent trend concerning boiling time. The interaction effect between treatment and boiling time was highly significant ($p < 0.0001$), indicating that the combination of treatment and boiling time influenced the red-green color component. Positive values indicate more yellowness, while negative values indicate more blueness. Similar to the a* parameter, the b* values also varied across treatments and boiling time. The b* parameter (yellow-blue color) registered the highest value on CV2%, followed by SP2% ($p < 0.0001$), compared to CON. The interaction effect between treatment and boiling time was highly significant ($p < 0.0001$), suggesting that the combination of treatment and boiling time affects the yellow-blue color component.

Table 7. Fatty acid composition in total lipids of egg yolks (average values/group).

	Experimental Groups									
Parameters	**CON**	**CV2%**	**SP2%**	**SEM**	***p*-Value**	**CON**	**CV2%**	**SP2%**	**SEM**	***p*-Value**
Yolk fat (% DM)	29.454	29.570	30.412	0.771	0.640					
Cholesterol (mg col/egg)	224.00	173.00	192.00	0.013	0.061					
Fatty acid content	g FAME/100 g Total FAME					mg fatty acid/egg				
Miristic C14:0	0.308 ^{b}	0.346 ^{a}	0.346 ^{a}	0.008	0.005	4.716	5.063	5.266	0.196	0.167
Pentadecanoic C15:0	0.060	0.140	0.060	0.052	0.399	0.932	2.182	0.918	0.760	0.423
Palmitic C16:0	25.44	26.14	26.05	0.242	0.121	389.9	380.2	395.0	13.500	0.738
Heptadecanoic C17:0	0.128 ^{b}	0.128 ^{b}	0.145 ^{a}	0.003	0.001	1.974 ^{ab}	1.869 ^{b}	2.227 ^{a}	0.093	0.043
Stearic C18:0	10.09	10.11	9.61	0.232	0.256	154.71	146.93	145.79	5.990	0.533
∑SFA	36.03	36.87	36.22	0.409	0.339	552.2	536.3	549.2	19.20	0.825
Miristioleic C14:1	0.079 ^{c}	0.092 ^{b}	0.108 ^{a}	0.003	<0.0001	1.217 ^{b}	1.345 ^{b}	1.645 ^{a}	0.066	0.001
Pentadecenoic C15:1	0.100	0.101	0.417	0.184	0.394	1.53	1.46	6.35	2.810	0.392
Palmitoleic C16:1	4.37	3.98	4.34	0.421	0.765	65.58	57.95	65.86	5.170	0.488
Heptadecenoic C17:1	0.069 ^{b}	0.075 ^{b}	0.089 ^{c}	0.003	0.002	1.056 ^{b}	1.095 ^{b}	1.358 ^{a}	0.070	0.016
Oleic C18:1	36.09 ^{b}	36.80 ^{ab}	37.59 ^{a}	0.406	0.050	552.8	535.8	570.1	20.100	0.499
Erucic C22 (1n9)	0.121 ^{a}	0.120 ^{a}	0.100 ^{b}	0.004	0.002	1.854 ^{a}	1.456 ^{b}	1.819 ^{a}	0.077	0.004
Nervonic C24 (1n9)	0.277	0.279	0.272	0.006	0.640	4.246	4.063	4.128	0.161	0.724
∑MUFA	41.11	41.43	42.94	0.508	0.050	628.3	603.2	651.2	20.4	0.280
Linoleic C18:2	15.66 ^{a}	14.27 ^{b}	13.97 ^{b}	0.108	<0.0001	239.94 ^{a}	207.67 ^{b}	211.94 ^{ab}	8.320	0.031
Linolenic γ C18:3n6	0.126 ^{b}	0.134 ^{ab}	0.138 ^{a}	0.002	0.0007	1.931	1.944	2.089	0.080	0.326
Linolenic α C18:3n3	0.218 ^{b}	0.319 ^{a}	0.233 ^{b}	0.017	0.001	3.343 ^{b}	4.651 ^{a}	3.541 ^{b}	0.283	0.011
Eicosadienoic C20 (2n6)	0.229 ^{b}	0.232 ^{b}	0.269 ^{a}	0.007	0.002	3.515	3.389	4.084	0.190	0.046
Eicosatrienoic C20 (3n6)	0.290 ^{a}	0.257 ^{b}	0.289 ^{a}	0.005	<0.0001	4.451 ^{a}	3.743 ^{b}	4.393 ^{ab}	0.178	0.023
Eicosatrienoic C20 (3n3)	0.252 ^{b}	0.234 ^{b}	0.274 ^{a}	0.006	0.001	3.866 ^{ab}	3.401 ^{b}	4.154 ^{a}	0.172	0.023
Arachidonic C20 (4n6)	3.613 ^{a}	3.548 ^{a}	3.178 ^{b}	0.093	0.010	55.420	51.570	48.230	2.34	0.129
Docosatetraenoic C22 (4n6)	1.501 ^{a}	1.318 ^{b}	1.537 ^{a}	0.043	0.005	23.01 ^{a}	19.16 ^{b}	23.32 ^{a}	1.01	0.019
Docosapentaenoic C22 (5n3)	0.071 ^{b}	0.109 ^{a}	0.071 ^{b}	0.003	<0.0001	1.087 ^{b}	1.593 ^{a}	1.073 ^{b}	0.054	<0.0001
Docosahexaenoic C22 (6n3)	0.691 ^{a}	1.056 ^{b}	0.632 ^{b}	0.021	<0.0001	10.591 ^{b}	15.357 ^{a}	9.595 ^{b}	0.483	<0.0001
∑PUFA	22.65 ^{a}	21.45 ^{b}	20.59 ^{c}	0.183	<0.0001	347.2	312.5	312.4	12.30	0.102
∑Ω3	1.23 ^{b}	1.72 ^{a}	1.21 ^{b}	0.024	<0.0001	18.887 ^{b}	25.001 ^{a}	18.364 ^{b}	0.784	<0.0001
∑Ω6	21.42 ^{a}	19.76 ^{b}	19.39 ^{b}	0.168	<0.0001	328.270	294.059	287.475	11.5	0.053
∑Ω6/Ω3	17.40 ^{a}	11.51 ^{c}	16.06 ^{b}	0.228	<0.0001	266.27 ^{a}	167.44 ^{b}	243.36 ^{a}	8.59	<0.0001
Other fatty acids	0.207	0.214	0.245	0.025	0.517	3.164	3.108	3.749	0.393	0.460

Where: CON, conventional diet; CV2%, conventional diet supplemented with 2% chlorella powder; SP (2%), conventional diet supplemented with 2% spirulina powder; $n = 6$ samples per group (3 eggs/sample); SEM, standard error of the mean; ^{abc} Mean values within a row with different letters are significantly different at $p < 0.05$.

Table 8. Effects of chlorella and spirulina powder on egg yolk color in fresh and boiled eggs.

Yolk Color Parameter		L*	a*	b*
CON	fresh yolk	45.76 ^{f}	1.321 ^{e}	14.03 ^{f}
	10 min boiling time	79.94 ^{e}	−1.389 ^{g}	19.32 ^{ef}
	20 min boiling time	104.33 ^{c}	−3.836 ^{h}	23.46 ^{de}
	30 min boiling time	125.97 ^{a}	−5.514 ^{i}	25.65 ^{d}
CV2%	fresh yolk	43.82 ^{f}	3.156 ^{d}	17.74 ^{ef}
	10 min boiling time	76.56 ^{e}	0.883 ^{e}	33.14 ^{c}
	20 min boiling time	100.05 ^{cd}	0.073 ^{ef}	51.60 ^{a}
	30 min boiling time	122.15 ^{a}	−1.070 ^{fg}	48.21 ^{a}
SP2%	fresh yolk	40.49 ^{f}	4.924 ^{c}	14.18 ^{f}
	10 min boiling time	74.86 ^{e}	6.414 ^{ab}	32.16 ^{c}
	20 min boiling time	95.11 ^{d}	7.601 ^{a}	39.83 ^{b}
	30 min boiling time	116.46 ^{b}	5.591 ^{bc}	49.31 ^{a}
	Main effect			
Treatment	CON	88.99 ^{a}	−2.354 ^{c}	20.61 ^{c}
	CV2%	85.65 ^{b}	0.761 ^{b}	37.68 ^{a}
	SP2%	81.73 ^{c}	6.132 ^{a}	33.87 ^{b}
	$SEM_{treatment}$	0.583	0.151	0.653
Boiling time	fresh yolk	43.36 ^{d}	3.134 ^{a}	15.318 ^{d}
	10 min boiling time	77.12 ^{c}	1.969 ^{b}	28.205 ^{c}
	20 min boiling time	99.83 ^{b}	1.279 ^{c}	37.897 ^{b}
	30 min boiling time	121.53 ^{a}	−0.331 ^{d}	41.457 ^{a}
	$SEM_{boiling\ time}$	0.682	0.177	0.764
	p-Value			
	Treatment	<0.0001	<0.0001	<0.0001
	Boiling time	<0.0001	<0.0001	<0.0001
	Treatment × Boiling time	0.293	<0.0001	<0.0001

Where: CON, conventional diet; CV2%, conventional diet supplemented with 2% chlorella powder; SP 2%, conventional diet supplemented with 2% spirulina powder; SEM, standard error of the mean. $^{a–i}$ Mean values within a column with different letters are significantly different at $p \leq 0.05$.

4. Discussion

From a nutritional point of view, the two microalgae (chlorella and spirulina) are considered food additives with high biological value due to their nutrient concentration. The results of our study analyses strongly indicate that spirulina shows higher antioxidant properties, carotenoid levels, polyphenols, and a superior DPPH inhibition, when compared to chlorella, which suggests that spirulina has a greater capacity to combat oxidative stress. On the other hand, the microalgae proximal composition showed that chlorella had a higher concentration of PUFA, particularly omega-3 fatty acids and a higher omega-3 content, and lower $\sum$ PUFA n-6/$\sum$ PUFA n-3 ratio compared to spirulina.

Other authors confirm that the microalgae contain the highest protein value with an excellent essential amino acid profile [33,34], bioactive compounds, PUFA fatty acids, polysaccharides, volatile and phenolic compounds, vitamins, sterols, and natural pigments [35]. The high levels of carotenoids and fatty acids, especially α-linolenic, are associated with health benefits and nutrition [36]. The microalgae utilization in animal feed improves productive performance, the immune system, antioxidant activity, and tissue regeneration [35]. Other authors [37] found a concentration of 3.291 mg/L chlorophyll *a*, 1.174 mg/L chlorophyll *b*, 4.466 mg/L total chlorophyll, and 0.919 mg/L carotenoids in blue-green algae spirulina. Abou-El-Souod et al. [38] stated that chlorella possesses chloroplasts that contain green photosynthetic pigments called chlorophylls *a* and *b*. Utilizing the process of photosynthesis, it exhibits rapid growth and multiplication by utilizing carbon dioxide, water, sunlight, and a minimal amount of minerals. Similar findings to our results have been reported in other studies investigating the antioxidant activity and fatty acid composition of spirulina and chlorella. Khan et al. [39] found that spirulina exhibited significantly higher antioxidant activity compared to the control group. The presence of active compounds such as phycocyanin and beta-carotene in spirulina contributed to its strong antioxidant potential. In a study, Stunda-Zujeva et al. [40] stated that phycocyanin is the main antioxidant of spirulina, offering various uses for health benefits, although care should be taken in terms of the antioxidant activity, which fluctuates. Numerous studies have highlighted the higher antioxidant capacity and beneficial fatty acid profiles, including higher concentrations of omega-3 polyunsaturated fatty acids, in both microalgae species compared to control groups. These fatty acids are known for their beneficial effects on human health, including cardiovascular health and anti-inflammatory properties. Another study by [41] investigated the fatty acid profiles of microalgae species and found that both spirulina and chlorella exhibited higher concentrations of omega-3 fatty acids, particularly ALA, compared to the control group. They also noted that these microalgae species had lower levels of saturated fatty acids, contributing to a more desirable fatty acid composition. Other researchers [42] evaluated the fatty acid composition of spirulina and highlighted its high content of gamma-linolenic acid (GLA), an omega-6 fatty acid with anti-inflammatory properties.

Our research revealed that adding chlorella and spirulina to the laying hens' diet at a 2% inclusion rate did not have a significant impact on initial or final body weight. Nevertheless, the group supplemented with spirulina demonstrated enhanced feed conversion efficiency, larger eggs, and higher rates of egg production compared to the control and chlorella groups. This suggests that dietary supplementation with spirulina could have more pronounced positive effects on egg production efficiency and size uniformity, with practical benefits for egg producers and consumer health. These findings are similar to those of other studies which studied different microalgae sources and inclusion levels and noticed an improved production parameter when including microalgae in poultry diets due to the high protein content, essential amino acids, vitamins, and minerals present in spirulina and chlorella. Additionally, the presence of certain bioactive compounds and antioxidants in microalgae may have positive effects on production performance. Mariey et al. [43] included four levels of spirulina powder (0, 0.10, 0.15, or 0.20%) in laying hens' diet and registered an improved egg production rate, daily egg mass, and feed conversion ratio compared to those of the control group. Shanmugapriya and Saravanababu [44] tested spirulina on broilers and found a significant increase in body weight.

Other studies have [45] supplemented the basal diet of laying hens raised under a chronic hot ambient temperature with spirulina powder (0.15 mg/kg diet) and seleno-methionine (0.10 mg/kg diet). The obtained results indicated that dietary spirulina and organic selenium showed improved productive performance under heat stress. In contrast, the chlorella supplementation at varying dosages of 2.5 g, 5.0 g, or 7.5 g per kg feed, in both spray-dried and bullet-milled/spray-dried forms, did not result in any impact on laying intensity, egg weight, daily egg mass production, or feed conversion. However, it was observed that the treatment groups exhibited an increase in yolk weight and an improvement in egg quality [46].

In a study conducted by Omri et al. [47], laying hens at 44 weeks of age were fed with diets containing 1.5% and 2.5% spirulina for a period of 6 weeks. The results indicated that the inclusion of 2.5% spirulina in the diet significantly increased egg weight. However, no significant effects were observed on other productive parameters, including dietary treatment, duration of the diet, or their interaction.

Concerning the antioxidant enzyme activity, the results obtained in our study showed that chlorella and spirulina dietary addition exhibited significant improvements in blood antioxidant enzyme activities (SOD, CAT, GSH) and total antioxidant capacity (TAC). Moreover, the increased serum levels of GSH and TAC in both experimental diets demonstrate and support the idea that the microalgae-supplemented diets positively influenced the hens' antioxidant status compared to the control group. The main antioxidant enzymes, such as SOD, CAT, and GSH, protect the organism against oxidative stress [48], improving the poultry immune system [49]. CAT is one of the most important antioxidant enzymes which mitigates oxidative stress via the catalysis of hydrogen peroxide [50]. Park et al. [51] obtained the same linearly increased GPx and SOD enzymes in broilers fed with spirulina and explained that this was due to the fact that spirulina contains antioxidants such a β-carotene, tocopherol, selenium, polypeptide pigment, or phenolic acids. Wu et al. [52] suggested that spirulina has stronger antioxidant capabilities than chlorella, which is probably due to the higher content of phenolic compounds.

Utilization of microalgae in laying hens' diet had no effect concerning the apparent digestibility coefficients. Our results are similar to those of [53], who reported that the incorporation of green seaweed (*Ulwa spp.*) meal between 20 and 35 g/kg in Boschveld hens' diets did not alter apparent nutrient digestibility.

Additional research [19] indicated that the inclusion of brown seaweed meal derived from (*Ecklonia maxima*) into the diet of Boschveld cockerels did not have a significant impact on the digestibility of dry matter, organic matter, crude protein, and fiber. This result was observed despite the seaweed inclusion rate ranging from 2 to 8 g/kg.

In our experiment, we obtained a high β-carotene content and increased antioxidant capacity of the yolk, which represents indicators of an improved egg quality, with potential health-promoting effects for consumers. Omri et al. [47] observed no effect ($p > 0.05$) on total cholesterol concentration when using spirulina (1.5% and 2.5%) in laying hen diets.

The dietary microalgae supplementation had no influence on egg quality parameters (egg weight and its components). Similar results on egg weight were observed by [46] using chlorella supplementation in laying hens (26-week-old) diets. Other authors, such as [54], used chlorella supplementation in Hy-Line brown laying hens, aged 70 weeks, without any effects on egg weight, but registered the highest Haugh units when supplementing diets with 2.4% liquid chlorella in their study.

Our data indicate that the dietary treatments of chlorella and spirulina influenced the yolk coloration, spirulina having a more pronounced effect on enhancing red color. Additionally, longer boiling times result in darker and lower/more negative values for a* (greenish-gray ring) and higher/more positive values for b*.

In other studies [34,46,54], both chlorella- and spirulina-supplemented diets were confirmed to increase the color of yolk by lutein dosing.

The intensity of yolk color can vary depending on the types and concentrations of carotenoids consumed by the laying hens. Englmaierová et al. [55], using chlorella at

12.5 g/kg, noticed a significant intensification of the yellowness of fresh yolk. In the case of boiled eggs, a statistically significant increase in redness was observed. Conversely, an extension of the boiling duration to 10 min resulted in an increase in lightness and a concomitant reduction in yolk coloration.

The L* value for fresh yolk indicates that the color of the fresh yolk has a moderately bright appearance. As the boiling time increased, the L* values also increased. The L* value for the 10 min. boiling indicated that the boiled yolk became significantly brighter ($p \leq 0.0001$) compared to the fresh yolk. The L* value increased progressively for the 20 min. and 30 min. boiling times, respectively; yolks became lighter as they were boiled for longer durations. The differences in L* values between the boiling time highlight the effect of heat exposure on the lightness of the yolks. This change in lightness can be attributed to structural and chemical transformations that occur during the cooking process, causing the denaturation of the proteins and altering the protein molecules. As a result, the yolks appear brighter or lighter in color [56]. According to Muñoz-Miranda and Iñiguez-Moreno [57], marine biopigments can be categorized into three main groups: chlorophylls, carotenoids, and phycobiliproteins. The rich carotenoid concentration of the pigments zeaxanthin, xanthophylls, and β-carotene offer different greenish, green, golden, red, and brown colors of algae [58]. Other authors [59] tested, in a short-term study, the effects of 1% and 3% spirulina supplementation on color, nutritional value, and stability of yolk. A decreased luminosity and increased redness ($p = 0.0001$) and yellowness ($p = 0.0103$) were observed for 1% supplementation, after only 15 experimental days, meaning that the high carotenoid levels present in spirulina are efficiently absorbed by the laying hens' gastrointestinal tract [60].

Dietary supplementation with chlorella at 1% and 2% levels on Hisex Brown laying hens aged 56 weeks revealed a significant increase in total carotenoid deposition by 46% and 119% for the 1% and 2% chlorella groups, respectively. This increase was accompanied by a significant improvement in the yolk egg color, as evidenced by the Roche Fan Yolk Color grade, which registered 5.0 and 6.1 for the 1% and 2% chlorella groups, respectively, compared to 4% for the control group ($p < 0.001$). These findings suggest that chlorella dietary supplementation can enhance the carotenoid content and improve the color of yolks in laying hens.

Omri et al. [47] obtained increases in egg yolk redness from 1.33 (C) to 12.67 (1.5% spirulina) and 16.19 (2.5% spirulina), and a significant yellowness (b*) reduction parameter from 62.1 (C) to 58.17 (1.5% spirulina) and 55.87 (2.5% spirulina). Overall, the yolks from the experimental diets were highly significantly ($p < 0.0001$) darker, exhibited a stronger red color, and had reduced yellowness compared to the yolks from the control group. Other studies [43] tested 0.1%, 0.15%, and 0.2% spirulina in laying hens' diet and observed increasing yolk color scores (RYCF) of 6.3, 6.7, and 7.6, respectively. Similarly, supplementation levels of 1.5%, 2%, and 2.5% of spirulina were tested by [61] and obtained significant yolk intensifications of 10.55, 11.43, and 11.66, respectively, compared to the control.

The present study also investigated the effectiveness of microalgae in enhancing the fatty acid composition of eggs, specifically through the increased presence of docosahexaenoic acid (DHA). The process of enriching eggs with omega-3 polyunsaturated fatty acids (n-3 PUFA) from dietary sources is gradual and requires time. However, achieving sufficient enrichment of eggs with these beneficial fatty acids is economically significant for the industry. The n-6/n-3 PUFA ratio reflected diet composition, with the ratio being lower for the eggs of the hens fed microalgae. Some studies consider that many salt and fresh-water microalgae, including spirulina, contain high concentrations of n3-long-chain polyunsaturated fatty acids (PUFA) (25–38%), including α-linoleic acid (ALA), eicosapentaenoic acid (EPA), and docosahexaenoic acid (DHA), which are anti-inflammatory and cardiovascular- and brain-protective [39,62]. Microalgae, due to their high concentrations of n-3 PUFA, present an exceptional n-6/n-3 PUFA ratio [39,62]. Studies have demonstrated that laying hens fed with microalgae-enriched diets produce DHA-enriched eggs [63–65].

5. Conclusions

In conclusion, the supplementation of laying hens' diet with chlorella and spirulina at a concentration of 2% each has demonstrated several positive effects on egg quality and nutritional content. This study has revealed significant improvements in egg weight, size, yolk intensity color, beta carotene content, and antioxidant capacity. Furthermore, the incorporation of chlorella has led to a noteworthy increase in omega-3 polyunsaturated fatty acids, resulting in a substantial reduction in the omega-6/omega-3 ratio. As a lower omega-6/omega-3 ratio is widely recognized for its potential benefits to human health and overall well-being, these findings have important implications beyond poultry production. While these results showcase the promising potential of chlorella and spirulina as valuable dietary supplements for laying hens, it is important to acknowledge that further research is necessary to comprehensively evaluate their capabilities in partially substituting costly protein sources in laying hens' diets. By doing so, these microalgae could contribute to more sustainable and economically viable feed ingredients within poultry production systems.

Author Contributions: Conceptualization, T.D.P.; methodology, T.D.P., G.M.C., N.C.P., A.C., R.P.T., M.S. and C.S.; software, T.D.P. and G.M.C.; validation, T.D.P.; formal analysis, T.D.P., N.C.P. and A.C.; investigation, T.D.P. and G.M.C.; resources, T.D.P.; data curation, T.D.P., G.M.C. and A.C.; writing—original draft preparation, T.D.P.; writing—review and editing, N.C.P. and G.M.C.; project administration, T.D.P.; funding acquisition, T.D.P. and G.M.C. All authors have read and agreed to the published version of the manuscript.

Funding: This research was funded by the Ministry of Research, Innovation and Digitalization, Project PN 23-20 0101, and the National Research Development Project to Finance Excellence, Project 8PFE/2021.

Institutional Review Board Statement: All experimental procedures of the study were performed in accordance with Directive 2010/63/EU on the protection of animals used for scientific purposes and experimental procedures, according to an experimental protocol (No. 6252/27.10.2021) approved by the Research Ethics Commission of the National Research and Development Institute for Animal Biology and Nutrition.

Informed Consent Statement: Not applicable.

Data Availability Statement: All data are contained within the article.

Conflicts of Interest: The authors declare no conflict of interest.

Abbreviations

∑ PUFA n-3	sum of polyunsaturated fatty acids with omega-3 double bond
∑ PUFA n-6	sum of polyunsaturated fatty acids with omega-6 double bond
a*	red-green intensity (in egg color determination)
ALA	alpha-linolenic acid
b*	yellow-blue intensity (in egg color determination)
C*a*	chlorophyll *a*
CAT	catalase
C*b*	chlorophyll *b*
Cc	total carotenoids
CF	crude fiber
CON	conventional diet
CP	crude protein
CV2%	diet with 2% *Chlorella vulgaris*
DCP	in vitro digestibility of protein
DDM	in vitro digestibility of dry matter
DEE	in vitro digestibility of ether extract
DF	dilution factor
DFI	daily feed intake (g/day/layer)
DHA	docosahexaenoic acid

DM	dry matter
DNFE	in vitro digestibility of non-fermentable extractive substance
DOM	in vitro digestibility of organic matter
DPPH	2,2-diphenyl-1-picrylhydrazyl
DTNB	5,5′-dithio-bis-[2-nitrobenzoic acid] (Ellman's reagent)
EDTA	ethylendiaminotetraacetic acid
EE	ether extract
EPA	eicosapentaenoic acid
EW	egg weight (g)
FAME	fatty acid methyl esters (for fatty acid chromatography)
FCR	feed conversion ratio (g feed/g egg)
GAE	gallic acid equivalents (for polyphenols)
GLA	gamma-linolenic acid
GSH	reduced glutathione (peroxidase)
HDEP	hen day egg production (%)
L*	lightness (in egg color determination)
ME	metabolizable energy
MUFA	total monounsaturated fatty acids
NBT	nitro blue tetrazolium
NFE	non-fermentable extractive substance
OM	organic matter
PUFA	total polyunsaturated fatty acids
SFA	total saturated fatty acids
SOD	superoxide dismutase
SP2%	diet with 2% *Spirulina platensis*
TAC	total antioxidant capacity
TNB	5-thio-2-nitrobenzoic acid

References

1. Bryden, W.L.; Li, X.; Ruhnke, I.; Zhang, D.; Shini, S. Nutrition, feeding and laying hen welfare. *Anim. Prod. Sci.* **2021**, *61*, 893–914. [CrossRef]
2. Wang, X.; Zhang, H.; Wang, H.; Wang, J.; Wu, S.; Qi, G. Effect of dietary protein sources on production performance, egg quality, and plasma parameters of laying hens. *Asian-Australas. J. Anim. Sci.* **2017**, *30*, 400–409. [CrossRef] [PubMed]
3. Kudełka, W.; Kowalska, M.; Popis, M. Quality of Soybean Products in Terms of Essential Amino Acids Composition. *Molecules* **2021**, *26*, 5071. [CrossRef]
4. Qin, P.; Wang, T.; Luo, Y. A review on plant-based proteins from soybean: Health benefits and soy product development. *J. Agric. Food Res.* **2022**, *7*, 100265. [CrossRef]
5. Alshelmani, M.I.; Abdalla, E.A.; Kaka, U.; Basit, M.A. Nontraditional feedstuffs as an alternative in poultry feed. In *Advances in Poultry Nutrition Research*; IntechOpen: London, UK, 2021.
6. Wang, Y.; Tibbetts, S.M.; McGinn, P.J. Microalgae as Sources of High-Quality Protein for Human Food and Protein Supplements. *Foods* **2021**, *10*, 3002. [CrossRef] [PubMed]
7. Katiyar, R.; Arora, A. Health promoting functional lipids from microalgae pool: A review. *Algal Res.* **2020**, *46*, 101800. [CrossRef]
8. Herrera, A.; D'imporzano, G.; Fernandez, F.G.A.; Adani, F. Sustainable production of microalgae in raceways: Nutrients and water management as key factors influencing environmental impacts. *J. Clean. Prod.* **2021**, *287*, 125005. [CrossRef]
9. Zhou, L.; Li, K.; Duan, X.; Hill, D.; Barrow, C.; Dunshea, F.; Martin, G.; Suleria, H. Bioactive compounds in microalgae and their potential health benefits. *Food Biosci.* **2022**, *49*, 101932. [CrossRef]
10. Riccio, G.; Lauritano, C. Microalgae with Immunomodulatory Activities. *Mar. Drugs* **2019**, *18*, 2. [CrossRef]
11. Coelho, D.F.M.; Alfaia, C.M.R.P.M.; Assunção, J.M.P.; Costa, M.; Pinto, R.M.A.; de Andrade Fontes, C.M.G.; Lordelo, M.M.; Prates, J.A.M. Impact of dietary Chlorella vulgaris and carbohydrate-active enzymes incorporation on plasma metabolites and liver lipid composition of broilers. *BMC Veter-Res.* **2021**, *17*, 229. [CrossRef]
12. Zheng, L.; Oh, S.T.; Jeon, J.Y.; Moon, B.H.; Kwon, H.S.; Lim, S.U.; Kang, C.W. The dietary effects of fermented Chlorella vulgaris (CBT®) on production performance, liver lipids and intestinal microflora in laying hens. *Asian-Australas. J. Anim. Sci.* **2012**, *25*, 261. [CrossRef] [PubMed]
13. Hyrslova, I.; Krausova, G.; Smolova, J.; Stankova, B.; Branyik, T.; Malinska, H.; Huttl, M.; Kana, A.; Doskocil, I.; Curda, L. Prebiotic and Immunomodulatory Properties of the Microalga *Chlorella vulgaris* and Its Synergistic Triglyceride-Lowering Effect with Bifidobacteria. *Fermentation* **2021**, *7*, 125. [CrossRef]
14. Matos, J.; Cardoso, C.L.; Falé, P.; Afonso, C.M.; Bandarra, N.M. Investigation of nutraceutical potential of the microalgae Chlorella vulgaris and Arthrospira platensis. *Int. J. Food Sci. Technol.* **2020**, *55*, 303–312. [CrossRef]

15. El-Shall, N.A.; Jiang, S.; Farag, M.R.; Azzam, M.; Al-Abdullatif, A.A.; Alhotan, R.; Dhama, K.; Hassan, F.-U.; Alagawany, M. Potential of Spirulina platensis as a feed supplement for poultry to enhance growth performance and immune modulation. *Front. Immunol.* **2023**, *14*, 1072787. [CrossRef]
16. Altmann, B.A.; Rosenau, S. Spirulina as Animal Feed: Opportunities and Challenges. *Foods* **2022**, *11*, 965. [CrossRef]
17. Smetana, S.; Sandmann, M.; Rohn, S.; Pleissner, D.; Heinz, V. Autotrophic and heterotrophic microalgae and cyanobacteria cultivation for food and feed: Life cycle assessment. *Bioresour. Technol.* **2017**, *245*, 162–170. [CrossRef]
18. Carpenter, K.J.; Clegg, K.M. The metabolizable energy of poultry feeding stuffs in relation to their chemical composition. *J. Sci. Food Agric.* **1956**, *7*, 45–51. [CrossRef]
19. Mhlongo, G.; Mnisi, C. Effect of seaweed (*Ecklonia maxima*) on apparent nutrient digestibility, growth performance, and physiological and meat quality parameters in Boschveld cockerels. *Poult. Sci.* **2023**, *102*, 102361. [CrossRef]
20. National Research Council. *Nutrient Requirements of Poultry*, 9th ed.; National Academic Press: Washington, DC, USA, 1994.
21. Panaite, T.D.; Nour, V.; Saracila, M.; Turcu, R.P.; Untea, A.E.; Vlaicu, P.A. Effects of linseed meal and carotenoids from different sources on egg characteristics, yolk fatty acid and carotenoid profile and lipid peroxidation. *Foods* **2021**, *10*, 1246. [CrossRef]
22. Boisen, S.; Fernandez, J.A. Prediction of the apparent ileal digestibility of protein and amino acids in feedstuffs and feed mix-tures for pigs by in vitro analyses. *Anim. Feed Sci. Technol.* **1995**, *51*, 29–43. [CrossRef]
23. Panaite, T.D.; Vlaicu, P.A.; Saracila, M.; Cismileanu, A.; Varzaru, I.; Voicu, S.N.; Hermenean, A. Impact of watermelon rind and sea buckthorn meal on performance, blood parameters, and gut microbiota and morphology in laying hens. *Agriculture* **2022**, *12*, 177. [CrossRef]
24. Arslan, R.; Eroglu, E.C.; Aksay, S. Determination of bioactive properties of protein and pigments obtained from *Spirulina platensis*. *J. Food Process. Preserv.* **2021**, *45*, e15150. [CrossRef]
25. Beyer, W.F., Jr.; Fridovich, I. Assaying for superoxide dismutase activity: Some large consequences of minor changes in conditions. *Anal. Biochem.* **1987**, *161*, 559–566. [CrossRef]
26. Aebi, H. Catalase in vitro. *Methods Enzymol.* **1984**, *105*, 121–126. [CrossRef] [PubMed]
27. Rahman, I.; Kode, A.; Biswas, S.K. Assay for quantitative determination of glutathione and glutathione disulfide levels using enzymatic recycling method. *Nat. Protoc.* **2006**, *1*, 3159–3165. [CrossRef]
28. Chrzczanowicz, J.; Gawron, A.; Zwolinska, A.; de Graft-Johnson, J.; Krajewski, W.; Krol, M.; Markowski, J.; Kostka, T.; Nowak, D. Simple method for determining human serum 2,2-diphenyl-1-picryl-hydrazyl (DPPH) radical scavenging activity—Possible application in clinical studies on dietary antioxidants. *Clin. Chem. Lab. Med.* **2008**, *46*, 342–349. [CrossRef]
29. AOAC. AOAC final action method 17.002, colour of eggs. *J. Assoc. Off. Anal. Chem.* **1973**, *56*, 272.
30. Untea, A.; Lupu, A.; Saracila, M.; Panaite, T. Comparison of ABTS, DPPH, Phosphomolybdenum Assays for Estimating An-tioxidant Activity and Phenolic Compounds in Five Different Plant Extracts. *Bull. UASVM Anim. Sci. Biotech-Nologies* **2018**, *75*, 2.
31. Untea, A.E.; Varzaru, I.; Panaite, T.D.; Gavris, T.; Lupu, A.; Ropota, M. The Effects of Dietary Inclusion of Bilberry and Walnut Leaves in Laying Hens' Diets on the Antioxidant Properties of Eggs. *Animals* **2020**, *10*, 191. [CrossRef]
32. Panaite, T.D.; Mironeasa, S.; Iuga, M.; Vlaicu, P.A. Liquid egg products characterization during storage as a response of novel phyto-additives added in hens diet. *Emir. J. Food Agric.* **2019**, *31*, 304–314. [CrossRef]
33. Kotrbáček, V.; Skřivan, M.; Kopecký, J.; Pěnkava, O.; Hudečková, P.; Uhríková, I.; Doubek, J. Retention of carotenoids in egg yolks of laying hens supplemented with heterotrophic Chlorella. *Czech J. Anim. Sci.* **2013**, *58*, 193–200. [CrossRef]
34. An, B.K.; Kim, K.E.; Jeon, J.Y.; Lee, K.W. Effect of dried Chlorella vulgaris and Chlorella growth factor on growth perfor-mance, meat qualities and humoral immune responses in broiler chickens. *Springerplus* **2016**, *5*, 718. [CrossRef] [PubMed]
35. Andrade, L.M.; Andrade, C.J.; Dias, M.; Nascimento, C.A.O.; Mendes, M.A. Chlorella and spirulina microalgae as sources of functional foods, nutraceuticals, and food supplements; an overview. *MOJ Food Process. Technol.* **2018**, *6*, 00144. [CrossRef]
36. Kang, H.K.; Salim, H.M.; Akter, N.; Kim, D.W.; Kim, J.H.; Bang, H.T.; Kim, M.J.; Na, J.C.; Hwangbo, J.; Choi, H.C.; et al. Effect of various forms of dietary Chlorella supplementation on growth performance, immune characteristics, and intestinal microflora population of broiler chickens. *J. Appl. Poult. Res.* **2013**, *22*, 100–108. [CrossRef]
37. Abdul-Adel, E.; Saleh, M.M.; Salman, J.M. Biodiesel production by Spirulina platensis under different NaCl concen-trations. *Biochem. Cell. Arch.* **2020**, *20*, 1745–1749.
38. Abou-El-Souod, G.W.; Hassan, L.H.; Morsy, E.M. Comparison of different media formulations and the optimal growing conditions on growth, morphology and chlorophyll content of green alga, chlorella vulgaris. *J. Am. Sci.* **2016**, *12*, 86–95.
39. Khan, F.; Shuvo, A.A.S.; Islam, K.M.S. Effects of dietary inclusion of Spirulina platensis on egg yolk pigmentation. *Live Res. Rural. Dev.* **2021**, *33*, 101.
40. Stunda-Zujeva, A.; Berele, M.; Lece, A.; Šķesters, A. Comparison of antioxidant activity in various spirulina containing products and factors affecting it. *Sci. Rep.* **2023**, *13*, 4529. [CrossRef]
41. Chen, B.; Wan, C.; Mehmood, M.A.; Chang, J.-S.; Bai, F.; Zhao, X. Manipulating environmental stresses and stress tolerance of microalgae for enhanced production of lipids and value-added products—A review. *Bioresour. Technol.* **2017**, *244*, 1198–1206. [CrossRef]
42. Torres-Durán, P.V.; Ferreira-Hermosillo, A.; Ramos-Jiménez, A.; Hernández-Torres, R.P.; Juárez-Oropeza, M.A. Effect of Spirulina maxima on postprandial lipemia in young runners: A preliminary report. *J. Med. Food* **2012**, *15*, 753–757. [CrossRef]

43. Mariey, Y.A.; Samak, H.R.; Ibrahem, M.A. Effect of using Spirulina platensis algae as a feed additive for poultry diets: 1-productive and reproductive performances oflocal spirulina algae- organic selenium-egg production-heat stress- laying hens. *Egypt. Poult. Sci.* **2015**, *32*, 201–215.
44. Shanmugapriya, B.; Saravanababu, S. Supplementary effect of Spirulinaplatensison performance, hematology andcarcass yield of broiler chickens. *Indian Streams Res. J.* **2014**, *4*, 1–7.
45. Abd El Wahab, E.; Abaza, I.M.; Saad, A. Effect of dietary supplementation of Spirulina Platensis and organic selenium on some biological traits of local laying hen under heat stress conditions. *Egypt. Poult. Sci. J.* **2019**, *39*, 207–217. [CrossRef]
46. Halle, P.; Janczyk; Freyer, G.; Souffrant, W.B. Effect of microalgae chlorella vulgaris on laying hen performance. *Arch. Zootech.* **2009**, *12*, 5–13.
47. Omri, B.; Amraoui, M.; Tarek, A.; Lucarini, M.; Durazzo, A.; Cicero, N.; Santini, A.; Kamoun, M. Arthrospira platensis (spirulina) supplementation on laying hens' performance: Eggs physical, chemical, and sensorial qualities. *Foods* **2019**, *8*, 386. [CrossRef]
48. Mirzaie, S.; Sharifi, S.D.; Zirak-Khattab, F. The effect of a Chlorella by-product dietary supplement on immune response, antioxidant status, and intestinal mucosal morphology of broiler chickens. *J. Appl. Phycol.* **2020**, *32*, 1771–1777. [CrossRef]
49. Hosseini-Vashan, S.J.; Golian, A.; Yaghobfar, A.; Zarban, A.; Afzali, N.; Esmaeilinasab, P. Antioxidant status, immune system, blood metabolites and carcass characteristic of broiler chickens fed turmeric rhizome powder under heat stress. *Afr. J. Biotechnol.* **2012**, *11*, 16118–16125. [CrossRef]
50. Nandi, A.; Yan, L.-J.; Jana, C.K.; Das, N. Role of Catalase in Oxidative Stress- and Age-Associated Degenerative Diseases. *Oxidative Med. Cell. Longev.* **2019**, *2019*, 9613090 . [CrossRef]
51. Park, J.H.; Lee, S.I.; Kim, I.H. Effect of dietary Spirulina (Arthrospira) platensis on the growth performance, antioxidant enzyme activity, nutrient digestibility, cecal microflora, excreta noxious gas emission, and breast meat quality of broiler chickens. *Poult. Sci.* **2018**, *97*, 2451–2459. [CrossRef]
52. Wu, L.C.; Ho, J.A.A.; Shieh, M.C.; Lu, I.W. Antioxidant and antiproliferative activities of Spirulina and Chlorella water ex-tracts. *J. Agric. Food Chem.* **2005**, *53*, 4207–4212. [CrossRef]
53. Nhlane, L.T.; Mnisi, C.M.; Madibana, M.J.; Mlambo, V. Nutrient digestibility, growth performance and blood indices of Boschveld chickens fed seaweed-containing diets. *Animals* **2020**, *10*, 1296. [CrossRef] [PubMed]
54. Jeon, J.-Y.; Kim, K.-E.; Im, H.-J.; Oh, S.-T.; Lim, S.-U.; Kwon, H.-S.; Moon, B.-H.; Kim, J.-M.; An, B.-K.; Kang, C.-W. The production of lutein-enriched eggs with dietary chlorella. *Korean J. Food Sci. Anim. Resour.* **2012**, *32*, 13–17. [CrossRef]
55. Englmaierová, M.; Skřivan, M.; Bubancová, I. A comparison of lutein, spray-dried Chlorella, and synthetic carotenoids effects on yolk colour, oxidative stability, and reproductive performance of laying hens. *Czech J. Anim. Sci.* **2013**, *58*, 412–419. [CrossRef]
56. Llave, Y.; Fukuda, S.; Fukuoka, M.; Shibata-Ishiwatari, N.; Sakai, N. Analysis of color changes in chicken egg yolks and whites based on degree of thermal protein denaturation during ohmic heating and water bath treatment. *J. Food Eng.* **2017**, *222*, 151–161. [CrossRef]
57. Muñoz-Miranda, L.A.; Iñiguez-Moreno, M. An extensive review of marine pigments: Sources, biotechnological applications, and sustainability. *Aquat. Sci.* **2023**, *85*, 68. [CrossRef]
58. García, J.L.; de Vicente, M.; Galán, B. Microalgae, old sustainable food and fashion nutraceuticals. *Microb. Biotechnol.* **2017**, *10*, 1017–1024. [CrossRef]
59. Rey, A.I.; De-Cara, A.; Rebolé, A.; Arija, I. Short-Term Spirulina (*Spirulina platensis*) Supplementation and Laying Hen Strain Effects on Eggs' Lipid Profile and Stability. *Animals* **2021**, *29*, 1944. [CrossRef]
60. Yunitasari, F.; Jayanegara, A.; Ulupi, N. Performance, Egg Quality, and Immunity of Laying Hens due to Natural Carotenoid Supplementation: A Meta-Analysis. *Food Sci. Anim. Resour.* **2023**, *43*, 282–304. [CrossRef]
61. Zahroojian, N.; Moravej, H.; Shivazad, M. Effects of dietary marine algae (*Spirulina platensis*) on egg quality and production performance of laying hens. *J. Agr. Sci. Technol.* **2013**, *15*, 1353–1360.
62. Kumar, R.S.P.; Sibi, G. Spirulina as poultry feed supplement to enhance nutritional value of chicken meat and eggs: A review. *Int. J. Microbiol. Res.* **2020**, *11*, 67–71.
63. Bruneel, C.; Lemahieu, C.; Fraeye, I.; Ryckebosch, E.; Muylaert, K.; Buyse, J.; Foubert, I. Impact of microalgal feed supplementation on omega-3 fatty acid enrichment of hen eggs. *J. Funct. Foods* **2013**, *5*, 897–904. [CrossRef]
64. Neijat, M.; Ojekudo, O.; House, J. Effect of flaxseed oil and microalgae DHA on the production performance, fatty acids and total lipids of egg yolk and plasma in laying hens. *Prostaglandins Leukot. Essent. Fat. Acids* **2016**, *115*, 77–88. [CrossRef] [PubMed]
65. Mens, A.J.W.; van Krimpen, M.M.; Kar, S.K.; Guiscafre, F.J.; Sijtsma, L. Enriching table eggs with n-3 polyunsaturated fatty acids through dietary supplementation with the phototrophically grown green algae Nannochloropsislimnetica: Effects of microalgae on nutrient retention, performance, egg characteristics and health parameters. *Poult Sci.* **2022**, *101*, 101869. [PubMed]

applied sciences

Article

Genetic Diversity and Physicochemical Characteristics of Different Wheat Species (*Triticum aestivum* L., *Triticum monococcum* L., *Triticum spelta* L.) Cultivated in Romania

Camelia Maria Golea [1], Paula-Maria Galan [2,3], Livia-Ioana Leti [2,3] and Georgiana Gabriela Codină [1,*]

1 Faculty of Food Engineering, "Ştefan cel Mare" University, 720229 Suceava, Romania
2 Vegetal Genetic Resources Bank "Mihai Cristea", 720224 Suceava, Romania
3 Faculty of Biology, "Alexandru Ioan Cuza" University, 700505 Iasi, Romania
* Correspondence: codina@fia.usv.ro; Tel.: +40-7-4546-0727

Abstract: Thirty-one varieties of wheat cultivated in Romania were analyzed regarding the genetic diversity and physicochemical properties, including the following determinations: moisture, ash, protein, wet gluten, lipid, starch content, falling number and damaged starch, considering the particularity of each species, its biological status and origin. The physicochemical data showed that the wheat samples presented large variability. The physicochemical properties of wheat flour were assessed by multivariable data analysis, using principal component analysis (PCA). All wheat samples clustered together according to their physicochemical data showed an association between all wheat species. The protein and ash contents were more related to the ancient wheat species, while the amounts of starch and damaged starch were associated with the modern ones. Positive correlations were obtained between protein and wet gluten content and between lipid and ash content. ISSR markers were used to analyze and compare genetic diversity among selected wheat cultivars. The obtained data were analyzed using NTSYSpc software considering the coefficients of similarity (Jaccard) and dissimilarity (Neighbor joining). The Jaccard coefficients varied from 0.53 to 1, reflecting the high genetic diversity characteristic of all wheat varieties.

Keywords: physicochemical characteristics; genetic similarity and dissimilarity coefficients; ISSR markers; *Triticum*

1. Introduction

The continuous growth of the world population highlights one of the most delicate problems of our century: providing food for mankind. Wheat is one of the most important cereal crops in the world, originally cultivated approximately 12,000 years ago in the Fertile Crescent [1]. In a course of history, about 8000 years ago, wheat became the main food source of most communities, so today it is widely cultivated around the world, covering 17% of the world's cultivated area and feeding 40% of the world's population [2]. Nowadays, wheat is a crop with multiple and diverse genetic resources available because of the general evolution of people and particularly the evolution of the agricultural segment, as there are currently many species of wheat cultivated by thousands of populations all over the world. Known functional genetic studies regarding the wheat crop are mostly focused on identifying natural variations, therefore contributing to the assembly and enrichment of wheat genetic stock [3,4]. By implementing different breeding programs, cultivators aim to obtain wheat crops with an increased agronomic performance, characterized by an improved yield, better quality and increased resistance to disease [4]. Sustainable wheat cultivation involves the conservation of species and the evaluation and characterization of genetic resources, as genetic diversity can nowadays be considered the key for many agricultural challenges.

Romania, due to its geographical location and temperate–continental climate, offers favorable conditions for wheat cultivation. Romanian wheat culture has been known since ancient times, from the Upper Neolithic and the Bronze Age [5]. Due to the food-friendly chemical composition, consisting of proteins, carbohydrates, lipids, minerals and vitamins but also its bread-making properties, wheat is used in human nutrition in the form of bread, pasta, biscuits and pastries [6]. As these products are very common for most countries globally, wheat occupies the first place in worldwide cereal production. The main wheat producers are considered to be European Union countries, China, India, Russia, USA, Canada, Ukraine and Pakistan [7]. In the EU, Romania achieved the 4th rank in wheat production after France, Germany and Poland [6], being one of the most important producers from Europe and the world. Among cereals, it occupies one of the largest areas of the total arable land, being of great economic importance in Romania and contributing to the country's food security [8]. In 2000, the wheat production in Romania was about 4.4 million tons per year. In 2020, this amount increased to 52.27%, and it is estimated to grow even more by 2030, up to 138%, which is 16 million tons of wheat annually [9]. The milling and bakery sectors depend on wheat production and grain quality. Romanian wheat varieties, such as Izvor, Glosa, Miranda, Pitar, Putna and Arieșan, must be adapted to the climatic conditions and soils on which they are cultivated to produce grains of as high of a quality as possible [10].

In order to describe the genetic diversity within and between populations or groups of individuals, molecular markers are used. Molecular markers are capable of detecting high levels of polymorphism [11]. These markers are provided by the wheat seed protein complex, which is mostly associated with the bakery quality of a product [12]. Gliadin, as an important protein compound, plays a primordial role in creating the gluten network, which is decisive in terms of dough rheological properties, especially when it comes to its elasticity and viscosity [13]. Gliadin is mainly responsible for dough viscosity, whereas glutenin, another important gluten protein, is responsible for dough elasticity [14]. Although there are other grains that contain gluten-type proteins, wheat is the only grain that can form dough when mixed. Chemically, wheat flour contains gluten proteins in proportions of 75–85% of the total protein content and there are also more proteins in the wheat endosperm [15]. Gliadin and glutenine have a major impact on the bread-making process and therefore on the technological properties and bread quality [14]. During the mixing process, gluten-type proteins influence dough rheological properties, such as strength, extensibility, elasticity and consistency [14,16]. However, during fermentation, they influence the porosity of the bakery products by keeping the resulting gases in the dough [17]. Moreover, during baking, gluten plays an important role in forming the shape and volume of the bread. They also contribute to the formation of flavoring substances, increase the shelf life of the bread and affect the crust color [17,18]. Wheat grains contain 8–20% of proteins, which is an average content of 11–12% [19]. In addition, wheat grains are also a rich source of carbohydrates, mainly related to the starch molecules but also to other compounds, such as lipids, minerals, etc. [1]. Starch has important properties for bread-making. It plays a significant role during the fermentation process, where it is hydrolyzed by the action of amylolytic enzymes, forming maltose, the main fermentable carbohydrate presented in the dough [20]. Maltose forms during the enzymatic hydrolysis of starch grains and participates in the formation of crust color and flavor substances in bread [21]. In addition, starch has an essential role in the baking process due to its ability to gelatinize, as well as in the bread retrogradation process [17]. When mixing the dough, the starch also hydrates the dough; this important role is attributed to the mechanically damaged starch granules. The higher the degree of mechanical damage, the more the amylolytic hydrolysis of starch increases [22]. The most commonly used methods used to measure wheat flour damaged starch granules and the activity of α-amylase are SDmatic and Hagberg Falling Number methods, which have been used in this study [23,24]. In addition to these, the evaluation of fat, ash and moisture content of wheat flours gives us complete information regarding their quality for bread-making purposes [25]. Wheat flour consists of water, which represents

13–15% of its mass and dry matter [26], which includes proteins, carbohydrates, lipids, mineral substances, vitamins, pigments and enzymes [27]. Although there are only small amounts of lipids, these compounds play an important technological role in bread-making because they form complexes with proteins and starch granules in the dough, influencing its rheological properties, the bread quality itself and its freshness [28]. Ash content is an indicator of the mineral substances present in wheat flour [25]. Therefore, the higher the mineral content, the better the nutritional value. Wheat chemical composition can be genetically modified, so it may have a tremendous impact on human health and diet. According to Feldman et al., there are more than 25,000 cultivated forms of wheat in the whole world, and this number may be at least twice as high as estimated [29]. According to Shewry and Hey, modern wheat cultures have a different composition than ancient ones, being lower in bioactive compounds and higher in other components such as dietary fibers. Some wheat components show high heritability due to their genetic effects on modern species, such as arabinoxylan presented in flour, alkylresorcinols in whole meal, sterols, tocols, etc. [30]. In order to obtain wheat with an optimal/average chemical composition and to overcome the variation between certain cultures due to the genetic impact, different genotypes of wheat can be blended in one [31].

The existence of reliable genetic markers is beneficial for evaluating the diversity of wheat germplasm but also for highlighting its possible temporal changes, which could be caused by selection for its quality. Thus, the following molecular DNA markers: inter-simple sequence repeat (ISSR), amplified fragment length polymorphism (AFLP), restriction fragment length polymorphism (RFLP), microsatellites and single nucleotide polymorphisms (SNP) have been developed and used to assess the relationships and levels of genetic diversity in wheat germplasm [32].

ISSR (Inter-simple sequence repeat) is a molecular biology technique discovered in 1994 by Zietkiewicz et al., which aims at DNA fingerprinting, by amplifying, using the PCR technique, repetitive regions within the nuclear and organelle genomes (chloroplasts and mitochondria) [33]. ISSR primers have a dominant character, but occasionally, they show codominance [34]. Interest in using ISSR primers is caused by the reproducibility of the results and the low costs of development and usage [35].

In this study, we evaluated the genetic diversity by utilizing ISSR markers (Inter-Simple Sequence Repeats) of 31 wheat varieties cultivated in Romania, with different origins and kept in the active collection of the "Mihai Cristea" Suceava Plant Genetic Resources Bank. The vast majority of common wheat varieties, taken in this study, of Romanian origin, such as Izvor, Glosa, Andrada and Dumbrava, or foreign ones, such as Sosthene, Amicus, Apache and Anapurna, were listed in the Romanian Official Variety Catalog, published in 2021. The varieties listed in this catalog are admitted for certification and commercialization based on tests of distinction, uniformity and stability for agronomic value and use. These wheat varieties are representative for Romania, and in particular, for the intensive wheat cultures in the northeast region, which resulted from applying breeding programs that aimed to develop wheat with certain characteristics such as higher yield, superior quality for bread-making, and resistance to biotic and abiotic stress. Additionally, the use of ancient wheat species from the gene bank collection, such as *Triticum monococcum* and *Triticum spelta*, highlights important aspects regarding ancient wheat species. To our knowledge, no other study has been reported so far on the genetic diversity of wheat varieties cultivated in Romania, one of the most important wheat producers in Europe. Moreover, the physicochemical characteristics of wheat samples of different species (*Triticum aestivum* L., *Triticum monococcum* L., *Triticum spelta* L.) cultivated in Romania have been determined in order to allow us to compare them by using multivariate principal component analysis (PCA) and to analyze the differences between wheat species. The physicochemical characteristics analyzed in this study, namely moisture, ash, protein, wet gluten, lipid, starch, falling number, and damaged starch, are the most relevant ones for the bread-making industry. This allows us to recommend the best wheat species and varieties that may be used in the future by farmers in order to obtain raw materials of a high quality for bread-making.

2. Materials and Methods

2.1. Plant Material

The genetic material consisting of thirty-one wheat samples is shown in Table 1. Fourteen samples originated from Romania, nine from France, five from Austria, one from Switzerland, one from Germany and one from Russia. The seeds of foreign varieties of wheat, cultivated in Romania, come from the northern and central regions of France, from northwestern Switzerland, from eastern Austria for *Triticum aestivum* and from the areas of northwestern Austria, southern Germany and southern Russia for *Triticum spelta*. Additionally, all 31 wheat samples belonged to three different species: twenty-five were *Triticum aestivum* L., two were *Triticum monococcum* L. and four were *Triticum spelta* L. The analyzed samples had various biological statuses: modern variety, landrace and breeding line. All wheat samples were grown in the northeastern region of Romania in 2020 by different farmers. The direct collection of genetic material from them was carried out by organizing our own exploration/collection missions in 2021, focused on local varieties, modern and old varieties, which lend themselves very well to the pedo-climatic and socio-economic conditions characteristic of this region. It was considered that this region, in which we encountered a transitional temperate continental climate, presented similar pedo-climatic conditions. Thus, local climate differences in the northeast region are due to altitude and latitude, resulting in an average annual temperature that slightly decreased from 10.41 °C in 2020 to 9.96 °C in 2021. The annual amount of precipitation was much lower in 2021 (276.9 mm) than in 2020 (493.9 mm). The favorable agricultural period for cultivating and harvesting autumn wheat in Romania is September 2020–July 2021. After harvesting, the wheat samples were subjected to a drying process until the moisture content of the seeds reached a maximum of 7% and then they were stored before being analyzed in a cold room at a variable temperature between 3 and 5 °C, for more than 12 months. Afterwards, the samples were ground using a laboratory mill 3100 (Perten Instruments, Hagersten, Sweden) and prepared for further analysis. The moisture, ash, protein, total starch, damaged starch, fat content, wet gluten and falling number values were measured.

2.2. Wheat Physiochemical Characteristics

The wheat flour physicochemical characteristics were analyzed according to the international standard methods: moisture content according to ICC 110/1, ash content according to ICC 104/1, protein content according to ICC 105/2, wet gluten content according to ICC 137/1, lipid content according to ICC 136, starch content according to AACC 76-13.01, falling number according to ICC 107/1, and damaged starch according to AACC 76-33.

2.3. Genomic DNA Isolation

Genomic DNA isolation was performed using the CTAB method of Doyle and Doyle [36] from 200 mg of seeds of each sample. After grinding to a thin powder using a laboratory mill 3100 (Perten Instruments, Hagersten, Sweden), wheat samples were mixed with 1200 μL CTAB buffer (20 g/L CTAB, 7.44 g/L $EDTA \cdot NA_2 \cdot 2H_2O$, 81.82 g/L Sodium chloride, 12.11 g/L TRIS ultrapure, PanReac Applichem, A4150,0500). The samples were incubated for 1 h at 65 °C with periodic mixing, followed by a centrifugation process at 14,000 rpm for 10 min at room temperature (RT). The supernatant was mixed with 200 μL freshly prepared chloroform: isoamyl alcohol solution (24:1) and centrifuged at 14,000 rpm for 10 min at RT. The upper phase was transferred into new tubes, mixed with 200 μL isopropyl alcohol and centrifuged at 13,500 rpm, 10 min, RT. The supernatant was carefully removed, and the pellet washed with 70% ethanol, followed by a centrifugation process at 13,500 rpm for 10 min at RT. The pellet was therefore dried for 15 min at RT and then dissolved in 200 μL nuclease-free water. The samples were kept overnight at 4 °C and stored at −20 °C for further examination.

Table 1. Identification of genotypes regarding species, accession name, origin and biological status.

Genotype Number	Scientific Name	Accession Name	Country	Biological Status
TA1	*T. aestivum* L.	Izvor	Romania	Modern variety [1]
TA2	*T. aestivum* L.	Glosa	Romania	Modern variety [1]
TA3	*T. aestivum* L.	Miranda	Romania	Modern variety [1]
TA4	*T. aestivum* L.	Andrada	Romania	Modern variety [1]
TA5	*T. aestivum* L.	Dumbrava	Romania	Modern variety [1]
TA6	*T.aestivum* L.	Aurelius	Austria	Modern variety [1]
TA7	*T. aestivum* L.	Sofru	France	Modern variety [1]
TA8	*T. aestivum* L.	Sosthene	France	Modern variety [1]
TA9	*T. aestivum* L.	Amicus	Austria	Modern variety [1]
TA10	*T. aestivum* L.	Sothys	France	Modern variety [1]
TA11	*T. aestivum* L.	Flavor	France	Modern variety [1]
TA12	*T. aestivum* L.	Solindo	France	Modern variety [1]
TA13	*T. aestivum* L.	Izalco	France	Modern variety [1]
TA14	*T. aestivum* L.	Tonnage	Austria	Modern variety [1]
TA15	*T. aestivum* L.	Sophie	France	Modern variety [1]
TA16	*T. aestivum* L.	Apache	France	Modern variety [1]
TA17	*T. aestivum* L.	Anapurna	France	Modern variety [1]
TA18	*T. aestivum* L.	Illico	Switzerland	Modern variety [1]
TA19	*T. aestivum* L.	Sf. Ilie	Romania	Modern variety [1]
TA20	*T. aestivum* L.	Lucăcești	Romania	Modern variety [1]
TA21	*T. aestivum* L.	Udești 1	Romania	Modern variety [1]
TA22	*T. aestivum* L.	Udești 2	Romania	Modern variety [1]
TA23	*T. aestivum* L.	Udești 3	Romania	Modern variety [1]
TA24	*T. aestivum* L.	Frumoasa	Romania	Modern variety [1]
TA25	*T. aestivum* L.	Tișăuți	Romania	Modern variety [1]
TM26	*T. monococcum* L.	SVGB-11842	Romania	Landrace [2]
TM27	*T.monococcum* L.	SVGB-11861	Romania	Breeding line [3]
TS28	*T. spelta* L.	Ebners Rotkorn	Austria	Modern variety [1]
TS29	*T. spelta* L.	Frankenkorn	Austria	Modern variety [1]
TS30	*T. spelta* L.	Alkoran	Russia	Modern variety [1]
TS31	*T. spelta* L.	Oberkulmer Rotkorn	Germany	Modern variety [1]

[1] Currently cultivated variety, which is distinguished by superior characteristics of quality, productivity, uniformity and stability compared to a primitive variety and is widely used as a parent in the breeding program. [2] Local variety of a plant species that has been obtained under the action of natural and/or artificial empirical selection, in specific environmental conditions and that presents a series of distinct individual characteristics in order to be associated with a specific geographical region. [3] Biological material obtained by breeders through artificial selection, based on conscious selection schemes.

2.4. Spectrophotometric Analysis of DNA

The extracted DNA was measured both quantitatively and qualitatively using ThermoScientific NanoDrop One. The quantitative determination refers to the concentration of DNA obtained in 1 μL and the qualitative analysis measures the purity (A260/A2680 ratio) of the samples.

2.5. PCR Analysis

The PCR reaction was performed using according to GoTaq G2 Green Master Mix protocol (Promega, M7822). The DNA samples were amplified using 11 ISSR markers, as shown in Table 2. The primer's sequences were identified in the literature [37] and they were synthetized at Eurogenetec, Belgium. The PCR mix was prepared in a final volume of 25 μL. Each reaction was comprised of GoTaq G2 DNA Polymerase, 2X Green GoTaq® G2 Reaction Buffer (pH 8.5), 3 mM $MgCl_2$, 400 μM each dNTPs, 0.5 μM primer and 30 ng DNA. Amplifications were performed in an Eppendorf Mastercycler under the following conditions: 94 °C for 2 min, followed by 30 cycles, each cycle consisting of three steps: (1) 94 °C for 30 s, (2) 48–58 °C for 30 s (depending on primer's Tm), (3) 72 °C for 1 min, and a final step of 72 °C for 7 min.

Table 2. ISSR primer sequence used in the PCR analysis.

Primer	Sequence (5′-3′) [1]	Tm (°C)	(%) GC
UBC841	GAGAGAGAGAGAGAGAYC	58	50
UBC843	CTCTCTCTCTCTCTCTRA	56	44.4
UBC854	TCTCTCTCTCTCTCTCRG	58	50
UBC855	ACACACACACACACACYT	56	44.4
UBC857	ACACACACACACACACYG	58	50
UBC859	TGTGTGTGTGTGTGTGRC	58	50
UBC880	GGAGAGGAGAGGAGA	48	60
UBC808	AGAGAGAGAGAGAGAGC	52	52.9
UBC810	GAGAGAGAGAGAGAGAT	50	47.1
UBC834	AGAGAGAGAGAGAGAGYT	56	44.4
UBC890	VHVGTGTGTGTGTGTGT	52	41.2

[1] A = Adenine; T = Thymine; C = Cytosine; G = Guanine; H = (A, T or C); R = (A or G); V = (A, C or G) e Y = (C or T); Tm = melting temperature.

2.6. Agarose Gel Electrophoresis

Amplified DNA fragments were separated using 2% agarose gel. For electrophoresis, the following components were used: agarose (BioRad, 1613101), TBE (0.89 M Tris-Borate, 0.02 M EDTA, pH 8.3, Lonza, BE50843), SYBR Safe DNA Gel Stain (Invitrogen, S33102), Gene Ruler 100 bp (Thermoscientific, SM#0323) and a constant power supply (PowerPac Basic, BioRad). Gels were run at 70 V for 100 min and subsequently analyzed using the GelDoc Go Imaging System, BioRad.

2.7. Statistical Analysis

The obtained data were analyzed considering the presence/absence of the amplified fragments. The presence was marked as 1 and the absence as 0. The obscure DNA fragments under UV exposure were not used in the statistical interpretation. All data were analyzed using NTSYSpc software, considering similarity (Jaccard) and dissimilarity (Neighbor joining) coefficients. To analyze the correlations between wheat samples and their physicochemical characteristics, principal component analysis (PCA) was performed using the software XLSTAT (version 2020.3.1., Addinsoft, Paris, France).

3. Results

3.1. Wheat Sample Characteristics

The physicochemical characteristics of wheat samples are shown in Table 3. The physicochemical data of each cultivar and line can be seen in the supplementary material—Table S1. All data indicated large variability among the wheat samples.

Table 3. Physicochemical characteristics of wheat flour.

Chemical Data	Minimum	Maximum	Mean	Standard Deviation	Variance
Moisture (%)	10.80	12.90	11.86	0.55	0.31
Ash (%)	1.18	2.06	1.45	0.23	0.05
Protein (%)	9.90	16.90	13.09	1.67	2.80
Wet gluten (%)	21.70	39.70	28.74	3.47	12.05
Lipid (%)	1.50	2.42	1.73	0.24	0.06
Starch (%)	53.45	64.10	58.55	5.11	26.21
Falling Number (%)	83.0	404	265.43	8.32	69.42
Damaged starch (UCDc)	1.40	19.40	9.92	3.24	10.54

The moisture content presented a maximum value of 12.90% indicating the fact that all wheat samples presented good storage stability. From the protein content point of view of which values varied between 9.90 and 16.90%, all wheat samples were good for breadmaking. However, the wet gluten content varied between 21.70 and 39.70%, indicating that some wheat samples of which wet gluten values were less than 22% were difficult to be used by bakery producers [25]. The lowest protein and gluten content were recorded for the TA8 and TA5 samples, whereas the highest was recorded for the TS 31 genotype number. The falling number values varied between 83 and 404 s, the lowest one being recorded for the TA24 genotype number and the highest one being recorded for TA4. This data means that wheat flours varied from the falling number point of view, from low, normal and high α-amylase activity [38]. The large variation of starch, damaged starch, fat and ash content may be due to many factors, such as wheat vegetation period, climatic conditions and their species [30].

3.2. DNA Amplification

DNA migration patterns characteristic of all 31 wheat samples, as a result of agarose gel electrophoresis, indicate that only 6 ISSR primers out of 11 presented clear and well-defined bands (UBC808, UBC810, UBC855, UBC857, UBC859 and UBC880). The migration pattern for UBC 808 can be seen in Figure 1, and the results obtained for the other five primers can be found in Supplementary Materials—Figures S1–S5.

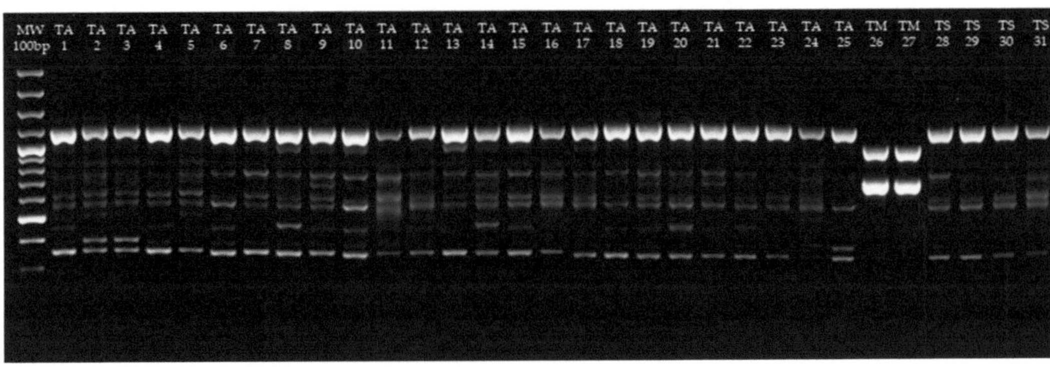

Figure 1. DNA migration pattern of the ISSR marker (UBC808) for all 31 samples of *Triticum* (2% agarose gel, migration for 130 min, 70 V).

The number of DNA bands per primer varied from 3 (for UBC859 and UBC880) to 8 (for UBC808), with a mean value of 4.83 bands/primer. Most of the primers had a number of polymorphic fragments equal to the number of amplified fragments, except UBC808, which had 7 polymorphic fragments from a total of 8. For each ISSR primer, we calculated the polymorphic information content (PIC) as proposed by Roldan-Ruiz et al. [39]. The formula was $PIC_i = 2f_i\,(1 - f_i)$, where f_i is the number of amplified fragments and $1 - f_i$ is the frequency of absent fragments. The polymorphic information content varied from 0.32 for UBC859 to 0.42 for UBC857, with a mean of 0.36. The total number of amplified fragments, fragment length, polymorphism percentage and PIC value can be found in Table 4.

The analysis of 31 wheat accessions using 6 ISSR primers showed that the mean allele number was 4.83; the minimum number of alleles was obtained for UBC859 and UBC880 (3 alleles), and the maximum number was 8 alleles for UBC808.

Considering the allele frequency, the average was 0.6, with the highest value for UBC880 (0.77) and the lowest for UBC808 (0.45). The other primers registered the following allele frequencies: UBC810—0.55, UBC855—0.6, UBC857—0.64 and UBC859—0.72. Primers with the lowest number of alleles (3 for UBC859 and UBC880) showed the highest allele frequency (0.72, respectively 0.77). Information about allele number and frequency can be found in Figure 2, and the exact values are shown in Supplementary Materials, Table S2.

Table 4. Number of amplified fragments, number of polymorphic bands and polymorphism percentage of each primer used in ISSR analysis.

Primer	Number of Amplified Fragments	Number of Polymorphic Fragments	Fragment Length (bp)	Polymorphism Percentage	Polymporphic Information Content (PIC)
UBC808	8	7	280–2250	87.5	0.34
UBC810	4	4	400–1000	100	0.39
UBC855	5	5	300–2500	100	0.37
UBC857	6	6	200–1500	100	0.42
UBC859	3	3	250–1000	100	0.32
UBC880	3	3	650–2000	100	0.33

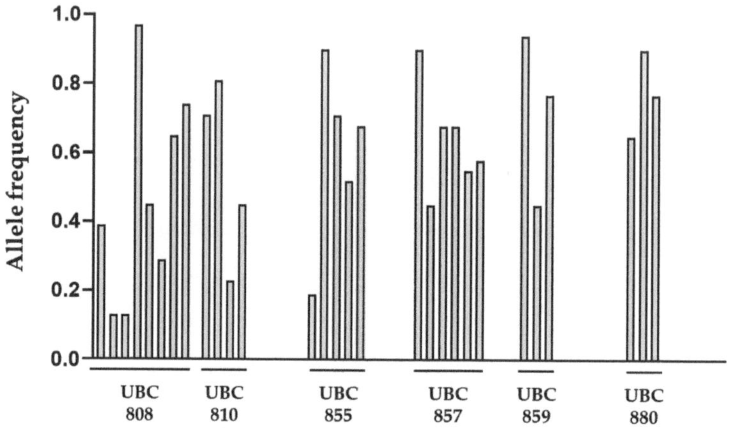

Figure 2. Allele number and frequency for all six ISSR primers.

3.3. Clustering Using the UPGMA Method

The migration pattern of each genotype was converted in a binary system where 0 means the absence of a certain DNA fragment, and 1 means the presence of it. The obtained data were analyzed in the NTSYS 2.21w statistical package (Applied Biostat LLC, Albany, NY, USA) using the UPGMA method. We calculated the Jaccard similarity coefficient, and the results can be found below.

According to the Jaccard similarity coefficient, the genotypes were divided into five clusters (C1–C5) with values between 0.53 and 1. C5 was the most extended cluster and contained 16 genotypes: 5 from Romania, 5 from France, 4 from Austria, 1 from Russia and 1 from Germany. Additionally, within C5, there was a subcluster formed by the *Triticum spelta* genotypes (TS28, TS29, TS30, TS31). C4 contained 6 genotypes, 5 from Romania and 1 from Switzerland, similar to C3, which also included 6 genotypes from different countries (2 from Romania, 3 from France and 1 from Austria). In C2, there was only one genotype from France, and in C1, there were two genotypes from Romania, as may be seen in Figure 3.

3.4. Clustering Using the Neighbor Joining Method

The Neighbor joining coefficient is related to the molecular distances between the analyzed samples and the results vary between 0 and 1, where values close to 1 show a higher genetic distance [40]. Table 5 shows the recorded values for all accessions. The most significant results were registered between samples 22 and 23 (coefficient = 0), where the genotypes showed molecular similarity. Additionally, the maximum value for genetic distance was obtained between samples 25 and 29 (coefficient = 0.247) due to a different genetic profile.

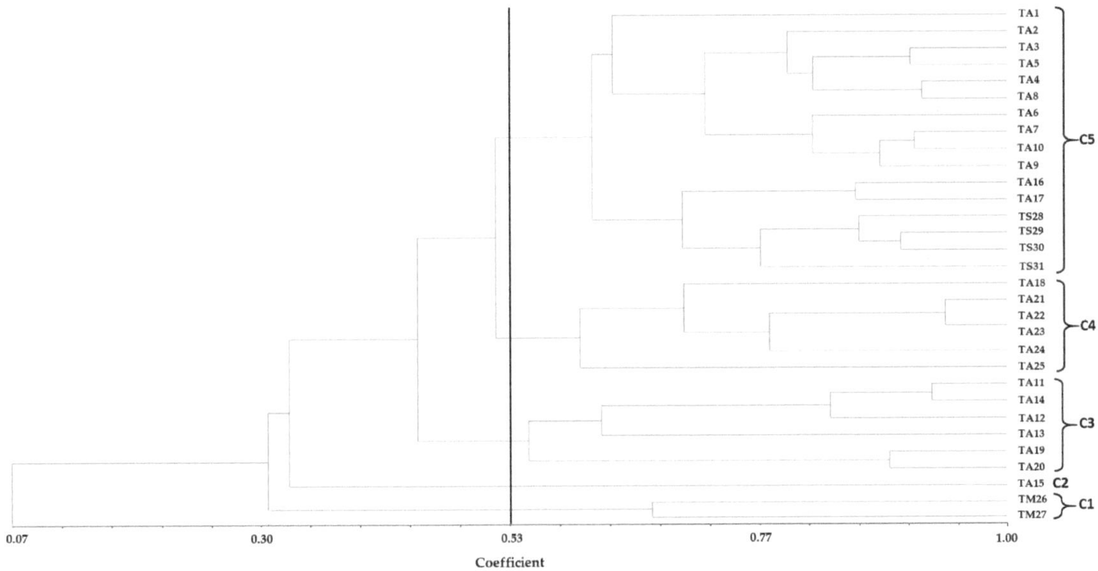

Figure 3. Dendrogram showing the genetic similarity among the 31 genotypes of *Triticum*, obtained by the UPGMA method and Jaccard coefficient, forming 5 clusters. Cut-off point of approximately 50% (dotted line).

3.5. Principal Component Analysis of the Physicochemical Characteristics and Scores of the Wheat Samples

Principal Component Analysis (Figure 4) was carried out to underline the correlations between the physicochemical characteristics of wheat samples and wheat varieties and how these varieties were influenced by their physicochemical characteristics. The first two principal components explained 54.55% of the total variance (PC1 = 19.79% and PC2 = 34.77%). The two plots underlined a good correlation between wheat varieties *Triticum spelta, Triticum monococcum* and *Triticum aestivum,* which were grouped close to each other. Generally, the ancient species *Triticum spelta* and *Triticum monococcum* were placed on the right part of the graph, whereas the *Triticum aestivum* was placed on the left. On the right part of the PCA, *Triticum monococcum* clustered together on the top, whereas *Triticum spelta* was on the bottom. However, some wheat samples of *Triticum monococcum* were placed in the same area as *Triticum monococcum* and *Triticum spelta,* indicating that some of those show some similarities with these varieties. The quality of ancient varieties was strongly characterized by the wheat physicochemical characteristics wet gluten, protein, lipid and ash content, all of them being placed alongside the PC2 component. However, *Triticum monococcum* was more related to the wet gluten and protein content, whereas the ash and lipid content of *Triticum spelta* were underlined by the first component PC1. *Triticum monococcum* was more related to the wheat physicochemical characteristics of damaged starch, falling number and starch, all of which were alongside the PC2 component. The physicochemical variable lipid content was positively correlated with the ash content ($r = 0.970$, $p < 0.05$) and protein with wet gluten ($r = 0.914$; $p < 0.05$), and there was also a positive correlation between damaged starch and falling number ($r = 0.672$; $p < 0.05$). These data are in agreement with those reported by others [25,41,42]. The position of starch in the space of main components indicated a negative correlation with wheat physicochemical characteristics lipid, fat, protein and wet gluten content. Similar data have also been reported by Golea et al. in their study [25].

Table 5. Dissimilarity Matrix among wheat samples using the NJ method and dissimilarity coefficient.

	TA1	TA2	TA3	TA4	TA5	TA6	TA7	TA8	TA9	TA10	TA11	TA12	TA13	TA14	TA15	TA16	TA17	TA18	TA19	TA20	TA21	TA22	TA23	TA24	TA25	TA26	TA27	TA28	TA29	TA30	TA31
TA1																															
TA2	0.042																														
TA3	0.026	0.016																													
TA4	0.02	0.022	0.006																												
TA5	0.004	0.038	0.022	0.016																											
TA6	0.01	0.052	0.036	0.03	0.014																										
TA7	0.033	0.009	0.007	0.013	0.029	0.043																									
TA8	0.004	0.038	0.022	0.016	0.018	0.014	0.029																								
TA9	0.064	0.022	0.038	0.044	0.06	0.074	0.031	0.06																							
TA10	0.024	0.018	0.002	0.004	0.02	0.034	0.09	0.02	0.04																						
TA11	0.016	0.058	0.042	0.036	0.02	0.006	0.049	0.02	0.08	0.04																					
TA12	0.014	0.028	0.012	0.006	0.01	0.024	0.019	0.01	0.05	0.01	0.03																				
TA13	0.056	0.098	0.082	0.076	0.06	0.046	0.089	0.06	0.12	0.08	0.04	0.07																			
TA14	0.006	0.036	0.02	0.014	0.002	0.016	0.027	0.002	0.058	0.018	0.022	0.008	0.062																		
TA15	0.036	0.078	0.062	0.056	0.04	0.026	0.069	0.04	0.1	0.06	0.02	0.05	0.02	0.042																	
TA16	0.088	0.046	0.062	0.068	0.084	0.098	0.055	0.084	0.024	0.064	0.104	0.074	0.144	0.082	0.124																
TA17	0.078	0.036	0.052	0.058	0.074	0.088	0.045	0.074	0.014	0.054	0.094	0.064	0.134	0.072	0.114	0.01															
TA18	0.069	0.027	0.043	0.049	0.065	0.079	0.036	0.065	0.005	0.045	0.085	0.055	0.125	0.063	0.105	0.019	0.009														
TA19	0.094	0.052	0.068	0.074	0.09	0.104	0.061	0.09	0.03	0.07	0.11	0.08	0.15	0.088	0.13	0.006	0.016	0.025													
TA20	0.008	0.034	0.018	0.012	0.004	0.018	0.025	0.004	0.056	0.016	0.024	0.006	0.064	0.002	0.044	0.08	0.07	0.061	0.086												
TA21	0.006	0.048	0.032	0.026	0.01	0.004	0.039	0.01	0.07	0.03	0.01	0.02	0.05	0.012	0.03	0.094	0.084	0.075	0.1	0.014											
TA22	0.009	0.033	0.017	0.011	0.005	0.019	0.024	0.005	0.055	0.015	0.025	0.005	0.065	0.003	0.045	0.079	0.069	0.06	0.085	0.001	0.015										
TA23	0.009	0.033	0.017	0.011	0.005	0.019	0.024	0.005	0.055	0.015	0.025	0.005	0.065	0.003	0.045	0.079	0.069	0.06	0.085	0.001	0.015	0									
TA24	0.001	0.031	0.015	0.009	0.007	0.021	0.022	0.007	0.053	0.013	0.027	0.003	0.067	0.005	0.047	0.077	0.067	0.058	0.083	0.003	0.017	0.002	0.002								
TA25	0.078	0.12	0.104	0.098	0.082	0.068	0.111	0.082	0.142	0.102	0.062	0.092	0.022	0.084	0.042	0.166	0.156	0.147	0.172	0.086	0.072	0.087	0.087	0.089							
TA26	0.074	0.116	0.1	0.094	0.078	0.064	0.107	0.078	0.138	0.098	0.058	0.088	0.018	0.08	0.038	0.162	0.152	0.143	0.168	0.082	0.068	0.083	0.083	0.085	0.004						
TA27	0.066	0.108	0.092	0.086	0.07	0.056	0.099	0.07	0.13	0.09	0.05	0.08	0.01	0.072	0.03	0.154	0.144	0.135	0.16	0.074	0.06	0.075	0.075	0.077	0.012	0.008					
TA28	0.122	0.08	0.096	0.102	0.118	0.132	0.089	0.118	0.058	0.098	0.138	0.108	0.178	0.116	0.158	0.034	0.044	0.053	0.028	0.114	0.128	0.113	0.113	0.111	0.2	0.196	0.188				
TA29	0.169	0.127	0.143	0.149	0.165	0.179	0.136	0.165	0.105	0.145	0.185	0.155	0.225	0.163	0.205	0.081	0.091	0.1	0.075	0.161	0.175	0.16	0.16	0.158	0.247	0.243	0.235	0.047			
TA30	0.104	0.68	0.084	0.1	0.104	0.11	0.07	0.098	0.075	0.078	0.12	0.1	0.159	0.101	0.18	0.02	0.026	0.035	0.061	0.1	0.109	0.104	0.104	0.095	0.182	0.178	0.161	0.06	0.052		
TA31	0.13	0.078	0.11	0.079	0.13	0.136	0.096	0.124	0.101	0.104	0.146	0.126	0.185	0.127	0.206	0.046	0.052	0.061	0.087	0.126	0.135	0.13	0.13	0.121	0.208	0.204	0.187	0.006	0.026	0.026	

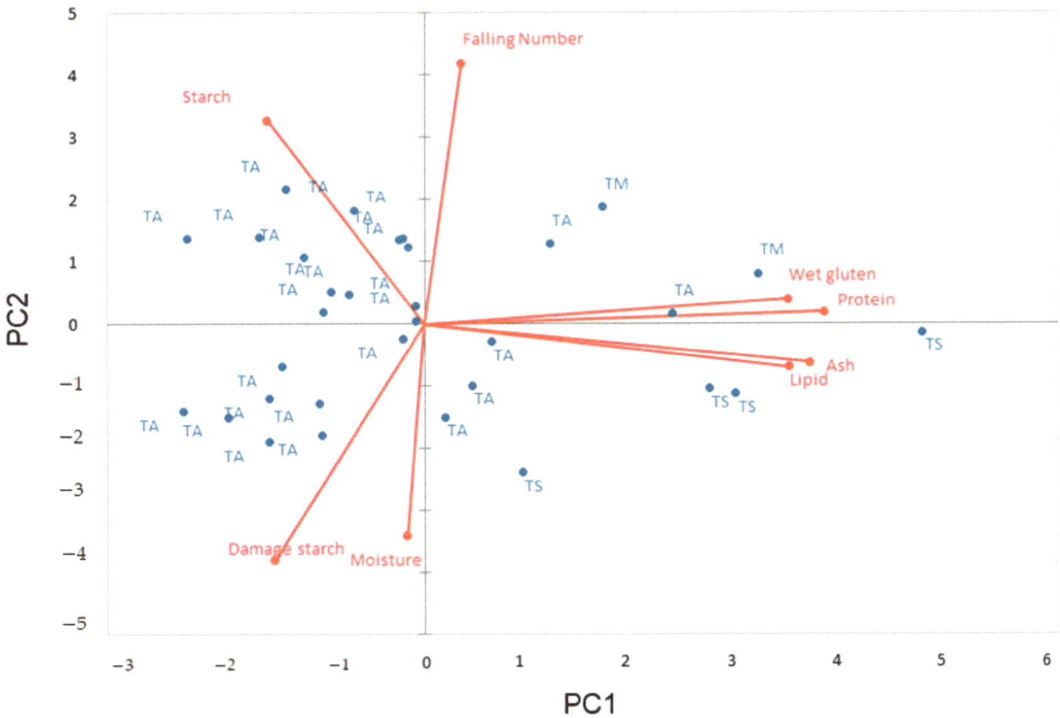

Figure 4. Principal component analysis of the physicochemical characteristics and scores of the wheat samples: TA—*Triticum aestivum*; TS—*Triticum spelta*; TM—*Triticum monococcum*.

4. Discussion

According to the physicochemical properties of wheat varieties, a large variability among samples was observed. All of them may be easily stored since their moisture value was less than 13%, characteristic for all wheat samples. This is an important criterion of wheat quality for bakery manufacturers, as a low moisture content means a long shelf during storage [25]. The quality content of wheat flour for bread-making is given especially by its gluten content. The technological quality of wheat flour used for bread-making is usually associated with the amount of gluten it consists of. Gluten is a protein obtained from glutenin and gliadin during mixing and is mainly responsible for dough rheological behavior during bread-making [13]. A high gluten content generally indicates flours of great quality that are suitable for bread-making [25]. According to our data, the mean value of gluten content is 28.74%, which indicates wheat samples with good bread-making quality [17]. However, the minimum gluten content value was 21.70%, which means that this type of wheat flour may be used in bread-making only after gluten addition or as a component of a mix with other kinds of flour with high gluten content [25]. Starch is the component that has the highest amount in wheat flour [20]. It is especially responsible for the formation of fermentable carbohydrates necessary during the fermentation process of the dough with baker's yeast for the formation of carbon dioxide that loosens the dough [43]. The large variability of starch, which varied between 55.30% and 64.10%, may be due to the other wheat compounds, such as protein, whose values are interconnected during grain development [44]. From a qualitative point of view, wheat flour presents intact and damaged starch granules. According to our study, the damaged starch varied in a high amount from low to high content in wheat flour. This can be degraded to fermentable sugars by amylases. Among the amylases, the most important is alpha amylase, an endoenzyme that attacks the starch granule inside [25]. The alpha amylase activity is suggested by falling number values, which varied according to our data from low, good and high ones for bread-making. The lipid and ash content varied in a high amount, probably due to the particularities of wheat samples [45].

From the genetic diversity point of view, the degree of polymorphism shown by ISSR markers is very high (almost 100%). Thus, we verified that the primers: UBC808 [(AG)8], UBC810 [(GA)8T], UBC855 [(AC)8 YT], UBC857 [(AC)8YG], UBC859 [(TG)8RC], UBC880 [(GGAGA)3] produced high levels of ISSR polymorphism in the common wheat, emmer and spelt genomes, reflecting the lifetime and abundance of the wheat genome. The unweighted pairwise group method with arithmetic means (UP-GMA) divided all cultivars by species/ploidy, reflecting a defined genetic structure. In addition, the Jaccard coefficient variation from 0.53 to 1 reflects the high genetic diversity of these wheat varieties [32].

Many scientists studied the suitability of the ISSR technique for genetic diversity studies of wheat, including some groups from Egypt [46], India [47], Iran [48] and Turkey [49]. ISSR primers were also used in the literature in order to study wheat accessions from Azerbaijan, and the results showed that they were superior to RAPD markers, producing more bands [50]. ISSR markers are widely used in genetic diversity assessment of distinct species. Mousavifard et al. [51] analyzed the potential of nine ISSR primers for eighty-nine diploid wheat samples. They reported a different polymorphic level from the present research, at 91.2%. At the same time, the number of migrated bands in agarose gel ranged between 16 for ISSR880 and 23 for UBC873. Regarding genetic similarity, the lowest value registered was 0.44, and the highest value was 0.9. These were reported for *Tr37a* of *Triticum boeoticum* subsp. *thaodar* and *Tr156* of *Triticum urartu*, respectively, and for *Tr37a* of *Triticum boeoticum* subsp. *thaodar* and *Tr156* of *Triticum urartu*. In this study, the authors highlighted an elevated level of genetic diversity within different species of wheat by using ISSR markers [51]. The same thing was reported by Du et al., who found different values for genetic distance, which ranged from 0.3115 to 0.3442 [52].

Lower levels of polymorphic bands were reported in another study, where some samples of *Triticum aestivum* were analyzed from a genetic diversity perspective. Using 16 ISSR primers, the percentage of polymorphic bands was 57.5%. In the same research, RAPD markers generated a higher value of polymorphic bands, 86.86% [53].

Molecular analysis can be efficiently used to identify plants with high adaptability to abiotic stress, such as drought [54,55] or salt tolerance [56]. Drought is a polygenic characteristic, and ISSR primers bind randomly in the genome, which might help to describe the trait-related regions. Deshmukh et al. [54] revealed that 3 of the 90 ISSR primers analyzed showed polymorphism related to tolerant and susceptible cultivars.

The PIC is an important tool for evaluating the quality of the marker and its ability to detect genetic diversity within and between species. ISSRs are dominant markers [39], and due to their biallelic nature, they tend to have a lower PIC. According to Botstein et al. [57]. PIC values can be characterized as follows: very informative (>0.5), medium informative (0.5 > PIC > 0.25) and less informative (<0.25). For wheat samples, the average PIC values for all primers range between 0.32 and 0.42, which corresponds to the medium informative category of PIC. This shows that the most informative primers were UBC857 (average PIC = 0.42) and UBC810 (average PIC = 0.39).

As there is a lot of known information nowadays regarding the nature and genetic diversity of different wheat accessions, there is an important fundament in terms of knowledge for creating superior wheat varieties with increased resistance to various biotic and abiotic stress factors [58]. An increased value of PIC (Polymorphism information content) for ISSR markers suggests a high genetic diversity, but a lower value can be the result of closely related genotypes. Kumar et al. showed that ISSR markers can reveal a significant value for PIC within the *Triticum aestivum* genotypes [59]. Despite the fact that ISSR primers are dominant and have a biallelic character, they are able to discriminate the variability between the wheat germplasm analyzed in this study. ISSR primers are often preferred in molecular biology studies due to their high reproducibility and the fact that there is no need for prior knowledge of the genome. Additionally, they have high transferability and accessibility, and they are low cost [60].

From the physicochemical point of view, the position of wheat samples on the PCA graph indicates differences between wheat species. These differences may be due to wheat physicochemical characteristics. The ancient species are more closely related to the protein, wet gluten, lipid and ash content, whereas the modern ones are related to the falling number, damaged starch and starch physicochemical characteristics. These data are similar to those reported by others [25,45,61]. They reported that ancient wheat varieties have a higher amount of protein, wet gluten, ash and lipid compared to the modern ones. However, some varieties (TA1, TA6, TA25) of *Triticum aestivum* presented similar characteristics, with ancient species being closely associated with those on the PCA graph, probably due to the fact that some of them were selected for their high protein and gluten content [61]. The close association between protein and wet gluten, which indicates a strong positive correlation ($p < 0.05$) between these variables, is explainable since wet gluten is formed from gluten proteins, which represent almost 75–85% of the total wheat protein content [14]. A significant high correlation ($p < 0.05$) between fat and ash, which is shown at the bottom right of the PCA graph, is explicable since the highest amount of minerals and fat presented in a wheat grain are located in wheat bran and germ. Therefore, higher amounts of germ and bran in wheat will lead to higher ash (minerals) and lipid content of the wheat flour [25]. The starch compound was negatively correlated with the rest of the physicochemical characteristics, such as protein, wet gluten, ash and lipid content according to its position in the space of the main components, as suggested by the PCA graph. This fact is explainable since the dry substance of wheat is made up of carbohydrates, proteins, lipids, minerals. Starch is the main compound of wheat, which represents almost 75–85% of the total dry substance of wheat [62]. Therefore, an increase in starch will lead to a decrease in proteins, the second in terms of weight in relation to dry matter [44]. Falling number values are indirectly correlated with damaged starch granules and directly correlated with protein content and the amount of wet gluten [25]. The falling number is a measure of the amylolytic activity of the wheat flour, which may be retained by glutenin in amounts that become higher as glutenin increased. Therefore, wheat flours that contain high amounts of proteins and wet gluten led to a decrease in amylolytic activity [42]. Falling number value is an expression of the wheat flour slurry viscosity heating to 100 °C and depends on damaged starch, which makes the starch more easily attacked by amylases [24]. Therefore, increased α-amylase activity is due to the high amount of damaged starch granules presented in wheat samples. Consequently, there is an inverse proportionality between damaged starch granules and falling number values [25].

5. Conclusions

The present study analyzed the physicochemical characteristics and genetic diversity of different wheat varieties cultivated in Romania of various species. Our data showed significant physicochemical differences between ancient and modern wheat species regarding ash, protein, wet gluten, lipid, starch, falling number and damaged starch values. All of the wheat samples clustered together according to their physicochemical data showed an association between wheat species. For all wheat samples, there were inverse correlations between the variables starch and physicochemical characteristics protein, wet gluten, lipid and ash and between falling number and damaged starch. The ISSR technique for genetic diversity showed that only 6 out of 11 ISSR primers had significant patterns of amplified fragments with clear and well-defined bands. The number of DNA bands per primer ranged from 3 to 8, with an average of 4.83 bands/primer. Most of the primers had a number of polymorphic fragments equal to the number of amplified fragments. The obtained data were analyzed using NTSYSpc software considering the coefficients of similarity (*Jaccard*) and dissimilarity (*Neighbor joining*). The genotypes obtained were divided into 5 groups (C1–C5). C5 was the most extensive cluster and contained 16 genotypes from different countries. C4 contained 6 genotypes, 5 from Romania and 1 from Switzerland. Additionally, C3 included 6 genotypes, also from different countries. In cluster C1, 2 genotypes were identified, both from Romania, and cluster C2 contained 1 genotype from France. Over time, many studies have been carried out that have demonstrated the utility and importance of ISSR-type molecular markers in genetic diversity research. The cultivars used in this study were selected due to their high wheat consumption and large cultivated areas of wheat grains in Romania. This research is of great importance worldwide considering that Romania occupies fourth place as a wheat producer at the European level. Knowing the genetic diversity of a germplasm collection and assessing the extent and nature of genetic variation in wheat is important for breeding programs and for the conservation of genetic resources. Except for a *Triticum aestivum* L. variety sample (TA5), which needs gluten addition to be used for bread-making, all analyzed wheat samples may be recommended to be cultivated by farmers. From the wheat samples analyzed, the ancient ones were of the highest quality for bread-making. They also presented the highest amounts of lipids and mineral nutrients, indicating that these species are nutritionally valuable for use in the bakery industry.

Supplementary Materials: The following supporting information can be downloaded at: https://www.mdpi.com/article/10.3390/app13084992/s1, Figure S1: DNA migration pattern of ISSR marker (UBC 810) for all 31 samples of *Triticum* (2% agarose gel, migration for 130 min, 70 V); Figure S2: DNA migration pattern of ISSR marker (UBC 855) for all 31 samples of *Triticum* (2% agarose gel, migration for 130 min, 70 V); Figure S3: DNA migration pattern of ISSR marker (UBC 857) for all 31 samples of *Triticum* (2% agarose gel, migration for 130 min, 70 V); Figure S4: DNA migration pattern of ISSR marker (UBC 859) for all 31 samples of *Triticum* (2% agarose gel, migration for 130 min, 70 V); Figure S5: DNA migration pattern of ISSR marker (UBC 880) for all 31 samples of *Triticum* (2% agarose gel, migration for 130 min, 70 V); Table S1: Physicochemical data of the analyzed wheat samples; Table S2: Number and frequency of alleles related to all six ISSR primers.

Author Contributions: Conceptualization, C.M.G. and G.G.C.; methodology, P.-M.G. and L.-I.L.; software, C.M.G. and G.G.C.; validation, C.M.G. and G.G.C.; formal analysis, P.-M.G. and L.-I.L.; investigation, C.M.G.; resources, C.M.G.; data curation, C.M.G.; writing—original draft preparation, C.M.G. and G.G.C.; writing—review and editing, G.G.C.; visualization, G.G.C.; supervision, G.G.C.; project administration, G.G.C.; funding acquisition, G.G.C. All authors have read and agreed to the published version of the manuscript.

Funding: This work was funded by the Ministry of Research, Innovation and Digitalization within Program 1—Development of national research and development system, Subprogram 1.2 —Institutional Performance—RDI excellence funding projects, under contract no. 10PFE/2021.

Institutional Review Board Statement: Not applicable.

Informed Consent Statement: Not applicable.

Data Availability Statement: Not applicable.

Conflicts of Interest: The authors declare no conflict of interest.

References

1. De Sousa, T.; Ribeiro, M.; Sabença, C.; Igrejas, G. The 10,000-Year Success Story of Wheat! *Foods* **2021**, *10*, 2124. [CrossRef]
2. Naushad, A.; Izhar, H.; Sardar, A.; Nagib, U.K.; Ijaz, H. Multivariate analysis for various quantitative traits in wheat advanced lines. *Saudi J. Biol. Sci.* **2021**, *28*, 347–352.
3. Jiang, H.; Gao, Q.; Li, L.; Kong, L.; Zhang, W.; Wu, A.; Yang, Y. Genetic Diversity of Recurrent Selection Populations with Ms2 Gene Assessed by Gliadins in Common Wheat (*Triticum aestivum L.*). *Agric. Sci. China* **2010**, *9*, 615–625. [CrossRef]
4. Eltaher, S.; Sallam, A.; Belamkar, V.; Emara, H.A.; Nower, A.A.; Salem, K.F.M.; Poland, J.; Baenziger, P.S. Genetic Diversity and Population Structure of F3:6 Nebraska Winter Wheat Genotypes Using Genotyping-By-Sequencing. *Front. Genet.* **2018**, *9*, 76. [CrossRef] [PubMed]
5. Popescu, A. Maize and Wheat—Top Agricultural Products Produced, Exported and Imported by Romania. Scientific Papers Series Management. *Econ. Eng. Agric. Rural Dev.* **2018**, *18*, 339–352.

6. Cappelli, A.; Cini, E. Challenges and Opportunities in Wheat Flour, Pasta, Bread, and Bakery Product Production Chains: A Systematic Review of Innovations and Improvement Strategies to Increase Sustainability, Productivity, and Product Quality. *Sustainability* **2021**, *13*, 2608. [CrossRef]
7. Júnior, R.d.S.N.; Ewert, F.; Webber, H.; Martre, P.; Hertel, T.W.; van Ittersum, M.K.; Asseng, S. Needed global wheat stock and crop management in response to the war in Ukraine. *Glob. Food Secur.* **2022**, *35*, 100662. [CrossRef]
8. Popescu, G.H.; Nicoale, I.; Nica, E.; Vasile, A.J.; Andreea, I.R. The influence of land-use change paradigm on Romania's agro-food trade competitiveness—An overview. *Land Use Policy* **2017**, *61*, 293–301. [CrossRef]
9. Cvijanovic, D.; Sterie, M.C.; Kovacevic, V.; Ion, R.A. Comparative analysis of wheat and sunflower seeds branches in Romania and Serbia. In Proceedings of the 5th International Conference on Economics and Social Sciences, Fostering Recovery through Metaverse Business Modelling, Bucharest, Romania, 16–17 June 2022.
10. Moroșan, E.; Secareanu, A.A.; Musuc, A.M.; Mititelu, M.; Ioniță, A.C.; Ozon, E.A.; Raducan, I.D.; Rusu, A.I.; Dărăban, A.M.; Karampelas, O. Comparative Quality Assessment of Five Bread Wheat and Five Barley Cultivars Grown in Romania. *Int. J. Environ. Res. Public Health* **2022**, *19*, 11114. [CrossRef]
11. Manifesto, M.M.; Feingold, S.; Hopp, H.E.; Schlattert, A.R.; Dubcoysky, J. Molecular Markers Associated with Differences in Bread-making Quality in a Cross Between Bread Wheat Cultivars with the Same High Mr Glutenins. *J. Cereal Sci.* **1998**, *27*, 217–227. [CrossRef]
12. Wang, M.; Wang, S.; Liang, Z.; Shi, W.; Gao, C.; Xia, G. From genetic stock to genome editing: Gene exploitation in wheat. *Trends Biotechnol.* **2018**, *36*, 160–172. [CrossRef] [PubMed]
13. Codină, G.G.; Bordei, D.; Pâslaru, V. The effects of different doses of gluten on rheological behavior of dough and bread quality. *Rom. Biotechnol. Lett.* **2008**, *13*, 37–42.
14. Wang, Y.; Chen, Y.H.; Zhou, Y.; Nirasawa, S.; Tatsumi, E.; Li, X.T.; Cheng, Y.Q. Effects of konjac glucomannan on heat-induced changes of wheat gluten structure. *Food Chem.* **2017**, *229*, 409–416. [CrossRef] [PubMed]
15. Kłosok, K.; Welc, R.; Fornal, E.; Nawrocka, A. Effects of Physical and Chemical Factors on the Structure of Gluten, Gliadins and Glutenins as Studied with Spectroscopic Methods. *Molecules* **2021**, *26*, 508. [CrossRef] [PubMed]
16. Attenburrow, G.; Barnes, D.J.; Davies, A.P.; Ingman, S.J. Rheological properties of wheat gluten. *J. Cereal Sci.* **1990**, *12*, 1–14. [CrossRef]
17. Hu, X.; Cheng, L.; Hong, Y.; Li, Z.; Li, C.; Gu, Z. An extensive review: How starch and gluten impact dough machinability and resultant bread qualities. *Crit. Rev. Food Sci. Nutr.* **2021**, 1–12. [CrossRef] [PubMed]
18. Ortolan, F.; Steel, C.J. Protein characteristics that affect the quality of vital wheat gluten to be used in baking: A review. *Compr. Rev. Food Sci. Food Saf.* **2017**, *16*, 369–381. [CrossRef]
19. Žilić, S.; Barać, M.; Pešić, M.; Dodig, D.; Ignjatović-Micić, D. Characterization of Proteins from Grain of Different Bread and Durum Wheat Genotypes. *Int. J. Mol. Sci.* **2011**, *12*, 5878–5894. [CrossRef]
20. Onyango, C. Starch and modified starch in bread making: A review. *Afr. J. Food Sci.* **2016**, *10*, 344–351.
21. Martínez-Anaya, M.A. Enzymes and bread flavor. *J. Agric. Food Chem.* **1996**, *44*, 2469–2480. [CrossRef]
22. Barrera, G.N.; Pérez, G.T.; Ribotta, P.D.; León, A.E. Influence of damaged starch on cookie and bread-making quality. *Eur. Food Res. Technol.* **2007**, *225*, 1–7. [CrossRef]
23. Wang, Q.; Li, L.; Zheng, X. A review of milling damaged starch: Generation, measurement, functionality and its effect on starch-based food systems. *Food Chem.* **2020**, *315*, 126267. [CrossRef]
24. Codină, G.G.; Mironeasa, S.; Mironeasa, C. Variability and relationship among Mixolab and Falling Number evaluation based on influence of fungal α-amylase addition. *J. Sci. Food Agric.* **2012**, *92*, 2162–2170. [CrossRef] [PubMed]
25. Golea, M.C.; Oroian, M.; Codină, G.G. Prediction of wheat flours composition using fourier transform infrared spectrometry (FT-IR). *Food Control* **2023**, *143*, 109318. [CrossRef]
26. Weidenbörner, M.; Wieczorek, C.; Appel, S.; Kunz, B. Whole wheat and white wheat flour—The mycobiota and potential mycotoxins. *Food Microbiol.* **2000**, *17*, 103–107. [CrossRef]
27. Gómez, M.; Gutkoski, L.C.; Bravo-Núñez, Á. Understanding whole-wheat flour and its effect in breads: A review. *Compr. Rev. Food Sci. Food Saf.* **2020**, *19*, 3241–3265. [CrossRef]
28. Pareyt, B.; Finnie, S.M.; Putseys, J.A.; Delcour, J.A. Lipids in bread making: Sources, interactions, and impact on bread quality. *J. Cereal Sci.* **2011**, *54*, 266–279. [CrossRef]
29. Feldman, M. Origin of cultivated wheat. In *The World Wheat Book. A History of Wheat Breeding*; Bonjean, A.P., Angus, W.J., Eds.; Lavoisier Publishing: Paris, France, 2001; pp. 3–56.
30. Shewry, P.R.; Hey, S. Do "ancient" wheat species differ from modern bread wheat in their contents of bioactive components? *J. Cereal Sci.* **2015**, *65*, 236–243. [CrossRef]
31. Migliorini, P.; Spagnolo, S.; Torri, L.; Arnoulet, M.; Lazzerini, G.; Ceccarelli, S. Agronomic and quality characteristics of old, modern and mixture wheat varieties and landraces for organic bread chain in diverse environments of northern Italy. *Eur. J. Agron.* **2016**, *79*, 131–141. [CrossRef]
32. Li, W.; Bian, C.-M.; Wei, Y.-M.; Liu, A.-J.; Chen, G.-Y.; Pu, Z.-E.; Liu, Y.-X.; Zheng, Y.-L. Evaluation of genetic diversity of sichuan common wheat landraces in China by SSR markers. *J. Integr. Agric.* **2013**, *12*, 1501–1511. [CrossRef]

33. Zietkiewicz, E.; Rafalski, A.; Labuda, D. Genome fingerprinting by simple sequence repeat (SSR)-anchored polymerase chain reaction amplification. *Genomics* **1994**, *20*, 176–183. [CrossRef] [PubMed]
34. Al-Turki, T.A.; Basahi, M.A. Assessment of ISSR based molecular genetic diversity of Hassawi rice in Saudi Arabia. *Saudi J. Biol. Sci.* **2015**, *22*, 591–599. [CrossRef]
35. Vieira, M.B.; Faustino, M.V.; Lourenço, T.F.; Oliveira, M.M. DNA-Based Tools to Certify Authenticity of Rice Varieties—An Overview. *Foods* **2022**, *11*, 258. [CrossRef] [PubMed]
36. Doyle, J.J.; Doyle, M. A rapid DNA isolation procedure for small quantities of fresh leaf tissue. *Phytochem. Bull.* **1987**, *19*, 11–15.
37. Cabral, P.D.S.; de Souza, L.C.; da Costa, G.F.; Silva, F.H.L.; Soares, T.C.B. Investigation of the Genetic Diversity of Common Bean (Phaseolus Vulgaris.) Cultivars Using Molecular Markers. *Genet. Mol. Res.* **2018**, *17*, gmr18106. [CrossRef]
38. Codină, G.G.; Dabija, A.; Oroian, M. Prediction of Pasting Properties of Dough from Mixolab Measurements Using Artificial Neuronal Networks. *Foods* **2019**, *8*, 447. [CrossRef]
39. Roldán-Ruiz, I.; Dendauw, J.; Van Bockstaele, E.; Depicker, A.; De Loose, M. AFLP Markers Reveal High Polymorphic Rates in Ryegrasses (*Lolium* spp.). *Mol. Breed.* **2000**, *6*, 125–134. [CrossRef]
40. Li, S.; Ramakrishnan, M.; Vinod, K.K.; Kalendar, R.; Yrjälä, K.; Zhou, M. Development and Deployment of High-Throughput Retrotransposon-Based Markers Reveal Genetic Diversity and Population Structure of Asian Bamboo. *Forests* **2020**, *11*, 31. [CrossRef]
41. Popa, N.C.; Tamba-Berehoiu, R.; Popescu, S.; Varga, M.; Codină, G.G. Predective model of the alveografic parameters in flours obtained from Romanian grains. *Rom. Biotechnol. Lett.* **2009**, *14*, 4234–4242.
42. Codină, G.G.; Mironeasa, S.; Bordei, D.; Leahu, A. Mixolab versus Alveograph and Falling Number. *Czech J. Food Sci.* **2010**, *28*, 185–191. [CrossRef]
43. Martínez, M.M.; Gómez, M. Rheological and microstructural evolution of the most common gluten-free flours and starches during bread fermentation and baking. *J. Food Eng.* **2017**, *197*, 78–86. [CrossRef]
44. Xie, Z.; Jiang, D.; Cao, W.; Dai, T.; Jing, Q. Relationships of endogenous plant hormones to accumulation of grain protein and starch in winter wheat under different post-anthesis soil water statusses. *Plant Growth Regul.* **2003**, *41*, 117–127. [CrossRef]
45. Kulathunga, J.; Reuhs, B.L.; Zwinger, S.; Simsek, S. Comparative Study on Kernel Quality and Chemical Composition of Ancient and Modern Wheat Species: Einkorn, Emmer, Spelt and Hard Red Spring Wheat. *Foods* **2021**, *10*, 761. [CrossRef]
46. Rizkalla, A.A.; Attia, S.A.A.; Abd El-Hady, E.A.A.; Hanna, N.S.; Nasseef, J.E. Genetic Diversity Based on ISSR and Protein Markers Associated with Earliness Trait in Wheat. *World Appl. Sci. J.* **2012**, *20*, 23–33. [CrossRef]
47. Rekha, M.; Sindhu, S.; Sushila, K.; Jag, S. The Use of SSR and ISSR Markers for Assessing DNA Polymorphism and Genetic Diversity among Indian Bread Wheat Cultivars. *Progress. Agric.* **2012**, *12*, 82–89.
48. Sofalian, O.; Chaparzadeh, N.; Javanmard, A.; Hejazi, M.S. Study the Genetic Diversity of Wheat Landraces from Northwest of Iran Based on ISSR Molecular Markers. *Int. J. Agric. Biol.* **2008**, *10*, 466–468.
49. Karaca, M.; Izbirak, A. Comparative Analysis of Genetic Diversity in Turkish Durum Wheat Cultivars Using RAPD and ISSR Markers. *J. Food Agric. Environ.* **2008**, *6*, 219–225.
50. Sadigova, S.; Sadigov, H.; Eshghi, R.; Salayeva, S.; Ojaghi, J. Application of Rapd and Issr Markers to Analyses Molecular Relationships in Azerbaijan Wheat Accessions (*Triticum aestivum* L.). *Bulg. J. Agric. Sci.* **2014**, *20*, 87–95.
51. Mousavifard, S.S.; Saeidi, H.; Rahiminejad, M.R.; Shamsadini, M. Molecular Analysis of Diversity of Diploid Triticum Species in Iran Using ISSR Markers. *Genet. Resour. Crop Evol.* **2015**, *62*, 387–394. [CrossRef]
52. Du, J.-K.; Yao, Y.-Y.; Ni, Z.-F.; Peng, H.-R.; Sun, Q.-X. Genetic diversity revealed by ISSR molecular marker in common wheat, spelt, compactum and progeny of recurrent selection. *Yi Chuan Xue Bao* **2002**, *29*, 445–452.
53. Nazarzadeh, Z.; Onsori, H.; Akrami, S. Genetic Diversity of Bread Wheat (*Triticum aestivum* L.) Genotypes Using RAPD and ISSR Molecular Markers. *J. Genet. Resour.* **2020**, *6*, 69–76. [CrossRef]
54. Deshmukh, R.; Tomar, N.S.; Tripathi, N.; Tiwari, S. Identification of RAPD and ISSR Markers for Drought Tolerance in Wheat (*Triticum aestivum* L.). *Physiol. Mol. Biol. Plants* **2012**, *18*, 101–104. [CrossRef]
55. Shokry, A.M.; Edris, S.; Ramadan, A.M.; Gadalla, N.O.; Bahieldin, A.; Arabia, S.; Engineering, G.; Division, B. Detection of Wheat (*Triticum aestivum*) Cultivars with Contrasting Performance under Abiotic Stresses. *Life Sci. J.* **2013**, *10*, 2746–2756.
56. Majeed, D.M.; Ismail, E.N.; Al-Mishhadani, I.I.; Sakran, N.M. Assessment of Genetic Diversity among Wheat Selected Genotypes and Local Varieties for Salt Tolerance by Using RAPD and ISSR Analysis. *Iraqi J. Sci.* **2018**, *59*, 278–286. [CrossRef]
57. Botstein, D.; White, R.L.; Skolnick, M.; Davis, R.W. Construction of a Genetic Linkage Map in Man Using Restriction Fragment Length Polymorphisms. *Am. J. Hum. Genet.* **1980**, *32*, 314–331. [PubMed]
58. Singh, K.; Singh, T.; Singh, V.; Verma, O.; Singh, S. Divergence Analysis in Certain Genotypes of Wheat (*Triticum aestivum* L. em. Thell). *J. Pharmacogn. Phytochem.* **2019**, *8*, 507–510.
59. Kumar, P.; Sharma, V.; Sanger, R.; Kumar, P.; Yadav, M.K. Analysis of Molecular Variation among Diverse Background Wheat (*Triticum aestivum* L.) Genotypes with the Help of ISSR Markers. *Int. J. Chem. Stud.* **2020**, *8*, 271–276. [CrossRef]
60. Ng, W.L.; Tan, S. Inter-Simple Sequence Repeat (ISSR) Markers: Are We Doing It Right? *ASM Sci. J.* **2015**, *9*, 30–39.

61. Spisni, E.; Imbesi, V.; Giovanardi, E.; Petrocelli, G.; Alvisi, P.; Valerii, M.C. Differential Physiological Responses Elicited by Ancient and Heritage Wheat Cultivars Compared to Modern Ones. *Nutrients* **2019**, *11*, 2879. [CrossRef]
62. Shewry, P.R.; Hawkesford, M.J.; Piironen, V.; Lampi, A.; Gebruers, K.; Boros, D.; Andersson, A.A.M.; Åman, P.; Rakszegi, M.; Bedo, Z.; et al. Natural variation in grain composition of wheat and related cereals. *J. Agric. Food Chem.* **2013**, *61*, 8295–8303. [CrossRef] [PubMed]

applied sciences

MDPI

Article

Enrichment of White Wheat Bread with Pistachio Hulls and Grape Seeds: Effect on Bread Quality Characteristics

Banu Koç * and Gamze Atar Kayabaşi

Gastronomy and Culinary Arts Department, Faculty of Tourism, Gaziantep University, Gaziantep 27310, Türkiye
* Correspondence: banukoc@gantep.edu.tr

Featured Application: Pistachio hulls and grape seeds are useful by-products of the pistachio and wine industries, respectively. Pistachio hulls and grape seeds can be used as alternative functional ingredients in wheat bread due to their high dietary fiber content and richness in antioxidants and phenolic substances. In this study, the effects of pistachio hulls (fresh and freeze-dried) and grape seeds (fresh and freeze-dried) on bread quality were compared in the 0.5–4% range of addition.

Abstract: In creating sustainable food systems, alternative uses of food waste and by-products as a source of phenolic compounds or dietary fiber in food formulations should be evaluated to reduce food losses and waste. In this study, wheat bread was fortified with agro-industrial by-products, namely, fresh pistachio hull (FPH), pistachio hull powder (PHP), fresh grape seeds (FGS), and grape seed powder (GSP), at different levels (0.5–4%). The effects of this enrichment on moisture content, crust and crumb color, specific volume, baking loss, total phenolic content, antioxidant activity, and textural properties were evaluated and compared with control bread. Fortification with pistachio hulls and grape seeds, especially in powdered form, improved the total phenolic content and antioxidant activity of the bread compared with the control bread. With the increase in the amount of PHP from 0 to 4.0%, the total phenolic content of the bread initially increased from 0.89 to 14.66 mg GAE/g dry weight and the specific volume decreased from 3.59 cm^3/g to 2.91 cm^3/g. Bread containing PHP showed a significant reduction in baking loss and specific volume, while the addition of FGS and GSP at a low level (0.5%) improved the specific volume. The brightness (L^*) of the crumb and crust gradually decreased with increasing levels of all additives. The crumbs of the breads with PHP and FPH were characterized by lower hardness, while GSP and especially FGS had higher hardness. All enriched breads (except those with FGS) were more elastic and easier to chew.

Keywords: pistachio hull; grape seeds; by-products; wheat bread; total phenolic content; antioxidant activity; texture

1. Introduction

Bread has always been a traditional staple food for people in many countries around the world. Thanks to its high carbohydrate content, bread is an important source of energy, but it also contains essential elements, such as protein, fat, and minerals. Bread has been the most widely used food for centuries due to its nutritional value and sensory properties [1]. The quality of bread is influenced by the dough composition [2]. However, refined wheat flour, which is commonly used for bread, has a lower nutritional quality because refining the flour results in a significant loss of dietary fiber, vitamins, minerals, and phytochemicals [3]. Therefore, bread is considered the best means of enrichment with bioactive compounds to develop health-promoting products. Since white flour bread is a food with low phenolic content, many scientists focused on fortifying wheat bread with food waste and by-products rich in phenolic antioxidants, such as grape seed powder [4,5], mango peel [6], dry onion skin [7,8], and pomegranate peel powder [9], which are often cheap, functional, and an excellent source of nutraceuticals and bioactive compounds [10].

The amount of waste generated in the food industry is significant, and it is critical to consider alternative recycling options. Grapes are the most widely produced fruit in the world and generate the most waste or by-products (pulp, peel, seeds, and stems). Grape seeds are rich in phenolic compounds, dietary fiber, proteins, vitamins, and organic acids [11]. Pistachio hulls are a major agro-industrial by-product of the pistachio industry and are accumulated in large quantities during industrial postharvest processing. Pistachio hulls are the soft mesocarp and epicarp of the pistachio, which are firmly attached to the hard inner shell. During ripening, it has a pinkish-yellow color that changes to a reddish-light yellow as it matures [12]. Researchers reported that pistachio hulls have antioxidant, antimicrobial, and antimutagenic effects [13–16]. Goli et al. [15] found that aqueous and methanolic extracts of pistachio hulls are rich in phenols, which effectively retard the deterioration of soybean oil at 60 °C, with efficacy comparable to that of the synthetic antioxidant BHA (butylated hydroxyanisole). Furthermore, pistachio hulls also contain polysaccharides, dietary fibers, essential oils, unsaturated fatty acids, phytosterols, carotenoids, chlorophylls, tocopherols, and triterpene acids [16–18]. Due to its antioxidant polyphenolic compounds (phloroglucinol, gallic acid, protocatechuic acid, galloyl-shikimic acid, 4-hydroxybenzoic acid, catechin, vanillic acid, eriodictyol-7-O-glucoside, naringin, and cinnamic acid) [19] and phytochemical contents, such as fatty acids (1500 mg/100 g dried hull), anacardic acids (355 mg/100 g dried hull), phytosterols (21.36 mg/100 g dried hull), carotenoids (0.55 mg/100 g dried hull), chlorophylls (1.14 mg/100 g dried hull), and tocopherols (0.98 mg/100 g dried hull) [16], pistachio hull can be used in bread formulation to improve bread quality. Due to their strong antioxidant activity, pistachio hulls might be an interesting additive to fortify cereal products. In addition, pistachio hulls are currently used as animal feed and have not been considered for food production. Therefore, pistachio hulls have the potential to be used for food upcycling, and there is a need to learn more about the use of pistachio hulls (in fresh or dried form) during this process. The enrichment of white wheat flour with these additives changes the quality of the bread. Therefore, when using non-traditional materials, special attention should be paid to the study of their functional and technological properties.

However, to our knowledge, there is no clear evidence to date for the use of pistachio hulls to enrich bread. In this work, the influence of the addition of fresh or freeze-dried pistachio hulls on the total phenolic content, antioxidant activity, moisture content, crust and crumb color, specific volume, baking loss, and textural and sensory properties of white wheat bread was specifically investigated. In addition, grape seeds, which have been used as an additive for many years, were also used as a fortifying agent and compared with the quality characteristics of breads with pistachio hulls.

2. Materials and Methods

2.1. Materials

Wheat flour with a moisture content of 14.3%, protein content of 11.9%, and ash content of 0.67% was supplied by Özmen Un San. ve Tic. A.Ş., Turkey. Salt and active instant dry yeast were purchased from the local market in Gaziantep, Turkey, while fresh pistachio hull and grape seeds were obtained as by-products from a company in Gaziantep, Turkey. Fresh hulls and grape seeds were stored in ALPE bags at −24 °C until use. Pistachio hulls and grape seeds that were added fresh were coarsely ground using a blender (Arzum, AR1056, İstanbul, Turkey), and those that were added as powder were freeze-dried at −55 °C for 48 h in a freeze-dryer (FreeZone 6, Labconco, Kansas City, MO, USA), and samples were ground after drying. All solvents and chemicals were purchased from Sigma Aldrich (St. Louis, MO, USA). The proximate composition (moisture, protein, oil, and ash) of PHP, FPH, GSP, and FGS was determined [20]. The proximate composition of the additives is shown in Table 1.

Table 1. Proximate compositions, color values, antioxidant activity, and total phenolic content of fresh and powdered pistachio hulls and grape seeds.

Properties	FPH	PHP	FGS	GSP
Moisture (%, wet basis)	71.40 ± 0.01	6.855 ± 0.09	21.90 ± 0.16	6.035 ± 0.04
Protein (%, dry basis)	10.51 ± 0.08	8.14 ± 0.02	5.36 ± 0.09	7.83 ± 0.026
Oil (%, dry basis)	6.35 ± 0.12	8.79 ± 0.04	9.10 ± 0.12	12.14 ± 0.01
Ash (%, dry basis)	7.88 ± 0.05	9.30 ± 0.18	1.62 ± 0.01	1.65 ± 0.06
L^*	48.74 ± 1.23	53.86 ± 1.29	24.26 ± 1.17	31.67 ± 2.10
a^*	20.30 ± 2.62	17.09 ± 0.49	15.92 ± 1.34	17.84 ± 0.53
b^*	21.61 ± 1.89	21.83 ± 0.52	14.50 ± 0.60	24.09 ± 0.59
Antioxidant activity (mg Trolox/g dry weight)	40.22 ± 0.03	59.39 ± 0.09	32.68 ± 0.01	54.74 ± 0.10
Total phenolic content (mg GAE/g dry weight)	104.9 ± 0.16	137.3 ± 0.25	66.48 ± 0.06	96.73 ± 0.15

FPH: fresh pistachio hull, PHP: pistachio hull powder, FGS: fresh grapeseed, GSP: grapeseed powder.

2.2. Preparation of Bread

Experiments were performed in a bread-baking machine (Tefal, Pain et Delice, Rumilly, France). The dough recipe was based on the bread standard TS 5000 [21]. For comparison, the control bread (control) was used, which was composed as follows: 400 g wheat flour, 240 mL water, 6 g salt, and 8 g yeast. Fresh pistachio hull (FPH), fresh grape seed (FGS), pistachio hull powder (PHP), and grape seed powder (GSP) were added separately to the bread formulation as additives. Wheat flour was replaced with FPH or FGS or PFP or GSP at five levels (0.5%, 1.38%, 2.25%, 3.13%, and 4% of wheat flour). The selected bread baking program included dough preparation, fermentation, and baking (total 3:30 h). The bread was left to rest for 2 h at room temperature before being prepared for analysis.

2.3. Moisture Content

The moisture content of the bread samples was determined using an infrared moisture analyzer (Daihan Scientific, MA10, Gangwon-do, Korea) at 105 °C.

2.4. Color

The CIE color values L^*, a^*, and b^* were measured at five different points within the crumb and crust regions using a colorimeter (3NH, China). The total color difference (ΔE) of the fortified bread relative to the control bread was calculated for each additive according to Hunter [22].

2.5. Specific Volume and Baking Loss

Baking loss (%) was determined by weighing the bread and the weight of the dough before baking. The specific volume of the bread (cm^3/g) was determined by dividing the volume of the bread by its weight. The volume of the bread was determined using the rapeseed displacement method according to the approved method AACC International 10-05.01 [23].

2.6. Total Phenolic Content and Antioxidant Activity

Ground bread samples (1 g) were extracted with 80% methanol (10 mL) in a shaking incubator (Mikrotest, MSC-30, Ankara, Turkey) at 37 °C for 1 h. The mixtures were then centrifuged (PCE Instruments, CFE100, Meschede, Germany) at 6000 rpm for 10 min, and the supernatants were collected.

The total phenolic content (TPC) in the methanol extract of the samples was determined using the Folin–Ciocalteau method and the procedure reported by [24]. All spectrometric measurements were performed in triplicate. The calibration curve was prepared using gallic acid, and the results were expressed as gallic acid equivalents (mg GAE/g dry weight).

The antioxidant activity of the samples was measured using the DPPH (1,1-diphenyl-2-picrylhydrazyl) radical-scavenging method according to Brand-Williams et al. [25] with some modifications. The extracts (0.1 mL) were added to a 2.9 mL DPPH solution (100 ppm). The mixture was shaken vigorously and left in the dark at room temperature for 30 min. Then, the absorbance was measured at 517 nm. The antioxidant activity was expressed as Trolox equivalents (mg Trolox/g dry weight).

2.7. Texture Analysis

Texture analysis of the crumb of bread samples was performed using a texture analyzer (Texture Analyzer TA-XT Stable Micro Systems, Surrey, UK) according to approved method 74-09 [26]. A cylindrical compression probe with a diameter of 36 mm was used for the analysis. Bread slices of 2.5 cm thickness were compressed to 50% of their original thickness at a test speed of 2 mm/s with a 30 s delay between the first and second compressions. Hardness, springiness, cohesiveness, and chewiness properties were measured.

2.8. Sensory Evaluation

A panel of 14 semi-trained panelists consisting of staff and students evaluated the sensory characteristics of the bread samples. The bread was cut into slices (2 cm thick). The sensory evaluation of the samples involved giving grades using a 5-point hedonic scale according to the crust and crumb color, chewiness, taste, flavor, and overall liking.

2.9. Statistical Analysis

The results presented are the average of three replicated observations. The significance of the level and type of additive was determined via an analysis of variance (ANOVA) for each bread quality parameter using SPSS Statics 22.0 (SPSS Inc., Chicago, IL, USA). Tukey's HSD multiple range test was applied at a 95% significance level to determine the significant differences. In addition, color values were correlated with the total phenolic content and antioxidant activity, and correlation coefficients were calculated.

3. Results

3.1. Moisture Content

The observations of the moisture content of the control and enriched breads are shown in Figure 1a. The values of moisture content of the breads prepared with different additives ranged from 36.89–37.84% for the breads enriched with fresh pistachio hulls (FPH), 38.05–38.53% for the breads enriched with freeze-dried pistachio hulls (PHP), 35.90–36.95% for the breads enriched with fresh grape seeds (FGS), and 37.46–38.45% for the breads enriched with freeze-dried grape seeds (GSP). The minimum moisture content of the enriched bread (35.90%) was found for the bread with the 0.5% addition of FGS and the maximum moisture content (38.53%) for the bread with the 2.25% addition of PHP. A significant increase in moisture content was observed for all enriched bread samples compared with the control (33.70%). It was concluded that the weight loss of the bread decreased with the addition of FPH, FGS, PHP, and GSP. In agreement with our results, the moisture contents of bread samples enriched with broad bean hulls [27], cumin and caraway powder seeds and by-products [28], parsley leaf powder [29], and Plantago ovata husk [30] were increased. This may be attributed to the fact that flours with added dietary fiber prevent water evaporation. According to the ANOVA (Figure 1a), there were significant differences in moisture content between the control bread and all the enriched breads ($p < 0.05$). In addition, the statistical analysis showed that only the type of additive had a significant effect ($p < 0.05$) on the moisture content, while the level of additive was not significant ($p > 0.05$). Figure 1a also shows that the moisture contents of the breads fortified with FHP and FGS increased slightly with the increasing addition levels, while the moisture contents of breads fortified with PHP and GSP decreased slightly with the increasing addition levels. In addition, breads fortified with powdered additives had higher moisture contents than the breads fortified with fresh additives. This was due to the higher water-holding capacity of powder products. Higher moisture content is both economical and necessary to extend the shelf life of bread [31].

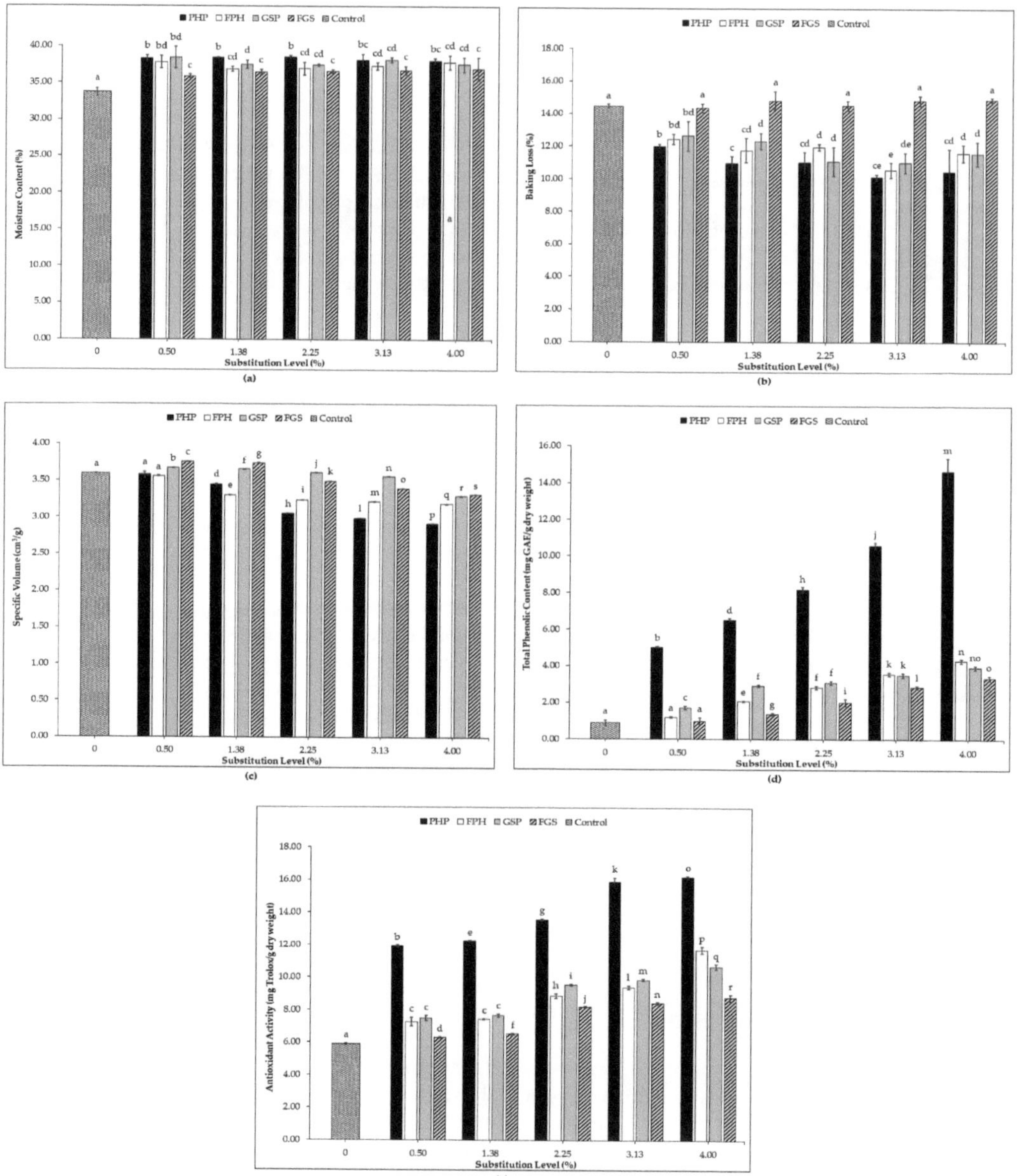

Figure 1. Moisture content (**a**), baking loss (**b**), specific volume (**c**), total phenolic content (**d**), and antioxidant activity (**e**) of control and enriched breads. Values of each parameter with the same letters were not significantly different ($p > 0.05$).

3.2. Baking Loss and Specific Volume

Baking loss and specific volume are important quality parameters of bread and are shown in Figure 1b and 1c, respectively. The lowest baking loss of the enriched breads

(10.11%) was found for the bread with the 3.13% addition of PHP, and the highest baking loss (14.87%) was found for the bread with the 4% addition of FGS. The baking loss values were lower for all the enriched breads (except for the breads with FGS) than for the control bread. This may be attributed to the fact that flours with fat, fiber, and gluten have free proteins that can bind free water in the dough, which prevents water evaporation and reduces baking loss [3,32]. The baking loss was influenced by the type and level of the additives ($p < 0.05$), and the control bread was significantly different from the enriched breads, except for the breads with FGS (Figure 1b). Furthermore, the baking loss values did not increase or decrease uniformly as a function of the level of additive; as the FGS content increased from 0.5% to 4%, the baking loss of the breads first decreased and then increased, with only the breads with FGS showing a statistically nonsignificant difference as a function of the change in additive level (Figure 1b). As mentioned above, the change in the other enriched breads was statistically significant ($p < 0.05$). The comparison of the type and level of additives showed that PHP was the most effective at reducing baking loss. Thus, pistachio hulls can be used to reduce baking loss. This result can be explained by the higher moisture content of the bread samples enriched with pistachio hulls.

The values of the specific volume of the breads enriched with different additives varied between 2.91–3.58 cm^3/g for PHP, 3.19–3.56 cm^3/g for FPH, 3.30–3.67 cm^3/g for GSP, and 3.32–3.76 cm^3/g for FGS. When comparing the specific volume of each enriched bread with that of the control bread (3.59 cm^3/g), it was found that the addition of pistachio hulls (both fresh and dry) resulted in a decrease in the specific volumes of the breads, while the addition of grape seeds in small amounts increased the specific volumes. There were also significant differences in the specific volume between the control bread and the enriched breads (except for the breads with the 0.5% additions of PHP and FPH) (Figure 1c, $p < 0.05$). The FGS breads had the highest specific volume, and the PHP breads had the lowest specific volume. Due to the higher water-holding capacity of the pistachio hull, which is a source of polysaccharides and dietary fiber [17], the water requirement of the dough increased, and the achievable volume was limited. In this process, the interaction between gluten and fiber resulted in a decrease in gas retention ability and a decrease in bread volume [33]. The specific volume always decreased as a result of fiber addition, which was previously reported by Ni et al. [27]. In addition, the results of the ANOVA showed that the type and level of additives had a significant effect on the specific volume, and it was clear that the specific volumes of all enriched breads decreased with an increasing level of additive (Figure 1c), and thus, low amounts of additives could be beneficial, while higher amounts could have negative effects on the specific volume.

3.3. *Total Phenolic Content and Antioxidant Activity*

The results on the effects of the different enrichment levels of FHP, PHP, FGS, and GSP on the total phenolic content and antioxidant activity of the bread samples are shown in Figure 1d and 1e, respectively. Significant increases in the total phenolic content and antioxidant activity of the breads were observed with increasing PHP levels. The total phenolic content and antioxidant activity of the breads initially increased from 0.89 to 14.66 mg GAE/g dry weight and from 5.90 to 16.19 mg Trolox/g dry weight, respectively, when the PHP level increased from 0.5% to 4.0%. When the total phenolic content and antioxidant activity of the control bread were compared, it was found that the addition of FHP, FGS, and GSP also increased them. There was no difference in the total phenolic content between the control bread and the breads with the 0.5% addition of FGS and GSP ($p > 0.05$), while the total phenolic content and antioxidant activity of all other enriched breads were different from those of the control bread ($p < 0.05$). Moreover, the total phenolic content and antioxidant activity of breads with powdered additives were higher than those of the breads enriched with fresh additives. This was due to the fact that the total phenolic content and antioxidant activity of powders are intrinsically higher due to the higher extraction yield resulting from the porous surfaces caused by freeze-drying (Table 1). In addition, according to the results from ANOVA, the type and level of additives had a significant effect on the total phenolic content and antioxidant activity, and it was clear that the total phenolic content and antioxidant activity increased with the increasing additive level in all the enriched breads (Figure 1d,e). In agreement with our results, the total phenolic contents and antioxidant activity of wheat bread samples enriched with Moldavian dragonhead [34] and parsley leaf powder [29] increased with increasing additive levels. Dziki et al. [29] found that the total phenolic content of bread to which parsley leaf powder was added at a concentration of 5% increased from 0.23 mg GAE/g (wheat flour) to 0.84 (wheat flour + 5% parsley leaf powder). Moreover, the total phenolic content of enriched bread increased linearly with the percentage (0 to 5%) increase in the addition of dragonhead leaves from 4.8 to 10.1 mg GAE/g dry matter, and the highest radical scavenging activity was obtained in bread with the highest percentage of dragonhead leaves [34]. Therefore, it was concluded in this study that the fortification of wheat flour with PHP significantly improved the functional quality of bread, which could have a positive impact on health.

3.4. Crumb and Crust Color of Bread

Figure 2 shows how the color parameters of the crumb and crust of the bread samples (L^*, a^*, and b^*) changed depending on the PHP, FPH, GSP, and FGS addition levels in the wheat flour. Since the color of bread correlates with its quality and consumer acceptance, the evaluation of color parameters is essential [29]. The L^* value of the crumb for all breads was higher than those of the crust. This was due to the fact that the color change of the bread crust was mainly influenced by the Maillard reaction. As a result of the Maillard reaction, brown melanoidins were formed, which darkened the crust. Furthermore, the color of the crumb was influenced by the color of the components added to the wheat flour, especially polyphenols and dietary fiber. Furthermore, statistically significant correlations ($p < 0.01$) were found between the crumb L^* and the total phenolic content and antioxidant activity (r = −0.944 and r = −0.925, respectively). The Maillard reaction had no effect on the formation of the crumb color. The reason for this was that the inside of the bread could not reach the same temperature as the crust [33]. The highest crumb L^* value was obtained in the control bread, and the presence of all additives (except the breads with the 0.5% and 1.38% addition of FGS) resulted in a significant reduction ($p < 0.05$) (Figure 2a). Similarly, the crust L^* value of the control bread was different from that of all the enriched breads ($p < 0.05$) (Figure 2d). The L^* value of the crumb and crust of the breads decreased with an increasing level of enrichment (Figure 2a,d). The results from the ANOVA also showed that the level and type of additives had an effect on the L^* value of the crumb and crust. In addition, the enrichment of the bread with PHP decreased the lightness of the crumb from 70.01 to 54.30 (bread with the 4% addition of PHP). The bread samples with the addition of pistachio hulls (both dry and fresh) had lower brightness in both the crust and crumb due to the color effects caused by the higher amount of phenolic compounds in pistachio hulls (Table 1).

The a^* values of the crumb and crust increased with the increasing levels of additives (Figure 2b,e). The most striking change in the crumb and crust a^* values was observed in the breads with PHP. The a^* values of the crumbs and crusts ranged from 1.84 and 5.43 (control bread) to 6.92 and 16.61 (for bread with the 4% addition of PHP), respectively. The increase in the redness of the crumb color of bread with PHP was probably due to the phenolic content and the natural color of the pistachio hull powder, which is red to light yellow and affects the color of the bread. In addition, statistically significant correlations ($p < 0.01$) were found between the crumb a^* value and the total phenolic content and antioxidant activity (r = 0.834 and r = 0.898, respectively). Moreover, both enzymatic and non-enzymatic browning could be the reason for higher redness values in the crust color of breads with pistachio hulls than in breads with grape seeds. Similarly, statistically significant correlations ($p < 0.01$) were found between the crust a^* value and the total phenolic content and antioxidant activity (r = 0.737 and r = 0.833, respectively). The a^* values of the crumb and crust of the control bread were significantly different from those of the enriched breads, except for the crumb a^* value of the breads with 0.5% and 1.38% FGS additions. The b^* value of the crumb and crust of all the enriched breads decreased with increasing levels of additives. The crumb b^* values decreased from 14.05 in the control bread to 8.87, 8.98, 8.20, and 8.36 when 4% of wheat flour was replaced with PHP, FPH, GSP, and FGS, respectively (Figure 2c,f). In addition, the breads containing pistachio hulls had lower crust b^* values than the breads with grape seeds. The analysis of the ANOVA also showed that the types and levels of additives had significant effects on the a^* and b^* values of the crumbs and crusts.

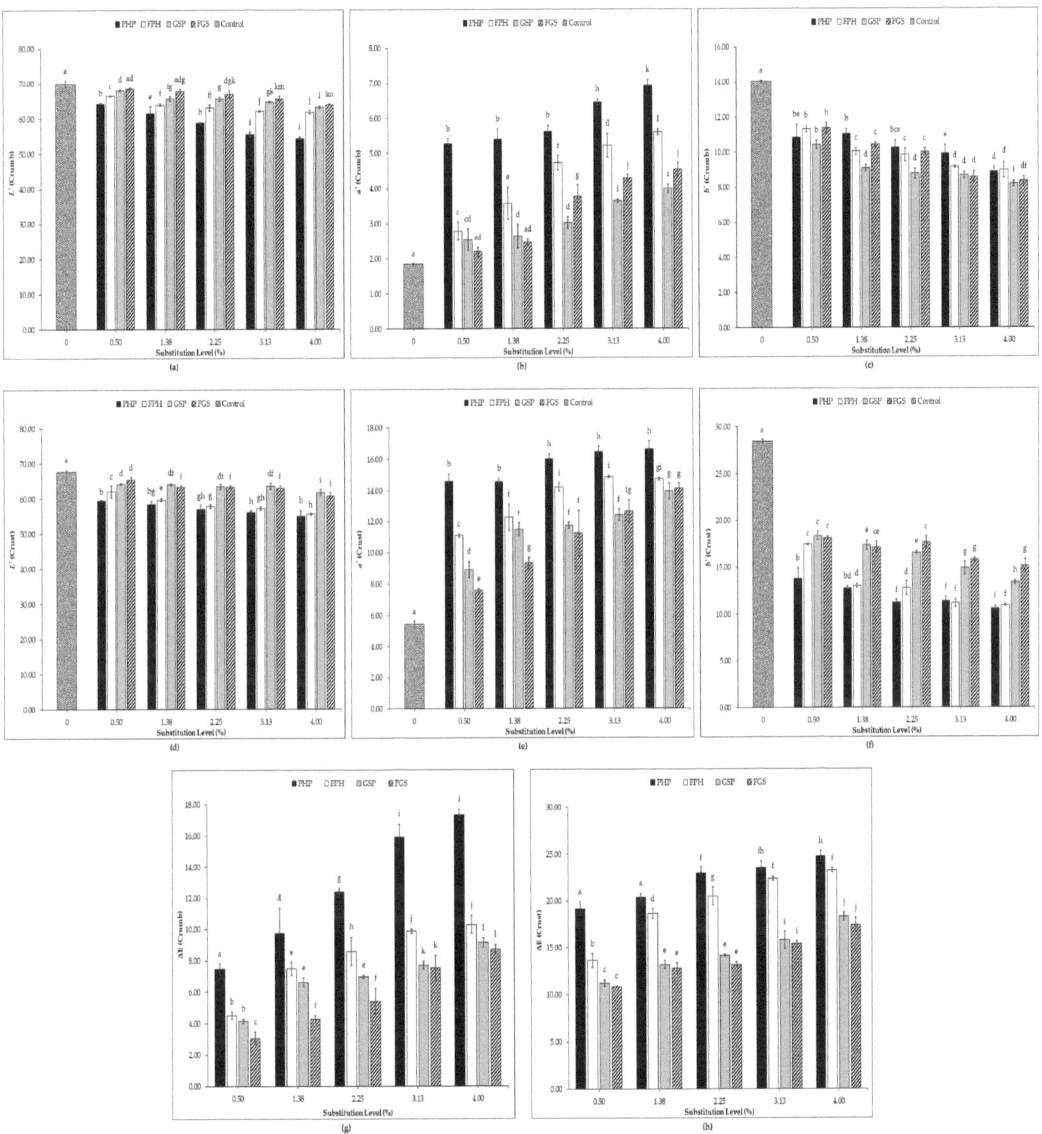

Figure 2. Crumb L^* (**a**), crumb a^* (**b**), crumb b^* (**c**), crust L^* (**d**), crust a^* (**e**), crust b^* (**f**), crumb color change (**g**), and crust color change (**h**) of the control and enriched breads. Values of each parameter with the same letters were not significantly different ($p > 0.05$).

The total color differences (ΔE) of the crumbs ranged from 7.45 to 17.31% for PHP, 4.52 to 10.29% for FPH, 4.16 to 9.13% for GSP, and 3.05 to 17.31% for FGS. This parameter (ΔE) indicated significant changes in color after the fortification compared with the control. As the amount of additive in the bread dough increased, the ΔE of the crumb increased (Figure 2g). The ΔE values of the crust also decreased with increasing additive levels (Figure 2h). In addition, the highest values for crumb and crust color changes were obtained for bread with PHP, while the lowest values were obtained for bread with FGS. The type of additive used had a significant effect on the color of the breads. The results from the ANOVA also clearly demonstrated that the ΔE values of the crumb and crust changed significantly depending on the type and level of additive (Figure 2g,h).

3.5. Textural Properties of Bread Crumb

The textural properties (hardness, springiness, cohesiveness, chewiness, and resilience) of the breads are shown in Figure 3; these are important quality factors because these properties are strongly related to consumer preferences. Hardness is defined as the loss of softness of the bread crumb and is measured in newtons (N) [35]. The hardness of the control bread was 32.8 N, indicating a firm and dense texture. The highest crumb hardness was found for the bread with the 4.0% addition of FGS (39.65 N), and the lowest was found for the bread with the 0.5% addition of FPH (26.37 N). As can be seen in Figure 3a, PHP and FPH showed softening effects on the crumb hardness, whereas FGS and bread with a higher level of 2.25% GSP had significantly equal and higher hardness values than the control bread. When the level of additives was increased from 0.5% to 4%, the hardness of the enriched bread crumbs (with the exception of FGS) initially decreased and then increased, and the hardness of the PHP and FPH breads to which 4% was added were similar to that of the control (Figure 3a). The hardness of the breads with GSP and FGS additions were higher than those of the breads with PHP and FPH additions. This result can be explained by the higher moisture contents of the breads with PHP and FPH (Figure 1a), the additions of dietary fiber (according to Akbari-Alavijeh et al. [17], pistachio hulls are a source of dietary fiber), and the increase in protein content (Table 1) decreasing the hardness [36–39]. In addition, the results of the ANOVA showed that the type and level of additives had significant effects on the hardness, and it was clear that the hardness in all enriched breads increased with increased levels of additives (Figure 3a). Similarly, Chen et al. [6], Korus et al. [40], and Dziki et al. [29] found that bread crumb hardness increased with increasing amounts of additives via fortification with mango peel powder, lyophilized kale, and parsley leaf powder, respectively.

Springiness describes the way the crumb of a product springs back following compression and is related to the elasticity of the bread crumb [34]. The lowest springiness of enriched bread (0.57) was found for the bread with the 1.38% addition of GSP and the highest springiness (1.68) was found for the bread with the 4% addition of FGS (Figure 3b). Like hardness, springiness is also strongly influenced by moisture content and water-holding capacity. Hong et al. [41] also reported that breads with mushroom powder were characterized by low specific volumes and springiness due to the high water-holding capacity of the mushroom powder. A similar trend to that observed for hardness was also observed for springiness, and the values for springiness of the bread with FGS were significantly higher than those of the control bread and the other enriched breads. It is also worth noting that consumers preferred bread with higher elasticity, as springiness is associated with freshness and loss of elasticity of the crumb, while bread with low springiness is associated with a crispy crumb [42,43]. The springiness of the PHP and FPH breads, as well as the bread with a higher GSP level of 3.13%, was similar to that of the control (Figure 3b). The results from the ANOVA showed that the springiness values of the bread crumbs changed significantly depending on the level of additive, except for the breads with PHP and FGS (Figure 3b). The results also showed that the type of additive affected the springiness.

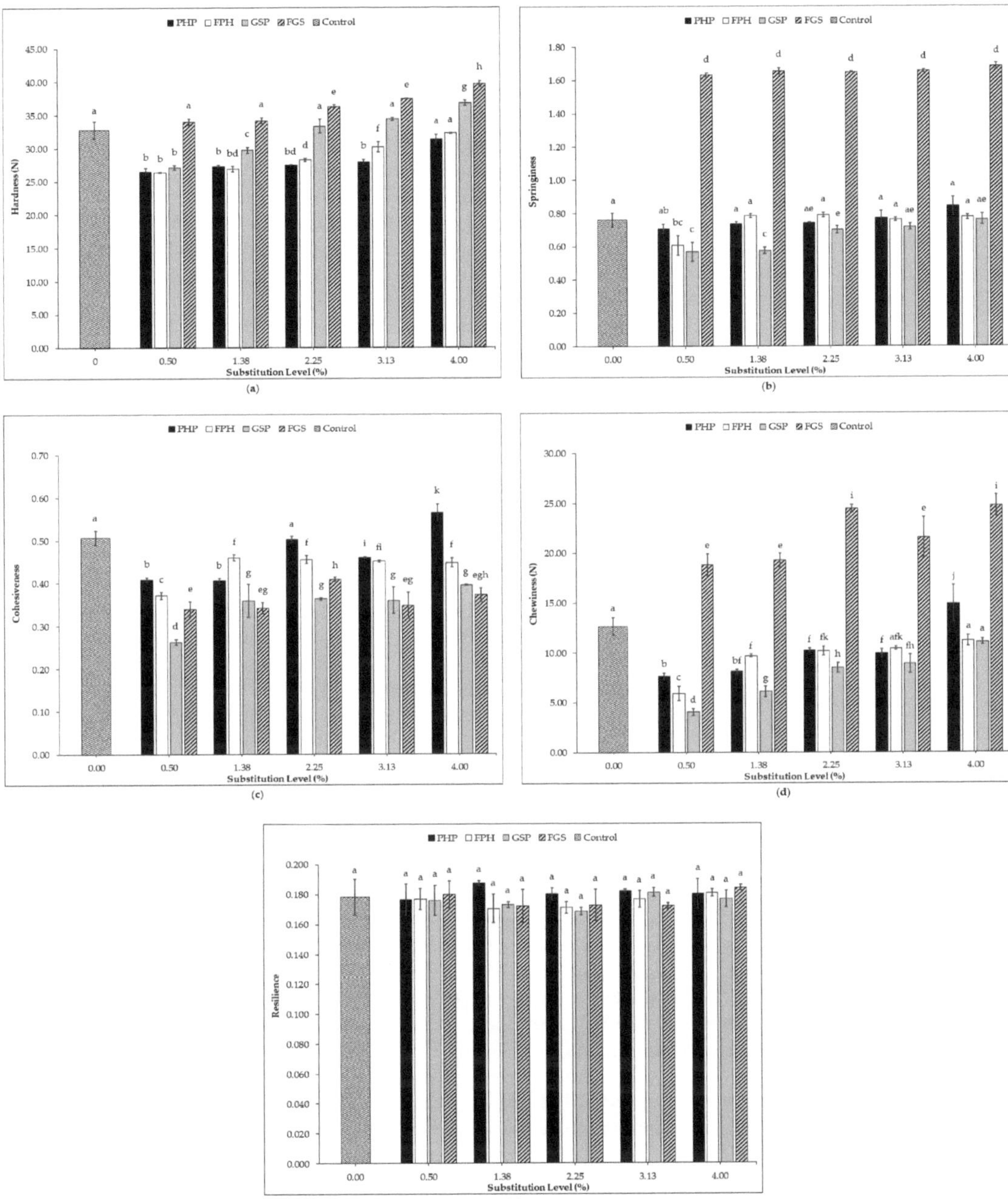

Figure 3. Hardness (**a**), springiness (**b**), cohesiveness (**c**), chewiness (**d**), and resilience (**e**) of the control and enriched breads. Values of each parameter with the same letters were not significantly different ($p > 0.05$).

Cohesiveness describes the amount of deformation a material can withstand before it breaks, the total strength of the internal bonds holding the product together, and the quantified internal resistance of the food structure [35]. The lowest cohesiveness of the enriched bread (0.263) was found for the bread with 0.5% GSP and the highest cohesiveness (0.565) was found for the bread with 4% PHP (Figure 3c). Based on the obtained results, the cohesiveness of the control bread was 0.507. After the addition of 0.5% PHP, FPH, GSP, and FGS, this value decreased to 0.409, 0.371, 0.263, and 0.340, respectively. When the level of the additive was increased from 0.5% to 4%, the cohesiveness of the enriched bread crumb increased. However, the cohesiveness was higher only in the bread with a PHP addition of 4.0% than in the control bread. In addition, there were no statistically significant differences between the control bread and the breads with a PHP addition of 3.18% ($p > 0.05$). When comparing the type of additive, it was found that the breads with grape seeds (both dry and fresh) had lower values for crumb cohesiveness than the breads with pistachio hulls (both dry and fresh). This indicates an inverse relationship between the hardness and cohesiveness values of the bread crumb. The loss of cohesiveness could have been due to lower moisture content and intramolecular bonds between bread components [44]. Bread with high cohesiveness is desirable because it is easy to form into a ball in the mouth instead of crumbling when chewed, while bread with low cohesiveness tends to break, which negatively affects consumer acceptance of the bread [45]. The addition of a higher level of 4% PHP improved the cohesiveness of the internal structure of the crumb. In addition, the results of the ANOVA showed that the type and level of the additives had a significant effect on the cohesiveness ($p < 0.05$).

Chewiness is an important parameter of bread texture that depends on springiness, hardness, and cohesiveness and gives the food the energy needed to require chewing before swallowing [35]. As shown in Figure 3d, the chewiness values of the control bread (12.65 N) decreased to 7.64 N, 5.94 N, and 4.04 N after the addition of PHP, FPH, and GSP, respectively. In contrast, an increase in chewiness (18.82 N) was observed after the addition of FGS. The results of the ANOVA showed that the type and level of additives had significant effects on chewiness, and it was clear that chewiness increased with increasing levels of additives in all the enriched breads (Figure 3d). The chewiness of the breads with FGS and the bread with 4.0% PHP were higher than that of the control bread. The bread with the 4.0% addition of GSP and a higher level of 3.13% FPH had significantly equal chewiness values to the control bread. Since the chewiness values of the fortified breads followed a similar trend to the hardness and springiness: as the hardness and springiness increased, the chewiness of the breads also increased, whereas chewiness was negatively related to the cohesiveness. While lower chewiness is desirable in bread, the current results showed that this parameter was improved by the addition of PHP, FPH, and GSP.

Resilience describes the rate at which the crumb recovers after compression [35] and is an important property related to bread freshness [46]. The resilience values of all enriched breads were not statistically different from those of the control ($p > 0.05$) and were quite consistent for all the enriched breads (Figure 3e). Thus, the addition of pistachio hulls and grape seeds could have a slight effect on the resilience of the breads, but this did not depend on the level and type of additives. Similar results were obtained by Tong et al. [47] and Dziki et al. [29], where the changes in the resilience of wheat bread due to enrichment with honey powder and parsley leaf powder, respectively, were not significant.

3.6. Sensory Quality of Bread

The appearance of the control bread and the enriched bread is shown in Figure 4. Sensory evaluation is an important parameter for assessing the quality of food products to meet consumer demands. The results of the sensory evaluation of the enriched breads and the control bread are shown in Figure 5. The data were analyzed in terms of the crumb and crust color, chewiness, flavor, taste, and overall liking.

Figure 4. Picture of control and enriched breads.

The highest score for the crust and crumb color was obtained by the control bread, and the lowest by the bread with PHP, which was consistent with the results of the color analysis. The results show that the breads that contained pistachio hulls had lower scores compared with the other breads. This may be attributed to the lower brightness of the bread and the higher redness, which may have had a negative effect on the color perception of the panelists. The crust color evaluation was not affected by the type and level of additives, while the crumb color evaluation was affected by the type and ratio of additives (Figure 5a,b). Moreover, it was found that the crust and crumb color scores for all enriched breads (except for bread with 0.5% and 4% additions of GSP and FGS) were different from those of the control bread ($p < 0.05$). No significant difference was found between the control sample and the enriched breads in terms of chewiness. The lowest chewiness score was found for the bread with 1.38% GSP and the highest was found for the bread with 4% PHP (Figure 5c). Flavor and taste scores were also not influenced by the type and level of additives (Figure 5d,e). The lowest flavor and taste scores were obtained for the breads with PHP; however, the control and all enriched breads did not differ significantly in their flavor and taste scores. Consequently, with a PHP level of 4% in the bread recipe, the flavor and taste scores of the bread were below 4 points. The highest score for overall liking was given to the control bread, and the lowest to the bread with PHP, as well as for all other sensory characteristics tested (with the exception of chewiness). Nevertheless, the control bread and all the enriched breads differed significantly in the overall liking score (except for the breads with 0.5% and 4% added GSP and FGS). The overall liking score was only affected by the type of additive (Figure 5f).

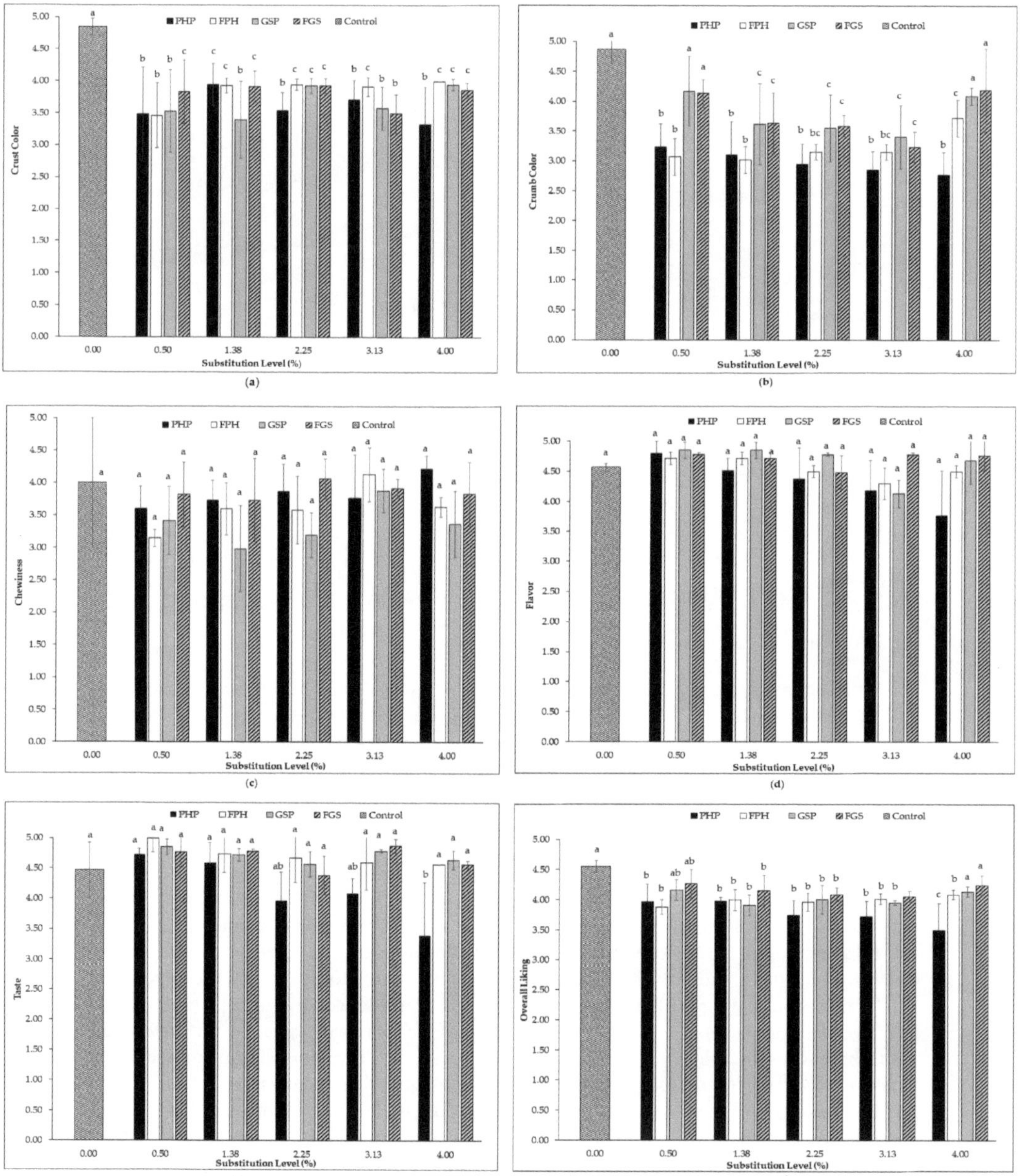

Figure 5. Results of the sensory analysis of the control and enriched breads. Crust color (**a**), crumb color (**b**), chewiness (**c**), flavor (**d**), taste (**e**), and overall liking (**f**). Values of each parameter with the same letters were not significantly different ($p > 0.05$).

4. Conclusions

Changes in moisture content, crust and crumb color, specific volume, baking loss, total phenolic content, antioxidant activity, and textural properties of bread were studied with respect to the addition of FPH, PHP, FGS, and GSP to wheat flour, and the results showed that the addition of by-products affected the quality characteristics of the bread.

The addition of PHP resulted in an increase in total phenolic content and antioxidant activity; however, the specific volume showed the lowest values among the other breads. Although PHP had a decreasing effect on the values of the specific volume of the bread samples, it was found that the overall quality characteristics were not negatively affected. The bread with FGS in the crust and crumb had the lowest value for color change for all additive levels. The lower hardness and chewiness and higher cohesiveness led to the PHP and FPH breads being of good quality. The current results show that most of the bread quality characteristics depended on the level of the additive used, and thus, it is important to control the amount of additives in the bread. The analysis of all these data suggests that pistachio hull powder is an alternative functional ingredient that can be used together with wheat flour for bread making due to its high fiber content and richness in antioxidant and phenolic substances. Although bread with pistachio hull powder caused an increase in total phenolic content and antioxidant capacity, and thus, value-added products were obtained, the sensory quality characteristics were lower than in the control and in bread with grape seeds. Therefore, considering the quality characteristics of bread, pistachio hull powder was found to be a good alternative additive to enrich bread.

Author Contributions: Conceptualization, B.K.; methodology, B.K.; software, B.K. and G.A.K.; validation, B.K. and G.A.K.; formal analysis, B.K. and G.A.K.; investigation, B.K. and G.A.K.; resources, B.K. and G.A.K.; data curation, B.K. and G.A.K.; writing—original draft preparation, data curation, B.K. and G.A.K.; writing—review and editing, B.K. and G.A.K.; visualization, B.K.; supervision, B.K.; project administration, B.K.; funding acquisition, B.K. All authors have read and agreed to the published version of the manuscript.

Funding: The authors acknowledge the financial support of Gaziantep University, Council of Scientific Research Projects (project no.: GSF.YLT.19.01) and Özmen Un Company, Gaziantep, Turkey.

Data Availability Statement: The data presented in this study are available on request from the corresponding author.

Conflicts of Interest: The authors declare no conflict of interest.

References

1. Sabanis, D.; Tzia, C.; Papadakis, S. Effect of different raisin juice preparations on selected properties of gluten-free bread. *Food Bioproc. Tech.* **2008**, *1*, 374–383. [CrossRef]
2. Rouille. J.; Le Bail, A.; Coucoux, P. Influence of formulation and mixing conditions on bread making qualities of French frozen dough. *J. Food Eng.* **2000**, *43*, 197–203. [CrossRef]
3. Zhao, C.C.; Lu, J.K.; Ameer, K. Effects of tofu whey powder on the quality attributes, isoflavones composition and antioxidant activity of wheat flour pan bread. *LWT* **2021**, *143*, 111166. [CrossRef]
4. Hoye, C., Jr.; Ross, C.F. Total phenolic content, consumer acceptance, and instrumental analysis of bread made with grape seed flour. *J. Food Sci.* **2011**, *76*, S428–S436. [CrossRef] [PubMed]
5. Majzoobi, M.; Azadmard-Damrichi, S.; Peighambardoust, S.H.; Aghamirzaei, M. Effects of grape seed powder as a functional ingredient on flour physicochemical characteristics and dough rheological properties. *J. Agric. Sci. Technol.* **2015**, *17*, 365–373.
6. Chen, Y.; Zhao, L.; He, T.; Ou, Z.; Hu, Z.; Wang, K. Effects of mango peel powder on starch digestion and quality characteristics of bread. *Int. J. Biol. Macromol.* **2019**, *140*, 647–652. [CrossRef]
7. Gawlik-Dziki, U.; Świeca, M.; Dziki, D.; Baraniak, B.; Tomiło, J.; Czyż, J. Quality and antioxidant properties of breads enriched with dry onion (*Allium cepa* L.) skin. *Food Chem.* **2013**, *138*, 1621–1628. [CrossRef]
8. Sagar, N.; Pareek, S. Fortification of multigrain flour with onion skin powder as a natural preservative: Effect on quality and shelf life of the bread. *Food Biosci.* **2021**, *41*, 100992. [CrossRef]
9. Altunkaya, A.; Hedegaard, R.; Brimer, L.; Gökmen, V.; Skibsted, L. Antioxidant capacity versus chemical safety of wheat bread enriched with pomegranate peel powder. *Food Funct.* **2013**, *4*, 722. [CrossRef]
10. Dziki, D.; Różyło, R.; Gawlik-Dziki, U.; Świeca, M. Current trends in the enhancement of antioxidant activity of wheat bread by the addition of plant materials rich in phenolic compounds. *Trends Food Sci. Technol.* **2014**, *40*, 48–61. [CrossRef]
11. Iuga, M.; Mironeasa, S. Potential of grape byproducts as functional ingredients in baked goods and pasta. *Compr. Rev. Food Sci.Food Saf.* **2020**, *19*, 2473–2505. [CrossRef] [PubMed]
12. Barreca, D.; Laganà, G.; Leuzzi, U.; Smeriglio, A.; Trombetta, D.; Bellocco, E. Evaluation of the nutraceutical, antioxidant and cytoprotective properties of ripe pistachio (*Pistacia vera* L., variety Bronte) hulls. *Food Chem.* **2016**, *196*, 493–502. [CrossRef]

13. Özbek, H.; Halahlih, F.; Göğüş, F.; Koçak Yanık, D.; Azaizeh, H. Pistachio (*Pistacia vera* L.) hull as a potential source of phenolic compounds: Evaluation of ethanol–water binary solvent extraction on antioxidant activity and phenolic content of pistachio hull extracts. *Waste Biomass Valor.* **2020**, *11*, 2101–2110. [CrossRef]
14. Özbek, H.; Yanık, D.; Fadıloğlu, S.; Göğüş, F. Optimization of microwave-assisted extraction of bioactive compounds from pistachio (*Pistacia vera* L.) hull. *Sep. Sci. Technol.* **2020**, *55*, 289–299. [CrossRef]
15. Goli, A.; Barzegar, M.; Sahari, M. Antioxidant activity and total phenolic compounds of pistachio (*Pistachia vera*) hull extracts. *Food Chem.* **2005**, *92*, 521–525. [CrossRef]
16. Grace, M.; Esposito, D.; Timmers, M.; Xiong, J.; Yousef, G.; Komarnytsky, S.; Lila, M. Chemical composition, antioxidant and anti-inflammatory properties of pistachio hull extracts. *Food Chem.* **2016**, *210*, 85–95. [CrossRef]
17. Akbari-Alavijeh, S.; Soleimanian-Zad, S.; Sheikh-Zeinoddin, M.; Hashmi, S. Pistachio hull water-soluble polysaccharides as a novel prebiotic agent. *Int. J. Biol. Macromol.* **2018**, *107*, 808–816. [CrossRef] [PubMed]
18. Arjeh, E.; Akhavan, H.; Barzegar, M.; Carbonell-Barrachina, Á. Bioactive compounds and functional properties of pistachio hull: A review. *Trends Food Sci. Technol.* **2020**, *97*, 55–64. [CrossRef]
19. Lalegani, S.; Gavlighi, H.A.; Azizi, M.H.; Sarteshnizi, R.A. Inhibitoryactivity of phenolic-rich pistachio green hull extract-enriched pasta on key type 2 diabetes relevant enzymes and glycemic index. *Food Res. Int.* **2018**, *105*, 94–101. [CrossRef]
20. AOAC. *Official Methods of Analysis of AOAC International*, 17th ed.; Horwitz, W., Ed.; Association of Official Analytical Chemists: Cincinnati, OH, USA, 2000.
21. *TSE 5000/T*; Bread. Turkish Standards Institute: Ankara, Türkiye, 2004.
22. Hunter, R.S. Scales for the measurements of color difference. In *The Measurement of Appearance*; John Wiley & Sons: New York, NY, USA, 1975.
23. AACC-I. *AACC International Approved Methods of Analysis*, 11th ed.; Method 10-05.01. Guidelines for measurement of volume by rapeseed displacement; American Association of Cereal Chemists: St. Paul, MN, USA, 2010.
24. Singleton, V.L.; Orthofer, R.; Lamuela-Raventós, R.M. Analysis of total phenols and other oxidation substrates and antioxidants by means of folin-ciocalteu reagent. *Oxid. Antioxid. Part A* **1999**, *299*, 152–178. [CrossRef]
25. Brand-Williams, W.; Cuvelier, M.; Berset, C. Use of a free radical method to evaluate antioxidant activity. *LWT* **1995**, *28*, 25–30. [CrossRef]
26. AACC. *AACC International Approved Methods of Analysis*; Method 74.09, Measurement Of Bread Firmness by Universal Testing Machine; AACC International Approved Methods: Arnold, MD, USA, 1995.
27. Ni, Q.; Ranawana, V.; Hayes, H.E.; Hayward, N.J.; Stead, D.; Raikos, V. Addition of Broad Bean Hull to Wheat Flour for the Development of High-Fiber Bread: Effects on Physical and Nutritional Properties. *Foods* **2020**, *9*, 1192. [CrossRef] [PubMed]
28. Sayed Ahmad, B.; Talou, T.; Straumite, E.; Sabovics, M.; Kruma, Z.; Saad, Z.; Hijazi, A.; Merah, O. Protein Bread Fortification with Cumin and Caraway Seeds and By-Product Flour. *Foods* **2018**, *7*, 28. [CrossRef] [PubMed]
29. Dziki, D.; Hassoon, W.H.; Biernacka, B.; Gawlik-Dziki, U. Dried and powdered leaves of parsley as a functional additive to wheat bread. *Appl. Sci.* **2022**, *12*, 7930. [CrossRef]
30. Pejcz, E.; Spychaj, R.; Wojciechowicz-Budzisz, A.; Gil, Z. The effect of Plantago seeds and husk on wheat dough and bread functional properties. *LWT* **2018**, *96*, 371–377. [CrossRef]
31. Vartolomei, N.; Turtoi, M. The influence of the addition of rosehip powder to wheat flour on the dough farinographic properties and bread physico-chemical characteristics. *Appl. Sci.* **2021**, *11*, 12035. [CrossRef]
32. Otegbayo, B.O.; Adebiyi, O.M.; Bolaji, O.A.; Olunlade, B.A. Effect of soy enrichment on bread quality. *Int. Food Res. J.* **2018**, *25*, 1120–1125.
33. Gómez, M.; Martinez, M.M. Fruit and vegetable by-products as novel ingredients to improve the nutritional quality of baked goods. *Crit. Rev. Food Sci. Nutr.* **2018**, *58*, 2119–2135. [CrossRef]
34. Dziki, D.; Cacak-Pietrzak, G.; Gawlik-Dziki, U.; Sułek, A.; Kocira, S.; Biernacka, B. Effect of Moldavian dragonhead (*Dracocephalum moldavica* L.) leaves on the baking properties of wheat flour and quality of bread. *CyTA J. Food* **2019**, *17*, 536–543. [CrossRef]
35. Szczesniak, A.S. Classification of textural characteristics. *J. Food Sci.* **1963**, *28*, 385–389. [CrossRef]
36. Martínez, M.M.; Díaz, Á.; Gómez, M. Effect of different microstructural features of soluble and insoluble fibres on gluten-free dough rheology and bread-making. *J. Food Eng.* **2014**, *142*, 49–56. [CrossRef]
37. Martínez-Cervera, S.; de la Hera, E.; Sanz, T.; Gómez, M.; Salvador, A. Effect of using erythritol as a sucrose replacer in making Spanish muffins incorporating Xanthan Gum. *Food Bioprocess Tech.* **2011**, *5*, 3203–3216. [CrossRef]
38. López, E.P.; Pérez, G.T.; de Erramouspe, P.L.J.; Cuevas, C.M. Effect of Brea Gum on the characteristics of wheat bread at different storage times. *Food Sci. Technol.* **2013**, *33*, 745–752. [CrossRef]
39. Różyło, R.; Hameed Hassoon, W.; Gawlik-Dziki, U.; Siastała, M.; Dziki, D. Study on the physical and antioxidant properties of gluten-free bread with brown algae. *CyTA J. Food* **2016**, *15*, 196–203. [CrossRef]
40. Korus, A.; Witczak, M.; Korus, J.; Juszczak, L. Dough rheological properties and characteristics of wheat bread with the addition of lyophilized kale (*Brassica oleracea* L. var. *sabellica*) powder. *Appl. Sci.* **2023**, *13*, 29. [CrossRef]
41. Hong, G.-H.; Kim, Y.-S.; Song, G.-S. Effect of oyster mushroom (*Pleurotus ostreatus*) powder on bread quality. *Prev. Nutr. Food Sci.* **2005**, *10*, 214–218. [CrossRef]
42. Martínez, M.M.; Román, L.; Gómez, M. Implications of hydration depletion in the in vitro starch digestibility of white bread crumb and crust. *Food Chem.* **2018**, *239*, 295–303. [CrossRef]

43. Tóth, M.; Kaszab, T.; Meretei, A. Texture profile analysis and sensory evaluation of commercially available gluten-free bread samples. *Eur. Food Res. Technol.* **2022**, *248*, 1447–1455. [CrossRef]
44. Ulziijargal, E.; Yang, J.-H.; Lin, L.-Y.; Chen, C.-P.; Mau, J.-L. Quality of bread supplemented with mushroom mycelia. *Food Chem.* **2013**, *138*, 70–76. [CrossRef]
45. Onyango, C.; Mutungi, C.; Unbehend, G.; Lindhauer, M.G. Modification of gluten-free sorghum batter and bread using maize, potato, cassava or rice starch. *LWT* **2011**, *44*, 681–686. [CrossRef]
46. Young, L.S. Applications of texture analysis to dough and bread. In *Breadmaking*; Woodhead Publishing: Cambridge, UK, 2012; pp. 562–579.
47. Tong, Q.; Zhang, X.; Wu, F.; Tong, J.; Zhang, P.; Zhang, J. Effect of honey powder on dough rheology and bread quality. *Food Res. Int.* **2010**, *43*, 2284–2288. [CrossRef]

applied sciences

MDPI

Article

Influence of the Protein-Based Emulsions on the Rheological, Thermo-Mechanical and Baking Performance of Muffin Formulations

Iuliana Banu [1], Livia Patrașcu [2], Ina Vasilean [1], Loredana Dumitrașcu [1] and Iuliana Aprodu [1,*]

[1] Faculty of Food Science and Engineering, Dunarea de Jos University of Galati, 111 Domneasca Str., 800008 Galati, Romania
[2] Cross-Border Faculty, Dunarea de Jos University of Galati, 111 Domneasca Str., 800008 Galati, Romania
* Correspondence: iuliana.aprodu@ugal.ro

Abstract: The impact of replacing the sunflower oil in a typical muffin formulation with different protein-based emulsions was investigated. Fundamental rheological measurements indicated significant differences between emulsions prepared with soy, lupin, and yeast proteins. The highest viscosity of 2.04 Pa·s was registered for the lupin protein-based emulsion, whereas the yeast protein-based emulsion exhibited the narrowest linear viscoelastic region. The influence of the protein-based emulsions on the thermo-mechanical properties of wheat flour dough was further investigated using the Mixolab device and Chopin+ protocol. Oil substitution with emulsion resulted in better starch gelatinization with the C3 torque of 0.46 Nm being registered for doughs with soy and lupin protein emulsions. Significant differences in terms of moisture, color, porosity, and texture were observed between muffins prepared with protein-based emulsions and control. The lower fat baked products retained higher amounts of water (25.05–26.00%) and exhibited slightly more vivid color (color intensity of 46.34–46.81) and harder texture (firmness of 5.64–5.86 N). The sensory analysis confirmed that soy, lupin, and yeast protein emulsions can be used for obtaining muffin samples with acceptable taste and flavor, and overall quality comparable to the control. These results indicate that the protein based-emulsions are promising oil replacers in muffin formulations.

Keywords: muffins; fat reduction; proteins; rheological properties

1. Introduction

Sweet baked food products, such as cakes, muffins, and biscuits, are convenient and affordable tasty snacks appreciated by many consumers of all ages. However, this category of food products is usually incriminated for its high sugar and fat contents. Nutrition related diseases, together with the increase in consumer concern and demand for healthier foods, led to a high interest in developing new formulations with less fat and/or sugar content. However, considering the contribution of the main ingredients to the technological behavior and sensorial perception, finding solutions for low calorie, sweet, baked products still represents a great challenge.

Fat is the energetic component with the highest calorific value and therefore, its reduction by substitution with other ingredients is highly appealing. Reducing the fat content of different types of baked foods represents a great challenge, as fats exert a high contribution to the sensorial characteristics like texture, softness, high volume, mouthfeel, and flavor [1–3]. Muffins are popular breakfast bakery products, mainly obtained from flour, sugar, fats, and eggs, and are usually known to contain up to 24% of fat in the final product or 40% fat relative to the flour content [3], both of animal or vegetable origin. There are many studies focusing on fat reduction in muffin cakes by substitution with different fat mimetics. The most tested in this respect are carbohydrate-based ingredients [2]. Different studies report the use of starch [4], chia seed mucilage [5], inulin [6], cocoa fibers [7], gums, polydextrose [2], and legume/fruit purees [1] for replacing various amounts of fats in typical muffin formulations. The use of oleogels is a rather new approach for fat replacement in muffins. In addition to reducing the total fat content, the use of oleogels represents a new healthier alternative for shortenings and saturated fats [8,9]. Regardless of the ingredient used to substitute fats in the muffin or cake formulations, different quality changes were observed, such as volume decrease or increase in crumb density and overall hardness, with negative consequences on the sensorial perception of the final product [6,10].

The use of protein-based ingredients for reducing the content of fats in bakery applications received less attention compared to other fat-based replacers. Protein-based fat replacers used so far for obtaining reduced fat baked products with improved porosity and pore area included animal protein sources, such as whey, milk, or egg white [11]. These protein sources are not considered sustainable ingredients to be used to feed the next generation of people, thus, alternative environmentally friendly sources need to be explored. Legume proteins are acknowledged as sustainable ingredients with high nutritional value and good functional properties [12].

The functional properties of proteins, such as high-water binding and retention capacity as well as foaming and emulsifying capacities, may contribute to the volume and mouthfeel of muffin cakes with a promising potential to imitate the fat function [3]. Azmoon et al. [3] incorporated a hydrocolloid-protein emulsion in a muffin cake recipe as an oil substitute. Authors obtained O/W emulsions using a mixture of guar and konjac gums as stabilizers together with soy protein isolate and achieved a 63% oil substitution rate in relation to the control sample's oil content, with no significant differences on quality characteristics of the final product. Using emulsions as fat substitutes in cakes was also studied by Grossi Bovi Karatay et al. [13] who substituted palm oil from pound cakes with emulsions obtained from canola oil and chickpea aquafaba (the viscous liquid formed when boiling legumes used by vegans as an egg replacement). The maximum rate of total fat reduction, in this case, was 25%, with no significant effect on quality characteristics of the final product.

Another interesting sustainable protein source, which was not previously considered as an ingredient for reducing the fat content of muffins, is the brewer's spent yeast extract. Although yeast extracts are rich in proteins, minerals, B vitamins, and many other biologically active compounds, their use in various food products is mostly related to flavoring properties [14,15]. However, taking into account their high protein content, the yeast extracts have good potential as emulsifiers and emulsion-stabilizing agents rich in essential amino acids for several applications in nutraceuticals and the food industry.

The objective of the study was to investigate the rheological properties of soy, lupin, and yeast protein-based emulsions and to evaluate the influence of oil substitution with these emulsions on the thermo-mechanical behavior of the dough and the baking performance of muffin formulations.

2. Materials and Methods

2.1. Materials

The following protein derivatives were considered in the experiment: soy protein isolate (Ubimedia S.R.L., Galati, Romania; 84.6% protein and 9.1% moisture), lupin concentrate (Lupipro 550, Luprodin; protein content of 57.8% and 9.4% moisture), and powdered yeast extract (69.1% protein and 4.9% moisture), obtained from *Saccharomyces cerevisiae* biomass through autolysis for 48 h at 55 °C, followed by centrifugation at 9000 rpm for 15 min to remove the yeast cell and freeze-drying (CHRIST Alpha 1–4 LD plus, Germany) the resulting liquid portion.

Superior white wheat flour (11.5% protein, 11.3% moisture and 0.4% ash) used for preparing muffin samples was purchased from Dobrogea Grup SA, Constanta, Romania. Other ingredients used for preparing the muffins, such as refined white sugar, fresh white eggs, sunflower oil (Spornic, Prutul SA, Galati, Romania), whole milk, and refined salt, were purchased from a local market (Galati, Romania).

2.2. Chemical Composition

The chemical compositions of the protein powders, wheat flour, and muffin samples were assessed using the following official methods: SR ISO 712:2005 [16] was used to determine the moisture content by air-drying the samples with a LabTech oven (Daihan Labtech Co., LTD, Kyungki-Do, Republic of Korea). The Kjeldahl method AACC 46–11.02 [17] (Raypa Trade, R. Espinar, S.L., Barcelona, Spain) and a nitrogen-to-protein conversion factor of 6.25 were used to determine the crude protein content and the SR ISO 2171:2002 [16] was employed to quantify the ash content.

2.3. Preparation of Emulsions

Suspensions with protein concentrations of 6% (*w*/*v*) were prepared by allowing the powders to well hydrate at room temperature, with over four hours of mixing at 600 rpm on a Cimarec i Poly magnetic stirrer (Thermo Fisher Scientific, Waltham, MA, USA). The protein suspensions were then mixed with sunflower oil (1:1 ratio, *w*/*v*) and further homogenized for 5 min at a speed of 15,000× *g*, using an Ultra Turrax ® IKA T18 basic homogenizer and the S18N-19G dispersing tool (IKA-Werke GmbH and Co. KG, Staufen, Germany).

2.4. Fundamental Rheological Measurements on Emulsions

The rheological characterization of the protein-based emulsions was performed by using a controlled-stress rheometer (AR2000ex, TA Instruments Ltd., New Castle, DE, USA). All measurements were performed at 20 °C with a cone–plate geometry (40 mm in diameter, 2° cone angle) and a closing gap of 1000 µm. The excess sample resulting when lowering the cone geometry towards the plate was trimmed off and the exposed edges of the sample were covered with a tinny layer of mineral oil, such as to limit the moisture loss during the measurements.

The emulsion samples were first subjected to a strain sweep test at a constant frequency of 1 Hz by increasing the oscillating strain in the 0.1–100% domain for checking the linear viscoelastic region (LVR).

Further frequency sweep testing was applied at constant strain within the LVR. The storage modulus (G′) and loss modulus (G″) of the protein-based emulsions were recorded while oscillating the frequency from 0.1 to 100 Hz.

The steady shear viscosities were finally recorded while running a stepped flow test at increasing shear rate values from 0.1 to 100 s^{-1}.

The TA Rheology Advantage Data Analysis Software V 4.8.3. (TA Instruments, New Castle, DE, USA) was used to analyze the results of the triplicate measurements.

2.5. Thermo-Mechanical Behavior

The Mixolab Chopin (Tripette & Renaud Chopin, Villeneuve La Garenne, France) device was used to estimate the influence of sunflower oil or proteins-based emulsion addition on the rheological properties of the white wheat flour dough. The amount of water required to prepare the wheat flour dough with standard torque value of 1.1 Nm (C1) at the initial mixing step was considered for all investigated samples. The empirical rheological measurements were performed on model doughs, prepared by well mixing in the Mixolab tank the white wheat flour with oil or protein-based emulsion (4% in respect to the flour) and water to reach water absorption (WA) of the flour. The volume fraction of the oil or protein-based emulsion added to prepare the model doughs was decided upon running preliminary tests (results not shown) in such a manner to allow for measuring of the torque while mixing at 30 °C in the first stage of the Mixolab protocol. At a higher volume oil fraction, the doughs were sticky, no appropriate contact with the mixing blades was ensured, and therefore, the dough resistance could not be correctly recorded by Mixolab. The torque of the resulting doughs was monitored over the 45 min long Chopin+ protocol, involving mixing coupled with thermal treatment. The typical Mixolab parameters were registered for estimating the influence of oil or emulsions addition on the properties of the main wheat flour constituents: C2 (Nm) torque providing information on the protein weakening under constant mixing and temperature (it was not possible to register the C2 for the doughs prepared with oil or soy and lupin protein emulsions), C3 (Nm) torque related to starch gelatinization, C4 (Nm) torque which is a measure of the starch gel stability at high temperature, and C5 (Nm) torque related to starch retrogradation upon cooling at 50 °C.

2.6. Preparation of the Muffins

Protein-based emulsions were further used to prepare muffins, following the procedure proposed by Martínez-Cervera et al. [18] with slight modifications. The muffin formulations included white wheat flour (100 g), sugar (100 g), fresh eggs (80 g), 50 mL emulsion, 50 mL of whole milk, 4 g of baking powder, and 1.5 g NaCl, as indicated in Table 1. The muffin samples prepared with emulsions based on soy protein, lupin protein, and yeast extract were coded M1, M2, and M3, respectively. The control sample (M) was prepared with sunflower oil instead of protein-based emulsion.

Table 1. Formulation of the muffin samples.

Ingredients	Muffin Samples			
	M	M1	M2	M3
Flour, g	100	100	100	100
Sugar, g	100	100	100	100
Egg, g	80	80	80	80
Milk, ml	50	50	50	50
Oil, ml	50	-	-	-
Soy protein emulsion, ml	-	50	-	-
Lupin protein emulsion, ml	-	-	50	-
Yeast protein emulsion, ml	-	-	-	50
Baking powder, g	4	4	4	4
NaCl, g	1.5	1.5	1.5	1.5

The batters were prepared by beating the ingredients with a Braun mixer (De'Longhi, Neu Isenburg, Hessen, Germany) for 3 min at the highest speed (4), followed by 2 min at a lower speed (2). The procedure for the preparation of the muffins is presented in Figure 1. For each type of emulsion, 12 paper cups were placed in the muffin baking trays and filled with 48 g of batter. The muffins were baked at 180 °C for 20 min using an electric oven (Electrolux, Stockholm, Sweden) with top and bottom heat. The oven was preliminarily heated to the desired temperature. The samples were then left for 1 h in the paper cups to cool down to room temperature.

Weighing out and mixing thoroughly the powdered ingredients (flour, baking powder and salt)
↓
Gradual addition of the eggs separately beaten with sugar
↓
Creaming in the oil/protein-based emulsion
↓
Gradual addition of the milk
↓
Mixing all ingredients, for 5 minutes, with a mixer
↓
Dispensing the batter (48 g) into muffin tins
↓
Baking at 180°C for 20 min using an electric oven
↓
Cooling down to room temperature

Figure 1. Flow diagram for muffin preparation.

2.7. Characterization of the Muffins

The muffins prepared with different types of protein-based emulsions to replace the oil were characterized by determining the moisture content, height, texture, color, and porosity. All measurements were performed within 12 h after baking.

Weight loss (WL) during baking was calculated, upon weighing the batter poured in the muffin cups (m_0) and the final baked products upon cooling (m_1), using the following equation:

$$WL = (m_0 - m_1) \times 100/m_0 \quad (1)$$

For each muffin formulation used in the study, five different samples were used to measure the highest point in respect to the flat base, using a caliper.

The firmness of the muffins' crumb was measured using the MLFTA apparatus (Guss, Strand, South Africa). The upper and bottom parts of the muffins were discarded and a probe with a diameter of 7.9 mm was used to penetrate the samples for 25 mm with a speed of 5 mm/s. The trigger threshold force was 20 g [19]. Each muffin sample was penetrated in three different points and the maximum force was registered as an indicator of crumb firmness.

The Chroma Meter CR-410 (Konica Minolta Business Solutions Europe GmbH) colorimeter was used to measure the brightness (L*), redness (a*), and yellowness (b*) of the muffin crumb. In order to estimate the impact of oil substitution by protein-based emulsion in the muffin formulation, the total color difference (ΔE) was calculated as:

$$\Delta E = ((L^*_{sample} - L^*_{control})^2 + (a^*_{sample} - a^*_{control})^2 + (b^*_{sample} - b^*_{control})^2)^{1/2} \quad (2)$$

Moreover, chroma (C*), which gives indications on color intensity, was calculated using Equation (3):

$$C^* = (a^{*2} + b^{*2})^{1/2} \quad (3)$$

The porosity of the muffin samples was assessed by image analysis, using the procedure described by Aprodu and Banu [20]. The muffins were prepared for pores' measurement through cutting them into halves. The scanned images with a resolution of 300 dpi (Canon Lide 210, Canon Inc., Tokyo, Japan) of the resulting sections were first gray-scaled and used for further analysis of porosity characteristics using the Image J software.

2.8. Sensory Analysis Test

The sensory analysis of the muffin samples was performed in agreement with the IFST Guidelines for Ethical and Professional Practices for the Sensory Analysis of Foods by 12 consumers. The consumer panel consisted of 7 women and 5 men aged between 23 and 50 years. The recruiting criteria of the consumer panel was the regular consumption of sweet baked food products, such as muffins, at least once every two days. Written informed consent was obtained from all participants to the sensory evaluation test, according to the decision of the Dunarea de Jos University Ethics Commission no. 28/19 October 2022. A pretest session was organized to instruct the members of the sensory panel on the use of the attributes selected for characterization of the muffins. The sensory evaluation was performed in a room that had the proper environmental controls and was sufficient in size to handle the panelists and products. Each consumer was asked to evaluate the sensory attributes of all muffin samples, placed on individual small white plates and codified with three-digit random numbers. The panelists were asked to use water for cleaning the palate before tasting each muffin sample. The products were submitted to a sensory acceptance test. The nine-point hedonic scale was used to rate the appearance, flavor, taste, color, texture, and overall quality of the muffins (1—"dislike extremely" to 9—"like extremely").

2.9. Statistical Analysis

The results are presented as average values ± standard deviation of triplicate measurements. The Minitab 19 (Minitab LLC, State College, PA, USA) software was used to perform statistical analysis of the results. Significant differences between samples were identified through the ANOVA method and the post hoc test, based on the Tukey method, when $p < 0.05$.

3. Results and Discussion

3.1. Rheological Properties of the Emulsions

Proteins are natural ingredients commonly used as emulsifying agents in the food industry [21]. Protein-based emulsions were used to reduce the fat content of the muffin formulations. Three different protein sources (soy, lupin, and spent brewery yeast) were used to stabilize the emulsions by forming a viscoelastic layer on the oil droplets [21]. Emulsions were prepared by mixing the 6% protein suspensions with sunflower oil in a 1:1 ratio (*w*/*v*), decided such as not to exceed the emulsifying capacity of the protein sources used in the study. The influence of the protein source on the rheological behavior of the emulsions was assessed by running low oscillatory and stepped flow tests.

The viscoelastic behavior of the emulsions was characterized through dynamic oscillation measurements within the LVR. The results of the frequency sweep test are presented in Figure 2. For all tested emulsions, both storage and loss moduli progressively increased with frequency. In agreement with the observations of Manoi & Rizvi [22] and Liu & Tang [23], these phenomena may be attributed to the solid-like behavior of the samples, exhibiting permanent interactions. Over the entire frequency domain considered in the experiment, the samples exhibited slightly higher G′ compared to G″ values, suggesting the formation of weak elastic gel-like emulsions.

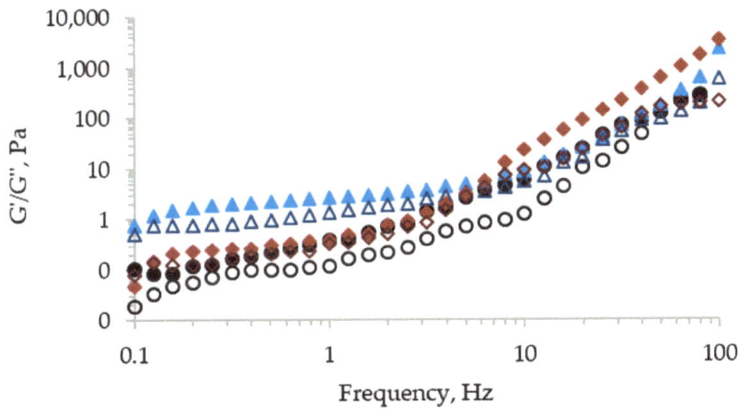

Figure 2. Evolution of the storage (G′—full symbol) and loss (G″—empty symbol) moduli with frequency for the emulsions prepared with soy proteins (circle), lupin proteins (triangle), and yeast proteins (diamonds).

The rheological behavior of the protein-based emulsions under flow conditions is presented in Figure 3. as apparent viscosity vs. shear rate rheograms. Regardless of the protein source, the emulsion samples presented shear-thinning behavior, which has been previously reported for protein-stabilized emulsions [22,23]. Liu and Tang [23] explained the shear-thinning behavior through the flocculation of the emulsion droplets. In addition, Berli et al. [24] suggested that the high shear-induced gradual disruption of the droplet aggregates might contribute to the shear-thinning behavior of the emulsions. The disruption of the droplets usually occurs at high flow rates when the hydrodynamic forces overcome the interfacial forces responsible for droplets aggregation [25].

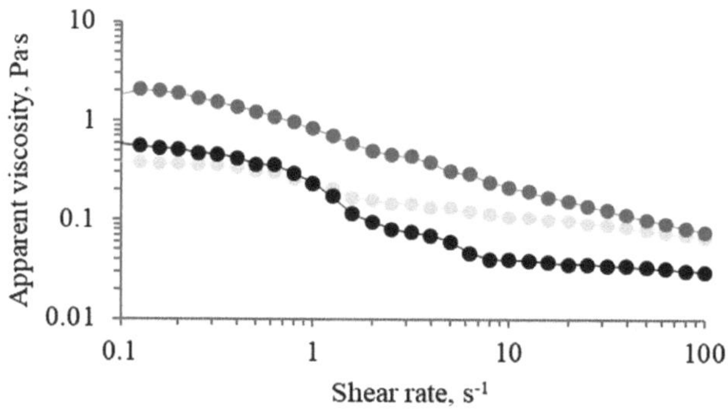

Figure 3. Evolution of the apparent viscosities as a function of shear rate for the emulsions prepared with soy proteins (light gray), lupin proteins (dark gray), and yeast proteins (black).

In the case of all tested samples, the shear stress increased, and the apparent viscosity decreased in the whole shear rate range of 0.1–100 s^{-1}. The highest values of apparent viscosity were registered for the lupin protein-based emulsion (Figure 3). At low shear rates, the soy protein-based emulsion had the lowest shear stress and viscosities values. For shear rates over 1 s^{-1}, the yeast protein-based emulsion was overtaken by the soy protein-based one (Figure 3). The slightly different shape of the yeast protein-based emulsion, with a more evident plateau of apparent viscosity values, might be due to the rather high content of minerals in the sample, which might weaken the attractive interaction between droplets [24].

3.2. Rheological Properties of the Doughs

The influence of oil addition and further substitution by emulsion on the thermo-mechanical behavior of the white wheat flour used to prepare the muffin samples was investigated by means of the Mixolab device, which allows for registering the torque of the dough while subjecting it to dual mixing and thermal constraints.

The typical Mixolab curve showing the thermo-mechanical behavior of the white wheat flour, used as a basis in the batter formulations, is presented in Figure 4a. The WA of 59.3% was used to get the target torque value C1 of 1.1 Nm. The addition of sunflower oil caused some important changes to the thermo-mechanical behavior of the dough (Figure 4b). The main consistency values were significantly lower ($p < 0.05$) compared to those measured for wheat flour with no oil addition (Table 2). The reduction of the torque values registered by Mixolab on the dough with the addition of sunflower oil was previously reported by Moreira et al. [26]. Moreover, they showed that oil addition resulted in lower WA values required to reach the target C1 of 1.1 Nm. In the case of our experiment, in order to actually measure the effect of oil and emulsion addition on the thermo-mechanical behavior of the dough, the optimum WA level of 59.3% measured for the white wheat flour was used for preparing all the dough samples. Thanks to the lubricating effect of the oil, softer consistency of the dough is obtained even at low water addition levels [27]. It is generally accepted that, when starch gelatinization occurs in the presence of lipids, the amylose-starch complexes have lower viscosity [28]. The maximum consistency registered for the oil-containing dough sample at 30 °C (C1) was 0.23 Nm, significantly lower compared with C1 of 1.08 Nm ($p < 0.05$) obtained for the white wheat flour sample. The increase of the temperature up to 45 °C resulted in the rapid drop of the torque measured on the oil-containing sample, indicating the impossibility to record the dough resistance by Mixolab, most probably occurring as a consequence of dough sliding on the mixing paddles when adhesion ceased. A similar rapid torque falling to zero was recently reported by Berceli et al. [29] when registering the Mixolab curve of the wheat dough with high oil content. This behavior of the dough suggests that oil addition might interfere with gluten proteins interaction, resulting in softer networks.

Further increase of the temperature up to 80 °C caused the rise of the dough consistency up to C3 of 0.42 Nm, which is five times lower compared with the C3 measured on the wheat flour, therefore suggesting limited gelatinization of the starch. Our observation is in agreement with the previous findings of Ai et al. [30] and Dun et al. [28], who showed that the formation of the starch-lipid complex might interfere with starch swelling, resulting in the reduction of peak viscosity. Finally, the C4 and C5 torque values registered on the oil-containing dough, when heating at 90 °C and further cooling to 50 °C, were significantly lower ($p < 0.05$) compared with the white wheat flour (Figure 4a,b). The lower values of the C3-C4 parameter suggest the improvement of the hot starch paste stability upon oil addition (Table 2). An additional distinctive characteristic of the oil-supplemented dough consists of the low C5-C4 value, suggesting increased resistance to starch retrogradation. Amylose-lipid complexes, formed when starch gelatinization occurs in the presence of lipids, are known to inhibit the retrogradation phenomenon [28].

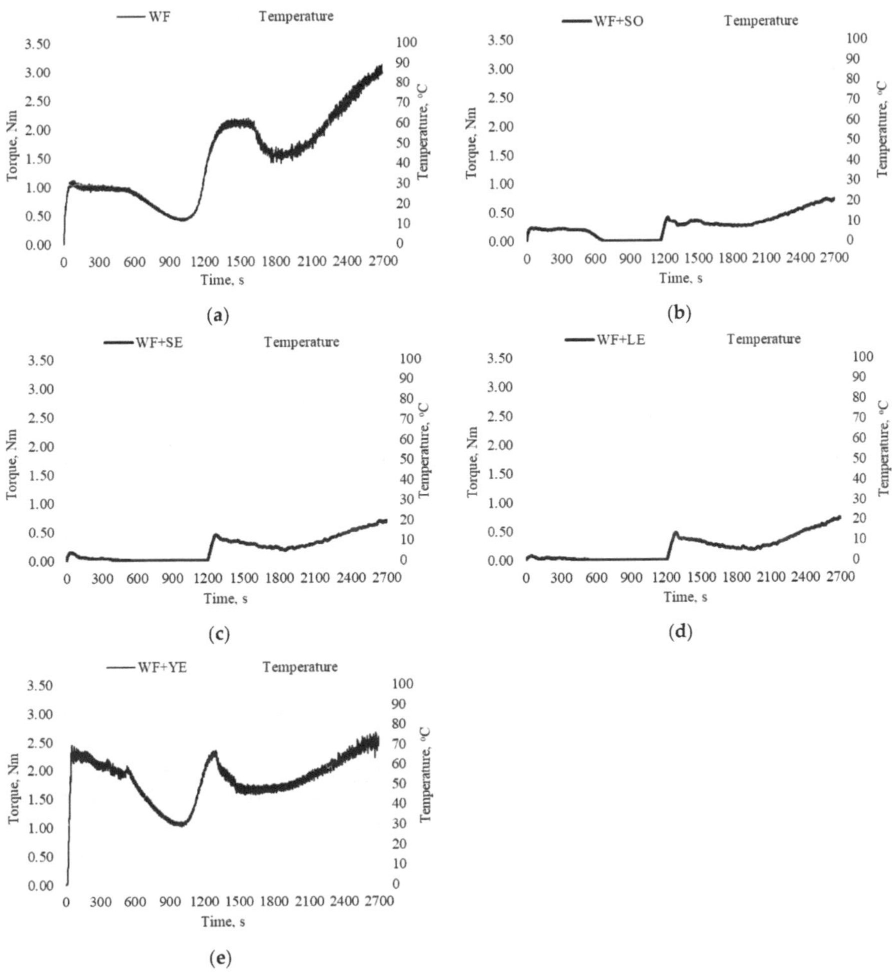

Figure 4. Mixolab curves of the dough samples prepared with: (**a**) white wheat flour (WF); (**b**) WF supplemented with sunflower oil (SO); (**c**) WF supplemented with soy protein-based emulsion (SE); (**d**) WF supplemented with lupin protein-based emulsion (LE); (**e**) WF supplemented with yeast protein-based emulsion (YE).

Table 2. Effect of supplementation with sunflower oil (SO), soy proteins (SE), lupin proteins (LE), and yeast proteins-based (YE) emulsion on the Mixolab parameters of the wheat flour dough (WF).

Doug Samples	C1, Nm	C2, Nm	C3, Nm	C4, Nm	C5, Nm	C3-C4, Nm	C5-C4, Nm
WF	1.08 ± 0.02 [b*]	0.42 ± 0.01 [b]	2.11 ± 0.01 [b]	1.52 ± 0.02 [b]	3.02 ± 0.01 [a]	0.59 ± 0.01 [b]	1.50 ± 0.02 [a]
WF + SO	0.23 ± 0.01 [c]	n.r.	0.42 ± 0.01 [d]	0.26 ± 0.01 [c]	0.73 ± 0.01 [c]	0.16 ± 0.02 [d]	0.47 ± 0.01 [d]
WF + SE	0.15 ± 0.00 [d]	n.r.	0.46 ± 0.01 [c]	0.17 ± 0.01 [d]	0.69 ± 0.02 [d]	0.29 ± 0.02 [c]	0.52 ± 0.03 [c]
WF + LE	0.06 ± 0.01 [e]	n.r.	0.46 ± 0.01 [c]	0.18 ± 0.01 [d]	0.71 ± 0.01 [c,d]	0.28 ± 0.01 [c]	0.53 ± 0.02 [c]
WF + YE	2.30 ± 0.02 [a]	1.05 ± 0.02 [a]	2.30 ± 0.01 [a]	1.64 ± 0.01 [a]	2.49 ± 0.01 [b]	0.66 ± 0.02 [a]	0.85 ± 0.02 [b]

n.r.—not recorded. * Significant differences among results on the same column, determined by ANOVA and Tukey post hoc test at $p < 0.05$, are shown by different superscript lowercase letters.

The thermo-mechanical behavior of the doughs prepared with emulsion instead of oil varied with the type of proteins used to prepare the emulsion. The maximum torque of the doughs prepared with soy and lupin proteins registered during mixing at a constant temperature of 30 °C were 0.15 Nm and 0.06 Nm, respectively. In the case of both samples, the gradual drop of the torque to zero was noticed during the initial mixing at 30 °C (Figure 4c,d). The more rapid decrease of the torque, compared with the dough sample prepared with sunflower oil, might be due to the fact that the lipids in the fine oil drops within the emulsion are more easily available for interacting with the macromolecular components of the wheat flour, resulting in faster termination of the adhesion between the mixing paddles and dough. A sharp increase of the torques was noticed when the temperature increased to 85–87 °C, the maximum C3 value of 0.46 Nm, registered for doughs with soy and lupin protein emulsions, being higher compared with the oil-based dough. These results suggest a better gelatinization of the starch as a consequence of a lower volume fraction of oil when emulsions are used to prepare the doughs. The lower C4 values registered for doughs prepared with soy and lupin proteins-based emulsions suggest the lower hot paste stability compared with the oil-based sample. Cooling the doughs to 50 °C in the final Mixolab stage resulted in the C5 values of 0.69–0.71 Nm, close to the one registered for the oil-based sample. These results indicate higher starch retrogradation in the case of the samples prepared with soy or lupin proteins emulsion (C5-C4 of 0.52–0.53 Nm) compared with the oil-based dough (C5-C4 of 0.47 Nm). In addition to oil, which exerts a major influence on the starch-related thermo-mechanical behavior of the doughs, the starch swelling and gelatinization are hampered by the interactions established by the proteins originating from the emulsion. Yang et al. [31] explained this phenomenon by the perturbation occurring in the hydrogen bonding network. Furthermore, the non-covalent interactions established between the gelatinized starch and the proteins are prone to alter the retrogradation behavior of the starch [28].

A different trend was observed in the case of the sample containing yeast proteins emulsion (Figure 4e). Analyzing the results presented in Table 2, one can see that, except for C5, all torque values were higher compared with the wheat flour dough. This particular rheological behavior is a consequence of the higher amount of water needed to prepare the dough of a specific work input (corresponding in this case to C1 of 1.1 Nm), which is due to the particular composition of the yeast extract used to prepare the emulsion. In addition to the high protein content, the yeast extract is particularly rich in minerals and B-complex vitamins. Vieira et al. [14] obtained lyophilized yeast extract with a protein content of 64% and high levels of macrominerals, trace elements (1.76 mg Fe/100 g dw and 11.9 mg Zn/100 g dw), and vitamins like nicotinic acid (77.2 mg/100 mg dw) and folic acid (3.01 mg/100 g dw). Moreover, Tomé [15] indicated that yeast extract is a good source of minerals (such as Ca over 130 mg/100 g dw and Fe over 5 mg/100 g dw) and group B vitamins (such as thiamine 10–12 mg/100 g dw, riboflavine 8–12 mg/100 g dw, niacine 90–110 mg/100 g dw, and pantothenic acid 12–20 mg/100 g dw). The human body needs all these micronutrients, and their adequate intake and availability are particularly important for a healthy lifestyle. Considering that over two billion people suffer from deficiencies of various nutrients worldwide, many programs and studies proposed the fortification of staple foods [32]. The influence of wheat flour supplementation with different levels of iron (30–90 ppm) or calcium (800–1600 ppm) alone or as micronutrient premix, additionally including different levels of vitamins like riboflavin (0.038–0.113 ppm), thiamine (0.033–0.098 ppm), niacin (0.25–0.75 ppm), and folic acid (0.008–0.023 ppm) on the rheological behavior of the dough and bread-making quality was investigated by Sudha and Leelavathi [32]. They indicated no alteration of the bread-making quality of the wheat flour, but the addition of iron resulted in the increase of WA by 1–2%, whereas the vitamins-minerals mixture raised the WA from 57.9 to 58.9%. Moreover, Aktar et al. [33] showed that wheat flour fortification with elemental iron and ZnO significantly increased the WA.

In the zones of the Mixolab curves where starch behavior prevails over proteins, smaller differences were found between C3 and C4 torques registered for the control dough and the samples with yeast proteins emulsion (Table 2). Finally, the lower C5 value suggests better starch retrogradation behavior of the dough upon yeast proteins emulsion addition.

3.3. Muffin Characterization

The influence of oil replacement by protein-based emulsion on the quality of muffins was established by assessing the physical properties of the muffins. Oil replacement by protein-based emulsion resulted in a significant increase of the protein and ash contents ($p < 0.05$) and significantly affected ($p < 0.05$) the height of the samples (Table 3). Our results are in line with the observations from previous studies reporting the height-reducing effect of partial oil replacement in muffin formulations with cocoa fiber [7], resistant dextrins obtained from wheat or maize starch [34], or cellulose ether emulsions [18]. Due to the higher amount of water used in the batter preparation when replacing the oil, the moisture content of the emulsion-based muffin samples was significantly higher ($p < 0.05$) compared with the control (Table 3). Among all investigated muffin samples, those prepared with soy proteins-based emulsion exhibited the lowest weight loss during baking, whereas no significant changes were noticed between muffins prepared with lupin or yeast protein emulsions and the control. These results indicate that the emulsion-based muffins have high water-holding capacity, and from an economic standpoint, the production process is advantageous.

Table 3. Quality characteristics of the muffin samples prepared with different types of emulsion (M—control prepared with sunflower oil, M1—muffins prepared with soy protein-based emulsion, M2—muffins prepared with lupin protein-based emulsion, M3—muffins prepared with yeast protein-based emulsion).

Muffin Sample	Weight Loss during Baking, g/100 g	Height, mm	Moisture, g/100 g	Protein, g/100 g d.w.	Ash, g/100 g d.w.
M	$7.43 \pm 0.86^{a,b*}$	4.68 ± 0.08^{a}	21.71 ± 0.24^{b}	8.31 ± 0.10^{b}	1.21 ± 0.02^{c}
M1	6.81 ± 0.29^{b}	4.46 ± 0.05^{b}	26.00 ± 0.54^{a}	9.31 ± 0.07^{a}	$1.29 \pm 0.04^{b,c}$
M2	8.84 ± 0.53^{a}	4.42 ± 0.08^{b}	25.05 ± 0.97^{a}	9.38 ± 0.04^{a}	$1.34 \pm 0.03^{a,b}$
M3	$8.32 \pm 0.50^{a,b}$	$4.54 \pm 0.09^{a,b}$	25.97 ± 0.98^{a}	9.45 ± 0.09^{a}	1.40 ± 0.04^{a}

* Significant differences among results on the same column, determined by ANOVA and Tukey post hoc test at $p < 0.05$, are shown by different superscript lowercase letters.

The characteristics of the pores of the four types of muffins (Figure 5) considered in the study were determined through image analysis and the results are presented in Table 4. The porosity-related characteristic provides indication of the ability of the batter to retain the CO_2 released by baking powder during heating. As one can see in Figure 5, the pore size and homogeneity of the pore structure varied with the muffin samples. The average pore size was significantly higher in the case of the muffin samples prepared with emulsions ($p < 0.05$). On the other hand, the control muffin sample had a higher number of uniformly distributed smaller pores (Figure 5). Considering that the same processing conditions were applied for preparing the batters and muffin samples, the observed differences could be attributed to the lower viscosity of the samples prepared with emulsions as the result of the higher amount of water in respect to the binding ability of the dry mixture. No important differences were noticed between the investigated samples in terms of the cell-total area ratio (Table 4).

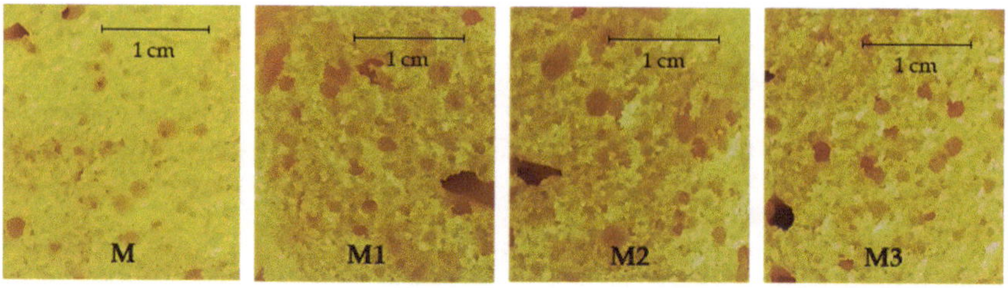

Figure 5. Structure of the crumb of muffin samples with different types of emulsion (M—control prepared with sunflower oil, M1—muffins prepared with soy protein-based emulsion, M2—muffins prepared with lupin protein-based emulsion, M3—muffins prepared with yeast protein-based emulsion). Images were captured using a Canon PowerShot G16 digital camera (Canon Inc., Tokyo, Japan).

Table 4. Porosity and texture characteristics of the muffin samples prepared with different types of emulsion (M—control prepared with sunflower oil, M1—muffins prepared with soy protein-based emulsion, M2—muffins prepared with lupin protein-based emulsion, M3—muffins prepared with yeast protein-based emulsion).

Muffin Sample	Porosity Characteristics		Firmness, N
	Average Size ($m^2 \cdot 10^{-6}$)	Cell-Total Area Ratio	
M	1.29 ± 0.01 [b*]	0.48 ± 0.01 [a]	5.00 ± 0.11 [b]
M1	1.76 ± 0.05 [a]	0.47 ± 0.01 [a]	5.64 ± 0.40 [a,b]
M2	1.74 ± 0.08 [a]	0.48 ± 0.04 [a]	5.86 ± 0.33 [a]
M3	1.69 ± 0.02 [a]	0.53 ± 0.04 [a]	5.79 ± 0.20 [a]

* Significant differences among results on the same column, determined by ANOVA and Tukey post hoc test at $p < 0.05$, are shown by different superscript lowercase letters.

The crumb cell structure exerts a major influence on how the texture and freshness of the product is perceived [20]. The firmness of the muffins was measured at 12 h after baking. Analyzing the results presented in Table 4, one can see that oil substitution by emulsion in the batter formulations resulted in a significant increase of firmness. Our results comply with the observations of Martinez-Cervera et al. [18] who reported a significant hardness increase when using cellulose ether emulsions in the muffins formulation instead of oil.

Color is an important attribute of food products which might influence the consumers' choice in a decisive manner. The color usually depends on the type and amount of ingredients used for the preparation of different foods, as well as on their interaction and transformation during processing. The difference in color (ΔE) between the control sample and muffins prepared with different types of protein-based emulsions is presented in Table 5. For all emulsion-based muffin samples, the ΔE fell within the 1.5–3.0 range, which, in agreement with Goyeneche et al. [35], suggests noticeable differences in respect to the sample prepared with sunflower oil. As all the samples exhibited similar a* values, the ΔE variations are mainly due to brightness (L*) and yellow (b*) chromaticity. The color intensity (C*) varied significantly with the source of fats ($p < 0.05$), being more vivid in the emulsion-based muffins (Table 5).

Table 5. Color characteristics of the muffin samples prepared with different types of emulsion (M—control prepared with sunflower oil, M1—muffins prepared with soy protein-based emulsion, M2—muffins prepared with lupin protein-based emulsion, M3—muffins prepared with yeast protein-based emulsion).

Muffin Sample	L*	a*	b*	ΔE	C*
M	79.35 ± 0.45 ^{a}	−2.79 ± 0.12 ^{a}	45.05 ± 0.17 ^{c}	-	45.13 ± 0.18 ^{c}
M1	77.48 ± 0.24 ^{b}	−2.75 ± 0.05 ^{a}	46.26 ± 0.11 ^{b}	2.24 ± 0.25 ^{a,b}	46.34 ± 0.11 ^{b}
M2	78.72 ± 0.17 ^{a}	−2.71 ± 0.13 ^{a}	46.73 ± 0.09 ^{a}	1.82 ± 0.09 ^{b}	46.81 ± 0.10 ^{a}
M3	77.23 ± 0.09 ^{b}	−2.76 ± 0.10 ^{a}	46.53 ± 0.08 ^{b}	2.62 ± 0.20 ^{a}	46.62 ± 0.09 ^{a,b}

* Significant differences among results on the same column, determined by ANOVA and Tukey post hoc test at $p < 0.05$, are shown by different superscript lowercase letters.

The scoring method was applied for the sensory analysis of the muffin samples and the sensory scores for the main attributes are shown in Table 6. Oil replacement by protein-based emulsions resulted in no significant differences in color, flavor, taste, and appearance of the muffin samples. In agreement with the results of the instrumental analysis, showing that among all investigated samples, those with lupin and yeast protein-based emulsions had the highest firmness values (Table 4), the textures of these muffins were rated with the lowest scores (Table 6). However, no important differences between control and emulsion-based muffins were observed in terms of overall quality.

Table 6. Sensory evaluation data of muffin samples prepared with different types of emulsion (M—control prepared with sunflower oil, M1—muffins prepared with soy protein-based emulsion, M2—muffins prepared with lupin protein-based emulsion, M3—muffins prepared with yeast protein-based emulsion).

Sensory Attributes	Muffin Samples			
	M	M1	M2	M3
Appearance	8.92 ± 0.29 ^{a*}	8.92 ± 0.29 ^{a}	8.75 ± 0.45 ^{a}	8.92 ± 0.29 ^{a}
Flavor	8.92 ± 0.29 ^{a}	8.67 ± 0.65 ^{a}	8.75 ± 0.45 ^{a}	8.83 ± 0.39 ^{a}
Taste	8.58 ± 0.52 ^{a}	8.42 ± 0.90 ^{a}	8.50 ± 0.52 ^{a}	8.58 ± 0.52 ^{a}
Color	8.83 ± 0.39 ^{a}	8.92 ± 0.29 ^{a}	8.92 ± 0.29 ^{a}	8.92 ± 0.29 ^{a}
Texture	8.67 ± 0.49 ^{a}	8.42 ± 0.67 ^{a,b}	8.08 ± 0.52 ^{a,b}	8.00 ± 0.60 ^{b}
Overall quality	8.83 ± 0.39 ^{a}	8.50 ± 0.52 ^{a}	8.50 ± 0.52 ^{a}	8.58 ± 0.52 ^{a}

* Significant differences among results on the same line, determined by ANOVA and Tukey post hoc test at $p < 0.05$, are shown by different superscript lowercase letters.

4. Conclusions

Fundamental and empirical rheological measurements were applied to assess the properties of the soy, lupin, and yeast protein-based emulsions and batters, respectively. The protein-based emulsions exhibited a weak elastic gel-like behavior with moderate frequency dependence on the storage and loss moduli. The stepped flow tests revealed the shear-thinning behavior of the samples. Among all investigated emulsions, the lupin proteins-based one exhibited the highest apparent viscosity over the entirety of the tested shear rate range. The results of the Mixolab test indicated that emulsion addition influences, in a different manner, the thermo-mechanical behavior of the white wheat flour doughs, depending on the protein source. The reduction of the volume fraction of fats added to the dough, when replacing sunflower oil with the soy or lupin proteins emulsion, ensured a better gelatinization of the starch. Higher torque values were registered for the dough prepared with yeast proteins emulsion in all stages of the Mixolab curve.

Oil substitution by emulsion resulted in muffins with higher moisture and firmness values. The structure of the muffins was assessed by analyzing the porosity of their cores. All samples exhibited a good ability to retain the CO_2 released during baking, the pore size being slightly higher in the case of the muffins with lower fat content. The chromatic parameters indicated that the samples prepared with protein emulsion have a more vivid color. In conclusion, the results reported herein confirm that protein-based emulsions considered in this study are sustainable ingredients for obtaining muffins with reduced fat content with acceptable quality characteristics.

Author Contributions: Conceptualization, I.A. and I.B.; methodology, L.D., I.V. and L.P.; software, I.A. and L.D.; validation, I.B., I.V. and L.P.; formal analysis, L.D. and L.P.; investigation, I.B., I.A., L.D., I.V. and L.P.; resources, I.A.; data curation, I.B.; writing—original draft preparation, I.A., I.B. and L.P.; writing—review and editing, I.A. and L.D.; supervision, I.A. All authors have read and agreed to the published version of the manuscript.

Funding: This work was supported by a grant of the Ministry of Research, Innovation and Digitization, CNCS—UEFISCDI, project number PN-III-P1-1.1-TE-2021-0459, within PNCDI III.

Institutional Review Board Statement: Not applicable.

Informed Consent Statement: Written informed consent was obtained from all subjects (patients) in this study, according to the decision of the Dunarea de Jos University Ethics Commission no. 28/19 October 2022.

Data Availability Statement: Not applicable.

Conflicts of Interest: The authors declare no conflict of interest.

References

1. Hussien, H.A. Using vegetable puree as a fat substitute in cakes. *Int. J. Nutr. Food Sci.* **2016**, *5*, 284–292. [CrossRef]
2. Colla, K.; Costanzo, A.; Gamlath, S. Fat replacers in baked food products. *Foods* **2018**, *7*, 192. [CrossRef] [PubMed]
3. Azmoon, E.; Saberi, F.; Kouhsari, F.; Akbari, M.; Kieliszek, M.; Vakilinezam, A. The effects of hydrocolloids-protein mixture as a fat replacer on physicochemical characteristics of sugar-free muffin cake: Modeling and optimization. *Foods* **2021**, *10*, 1549. [CrossRef]
4. Rodriguez-Sandoval, E.; Prasca-Sierra, I.; Hernandez, V. Effect of modified cassava starch as a fat replacer on the texture and quality characteristics of muffins. *J. Food Meas. Charact.* **2017**, *11*, 1630–1639. [CrossRef]
5. Fernandes, S.S.; de las Mercedes Salas-Mellado, M. Addition of chia seed mucilage for reduction of fat content in bread and cakes. *Food Chem.* **2017**, *227*, 237–244. [CrossRef]
6. Zahn, S.; Pepke, F.; Rohm, H. Effect of inulin as a fat replacer on texture and sensory properties of muffins. *Int. J. Food Sci. Technol.* **2010**, *45*, 2531–2537. [CrossRef]
7. Martínez-Cervera, S.; Salvador, A.; Muguerza, B.; Moulay, L.; Fiszman, S.M. Cocoa fibre and its application as a fat replacer in chocolate muffins. *LWT-Food Sci. Technol.* **2011**, *44*, 729–736. [CrossRef]
8. Jeong, S.; Lee, S.; Oh, I. Development of antioxidant-fortified oleogel and its application as a solid fat replacer to muffin. *Foods* **2021**, *10*, 3059. [CrossRef]
9. Manzoor, S.; Masoodi, F.A.; Naqash, F.; Rashid, R. Oleogels: Promising alternatives to solid fats for food applications. *Food Hydrocoll. Health* **2022**, *2*, 100058. [CrossRef]
10. Pehlivanoglu, H.; Ozulku, G.; Yildirim, R.M.; Demirci, M.; Toker, O.S.; Sagdic, O. Investigating the usage of unsaturated fatty acid-rich and low-calorie oleogels as a shortening mimetics in cake. *J. Food Process. Pres.* **2018**, *42*, e13621. [CrossRef]
11. Therdthai, N. Sugar, salt and fat reduction of bakery products. *Adv. Food Nutr. Res.* **2022**, *99*, 283–327.
12. Boye, J.I.; Zare, F.; Pletch, A. Pulse proteins: Processing, characterization, functional properties and applications in food and feed. *Food Res. Int.* **2010**, *43*, 414–431. [CrossRef]
13. Grossi Bovi Karatay, G.; Rebellato, A.P.; Joy Steel, C.; Dupas Hubinger, M. Chickpea aquafaba-based emulsions as a fat replacer in pound cake: Impact on cake properties and sensory analysis. *Foods* **2022**, *11*, 2484. [CrossRef]
14. Vieira, E.F.; Carvalho, J.; Pinto, E.; Cunha, S.; Almeida, A.A.; Ferreira, I.M. Nutritive value, antioxidant activity and phenolic compounds profile of brewer's spent yeast extract. *J. Food Compos. Anal.* **2016**, *52*, 44–51. [CrossRef]
15. Tomé, D. Yeast extracts: Nutritional and flavoring food ingredients. *ACS Food Sci. Technol.* **2021**, *1*, 487–494. [CrossRef]
16. *SR ISO 712:2005; SR ISO 2171:2002*; Romanian Standards Catalog for Cereal and Milling Products Analysis. ASRO: Bucharest, Romania, 2008.
17. AACC International. *Approved Methods of Analysis*, 11th ed.; Method 46-11.02; American Association of Cereal Chemists International: Saint Paul, MN, USA, 2000.
18. Martínez-Cervera, S.; Salvador, A.; Sanz, T. Cellulose ether emulsions as fat replacers in muffins: Rheological, thermal and textural properties. *LWT-Food Sci. Technol.* **2015**, *63*, 1083–1090. [CrossRef]
19. Banu, I.; Aprodu, I. Investigation on functional, thermo-mechanical and bread-making properties of some white and black rice flours. *Appl. Sci.* **2022**, *12*, 4544. [CrossRef]
20. Aprodu, I.; Banu, I. Rheological, thermo-mechanical, and baking properties of wheat-millet flour blends. *Food Sci. Technol. Int.* **2015**, *21*, 342–353. [CrossRef] [PubMed]

21. Wilde, P.; Mackie, A.; Husband, F.; Gunning, P.; Morris, V. Proteins and emulsifiers at liquid interfaces. *Adv. Colloid Interface Sci.* **2004**, *108*, 63–71. [CrossRef]
22. Manoi, K.; Rizvi, S.S. Emulsification mechanisms and characterizations of cold, gel-like emulsions produced from texturized whey protein concentrate. *Food Hydrocoll.* **2009**, *23*, 1837–1847. [CrossRef]
23. Liu, F.; Tang, C.H. Cold, gel-like whey protein emulsions by microfluidisation emulsification: Rheological properties and microstructures. *Food Chem.* **2011**, *127*, 1641–1647. [CrossRef]
24. Berli, C.L.; Quemada, D.; Parker, A. Modelling the viscosity of depletion flocculated emulsions. *Colloid. Surf. A-Physicochem. Eng. Asp.* **2002**, *203*, 11–20. [CrossRef]
25. McClements, D.J. Emulsion Rheology. In *Food Emulsions: Principles, Practices, and Techniques*, 2nd ed.; McClements, D.J., Ed.; CRC Press: Boca Raton, FL, USA, 2004.
26. Moreira, R.; Chenlo, F.; Torres, M.D. Effect of shortenings on the rheology of gluten-free doughs: Study of chestnut flour with chia flour, olive and sunflower oils. *J. Texture Stud.* **2012**, *43*, 375–383. [CrossRef]
27. Zoulias, E.I.; Oreopoulou, V.; Tzia, C. Textural properties of low-fat cookies containing carbohydrate-or protein-based fat replacers. *J. Food Eng.* **2002**, *55*, 337–342. [CrossRef]
28. Dun, H.; Liang, H.; Zhan, F.; Wei, X.; Chen, Y.; Wan, J.; Ren, Y.; Hu, L.; Li, B. Influence of O/W emulsion on gelatinization and retrogradation properties of rice starch. *Food Hydrocoll.* **2020**, *103*, 105652. [CrossRef]
29. Berceli, M.B.; Izsó, E.; Gergely, S.; Salgó, A. Effects of special additives in wheat dough system measured by Mixolab technique. *Czech J. Food Sci.* **2021**, *39*, 460–468. [CrossRef]
30. Ai, Y.; Hasjim, J.; Jane, J.L. Effects of lipids on enzymatic hydrolysis and physical properties of starch. *Carboh. Polym.* **2013**, *92*, 120–127. [CrossRef]
31. Yang, C.; Zhong, F.; Goff, H.D.; Li, Y. Study on starch-protein interactions and their effects on physicochemical and digestible properties of the blends. *Food Chem.* **2019**, *280*, 51–58. [CrossRef]
32. Sudha, M.L.; Leelavathi, K. Influence of micronutrients on rheological characteristics and bread-making quality of flour. *Int. J. Food Sci. Nutr.* **2008**, *59*, 105–115. [CrossRef]
33. Akhtar, S.; Anjum, F.M.; Rehman, S.U.; Sheikh, M.A. Effect of mineral fortification on rheological properties of whole wheat flour. *J. Texture Stud.* **2009**, *40*, 51–65. [CrossRef]
34. Martínez-Cervera, S.; de la Hera, E.; Sanz, T.; Gómez, M.; Salvador, A. Effect of nutriose on rheological, textural and sensorial characteristics of Spanish muffins. *Food Bioprocess Technol.* **2013**, *6*, 1990–1999. [CrossRef]
35. Goyeneche, R.; Agüero, M.V.; Roura, S.; Di Scala, K. Application of citric acid and mild heat shock to minimally processed sliced radish: Color evaluation. *Postharvest Biol. Technol.* **2014**, *93*, 106–113. [CrossRef]

applied sciences

MDPI

Article

A Comparative Study on the Development of Bioactive Films Based on β-glucan from Spent Brewer's Yeast and Pomegranate, Bilberry, or Cranberry Juices

Ionut Avramia * and Sonia Amariei

Faculty of Food Engineering, Stefan cel Mare University of Suceava, 720229 Suceava, Romania
* Correspondence: ionut.avramia@usm.ro

Abstract: This study provides new insight into developed bioactive films. The development of film-forming solutions from different fruit juices has demonstrated a major impact of bioactive compounds in film formulation, from smart packaging to bioactive packaging that releases the compounds from the oral solution at the same time as the packaged product. There were significant influences among independent parameters such as yeast β-glucan, gelling agent, fruit juice, or surfactant for each individual film. However, in this study, it was found that the amount of juice was the most significant factor in assigning their properties to all three types of films made of different juices (pomegranate, bilberry, and cranberry). Properties such as WVTR through the film varied within quite wide limits between 3.2562 and 32.1869 g/h·m^2, while their solubilization time started from a few seconds for a complete dissolution or ensured only partial dissolution after 10 min of stirring (in the case of films made of β-glucan and cranberry juice). Therefore, natural biopolymer-based films developed with excellent water vapor barrier properties and improved solubility have a huge potential for application as packaging materials for dry powdered such as pharmaceuticals.

Keywords: bioactive film; bilberry juice; cranberry juice; pomegranate juice; yeast β-glucan

1. Introduction

From antibacterial properties to UV radiation protection, bioactive films are used in a wide range of applications [1,2], including for the antioxidant properties by supplementing compounds such as naphtho-γ-pyrone in film-forming solutions (FFSs) based on yeast cell walls [3] to the use of edible films as carriers of bioactive compounds [4] or the incorporation of lactic acid bacteria (LAB) producing bacteriocins into edible whey protein films for antimicrobial properties [5] and up to product packaging materials that dissolves into the oral solution formed with the product [6,7].

A new possibility for the development of bioactive films is the incorporation of different types of fruit juices into the film-forming solution to impart antioxidant and antimicrobial properties [8,9]. Suitable matrices of FFS can be represented by yeast cell walls [3,10] or β-glucans extracted by purification from the cell walls, with immunostimulatory effect.

The bioactivity of films developed from juices, compared to edible films, consists mainly of the high content of tannins and polyphenols such as in the following: pomegranate juice (*Punica granatum* L.) ellagitannins between 2010 and 6420 mg/L (up to 103 mg/L ellagic acid) and gallic acid up to 8.55 mg/L [11–13]; up to a 444.5 mg/100 g of anthocyanins in bilberry juice (*Vaccinium myrtillus* L.) and 65 mg/100 mL flavonols [14,15]; or about 59 compounds identified by Vilkickyte et al., (2022) in lingonberry (*Vaccinium vitis-idaea* L.) including quercitin, catechins, flavonols, and a phenolic content up to a total of 760 mg/100 g fresh weight [16,17].

With the synergistic action of pomegranate, bilberry, and cranberry juices, along with β-glucans from yeast, it confers antioxidant and antidiabetic properties, as well as immunomodulatory activity [17–21]. For example, in bilberries, it has been found that due to anthocyanosides, in case of vascular complications the fruit is used instead of leaves. For a daily intake of 25% anthocyanosides from the fruit extract, doses of up to 240–480 mg per day have been established [22]. In the case of cranberries, a consumption of 240 mL cranberry juice (45.7 mg polyphenols) daily for a period of 12 weeks was beneficial in the control of type II diabetes [21]. Vascular functions can also be visibly improved after a month of consumption of cranberry due to the high content of polyphenols (especially proanthocyanidins type A, anthocyanins, flavonols, and phenolic acids) [23]. Besides bilberry-derived bioactive compounds with hypoglycemic activity, pomegranate juice is also particularly valuable and can be used in the management of diabetes complications from vasoprotective effect to reduction in serum glucose levels. In a communication on how pomegranate juice acts on glucose metabolism, Carlos et al. (2022) summarized in a table up to 10 case reports on the effect of pomegranate juice in type II diabetes. These includes blood glucose reduction after consuming 200 mL for 4 days while another report revealed that after 4 weeks the PON1, an HDL-associated enzyme increased, helping to reduce total cholesterol [24]. Films developed from pomegranate extract, in addition to the antioxidant and antimicrobial activity, were shown to improve mechanical properties, thermal stability, and inhibited the activity of *Penicillium digitatum* [25].

β-glucans are described as high molecular weight polymers of glucose that can be isolated from oats, plants, bacteria, fungi, and yeast [26]. Depending on the source of origin, they can play roles ranging from dietary fiber (as in the case of β-glucans from oats) to anti-cancer, anti-inflammatory, and immune system stimulating roles (β-glucans from mushrooms and yeast) [27]. The benefits of using β-glucans isolated from yeast cells can be attributed to the ability to stimulate both the innate and acquired immune systems [28,29]. The valorization of spent brewer's yeast through the extraction of β-glucans has a high economic potential, is well documented in the literature and promotes the circular economy [30,31]. Previously films made from yeast β-glucan were made for the improved mechanical properties given by the β-1,3/1,6 structure of glucans [32].

Fruit juices and β-glucan are incorporated into film-forming solution so it presents the active agents as components of films. Once they dissolve in the oral solution, it releases the active ingredients. Of course, some studies point out that the addition of bioactive compounds to the film-forming solution, without prior encapsulation of the active agent, will lead in a reduction in effectiveness and deterioration of bioactivities [33]. Regarding β-glucans, they are used especially in dry form in order to support the overall immune function [34] while studies on juices have shown that drying up to 80 °C maintains vitamins, antioxidant activity, and anthocyanins content in the same parameters as in the case of lyophilization [35,36].

The potential applications of β-glucan/juice-based films as packaging agents for pharmaceutical dry powders were investigated due to the antioxidant, immunostimulant, and antidiabetic properties. In this context, the present research was carried out to evaluate the bioactive films made of β-glucan and the three types of fruit juices in terms of uniformity of film thickness, moisture vapor transmission rate, water vapor permeability, and dissolution time.

2. Materials and Methods

2.1. Raw Materials and Chemicals

Pomegranate juice (BioAgros, Pella, Greece) (dw. 12.42%, pH = 3.03), bilberry juice (*Vaccinium myrtillus* L.) (dw. 11.5%, pH = 3.3), and cranberry juice (dw., 6.85%) were purchased from a local supermarket. β–glucan was extracted by the alkaline-acid method from spent brewer's yeast provided by the SC. Bermas SA. brewery (Suceava, Romania). The alkaline process involved treating the autolyzed cells in a ratio of 1 to 5 (*w*/*v*) with a 1.5 N NaOH solution (at 90 °C for 2 h). The pellet resulted after centrifugation was treated with 1 N hydrochloric acid at 75 °C for 2 h for the elimination of amorphous substances. The β-glucans obtained were centrifuged, washed with distilled water and stored in the refrigerator until use. Chemical reagents from film composition were: Sodium alginate, Product No. 9180.1 (Carl Roth, Karlsruhe, Germany), Glycerin, Product No. G7893 (Sigma-Aldrich, St. Louis, MO, USA, ACS reagent ≥99.5%), Soybean oil (oil of genetically unmodified soybeans, Dachim SRL, Cluj, Romania).

2.2. Film Preparation and Casting

The film-forming solution was made in the following order. First, β-glucans were added up to a total content of 1.5 g dry weight (dw.), then the gelling agent (sodium alginate) was added in an amount of maximum 1 g while the various fruit juices were calculated in order a total amount of solid substance to not exceed 6 g in FFS. To this composition glycerin was also added as a plasticizer as 25% of the total solids while the 2% soybean oil was added for the bilberry and cranberry juice films. The film-forming solution was made up with distilled water at volume of 150 mL. The stirring time to homogenize was 30 min at 80 °C, then FFS was casted onto petri dishes and subjected to drying process for 48 h at 40 °C [37,38]. The overall chart flow is shown in Figure 1.

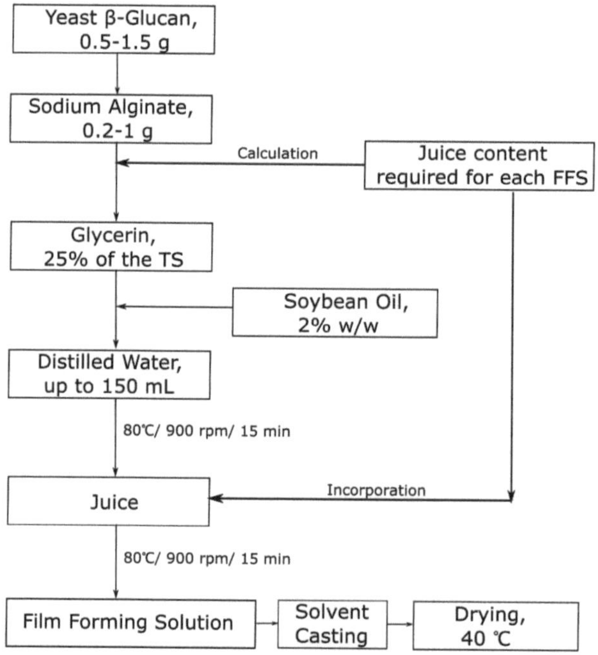

Figure 1. Flowchart of the production of bioactive β–glucan/juice films.

The development of β-glucan and pomegranate juice-based films was carried out according to an experimental design involving 15 experiments with three different levels for β-glucan, pomegranate juice, and sodium alginate (0.5–1.5 g; 10–30 mL and 0.2–0.6 g). In the case of films containing bilberry juice and cranberry juice, the incorporation of soybean oil was considered in order to observe changes in dissolution time. In the 7 samples of bilberry juice films, the sodium alginate remained constant at 0.8 g, while in the case of the cranberry juice films, the amount of SA was increased to 1 g due to increased juice content (between 36 and 43 g juice corresponding to 2.5–3 g dw.). The amount of yeast β-glucan did not exceed 1.5 g in all film-forming solutions.

2.3. Methods

2.3.1. Thickness

Thickness was determined with the thickness gauge PosiTector 6000 (DeFelsko, Ogdensburg, NY, USA). On each film, on the surface to be analyzed, such as WVTR or WVP, thickness was measured in ten different locations. The average of ten determinations (μm) was used in establishing the film properties.

2.3.2. Water Vapor Transmission Rate (WVTR)

WVTR tests ($g/h{\cdot}m^2$) were measured using the dry cup method according to the ASTM E96/96M method [39]. Circular pieces of the films were sealed horizontally on polystyrene petri dishes containing approximately 10 g of $CaCl_2$ as the desiccant. The cups (with an internal 0% RH) were placed in an environmental chamber with a saturated solution of NaCl at the bottom, providing a 75% RH. The WVTR of the films was calculated by dividing the slope (the weight gain versus time) to the area of exposed film using the following equation:

$$WVTR = \frac{\Delta W}{\Delta t \times A}\left(\mathrm{g/h{\cdot}m^2}\right) \quad (1)$$

where $\Delta W/\Delta t$ is the slope (g/h) and A represents the area exposed to the water vapor flux (m^2).

The weight of the cups was measured every 8 h within 72 h. A number of ten data points over the 72 h time was provided for accurate information about the amount of water gain.

2.3.3. Water Vapor Permeability (WVP)

WVP describes how easily a film is penetrated by water vapor and was calculated by dividing the WVTR to water vapor partial pressure across the film and multiplying by the thickness of the film. The water vapor permeability at the specific RH chamber the follows the next equation:

$$WVP = \frac{WVTR \times L}{\Delta p}\left(\mathrm{g{\cdot}mm/kPa{\cdot}h{\cdot}m^2}\right) \quad (2)$$

where $WVTR$ is the water vapor transmission rate, L is the film thickness (in mm), and Δp is the water vapor partial pressure across the film (kPa) which is calculated according to:

$$\Delta p = S{\cdot}(R_1 - R_2)(\mathrm{kPa}) \quad (3)$$

where S is the saturation vapor pressure of the water (3.1687 kPa), R_1 is the RH inside the desiccator, and R_2 is the RH inside the cup filled with $CaCl_2$.

2.3.4. Dry Weight Determination

The percentage dry weight content was determined gravimetrically by referring to the dry substances which has held in an oven at 105 ± 1 °C. Samples were dried for at least 24 h followed by cooling in desiccator prior to weighing [40].

2.3.5. Dissolution Time

The film samples were cut into squares of 2 × 2 cm, immersed in 50 mL of distilled water, and then vigorously shaken. The dissolution time was determined when the film dissolved and no visible fragments were found in suspension. The dissolution time (min) was recorded using a chronometer [41].

2.3.6. Scanning Electron Microscopy (SEM)

SEM images were recorded by using a scanning electron microscope (VEGA II LMU, Tescan, Brno, Czech Republic). For higher depth of field, a secondary electron detector was used at an accelerating voltage of 30 kV.

2.3.7. Optical Properties of the Film Samples

The absorption and transmission spectra in the 200–800 nm region of bioactive films was investigated by means of a UV-Vis-NIR Shimadzu 3600 spectrophotometer (Tokyo, Japan). Each sample (1 × 4 cm) was placed directly in the side of spectrophotometer cell and an empty cell was used as reference. The ability to block UV radiation is crucial for materials used in food packaging. UV penetration into food can cause unwanted chemical changes, such as the creation of unpleasant odors, by forming free radicals and their derivatives. By reducing the amount of light that passes through the packaging, the rate at which photo-oxidation occurs in food products can be slowed down [42].

2.3.8. Statistical Analysis of the Results

The data was processed by using XLSTAT software for Excel 2022 version (Addinsoft, New York, NY, USA). Significant differences between samples were considered at $p < 0.05$ by applying Tukey test.

3. Results

The order of development and examination of β-glucan-based films from yeast was as follows. First, 15 films were developed from β-glucan and pomegranate juice following a Box–Behnken optimization of response surface methodology. After identification of the optimal composition and validation of the data by laboratory tests on the developed films, acceptable properties for packaging products of powdery nature were observed [37]. The standard errors of the design are shown in Figure 2. As a result of the preceding analysis results, it was decided, in order to facilitate rapid dissolution, to use a surfactant at a concentration of 2% (soybean oil) in the development of the β-glucan and bilberry juice films [38]. For a comparative study, it was considered to make four control films without soybean oil. In the last stage of the research, the amount of solids in the FFS of films designed with β-glucan and cranberry juice was increased to 5 g, with the use of soybean oil in half of the samples, keeping the plasticizer at 25% of the total solids and increasing the gelling agent (sodium alginate) to 1 g in each film. For statistical analysis, data were determined in triplicate, the exception was the determination of the thickness, which was measured in 10 points for the accuracy of measurements.

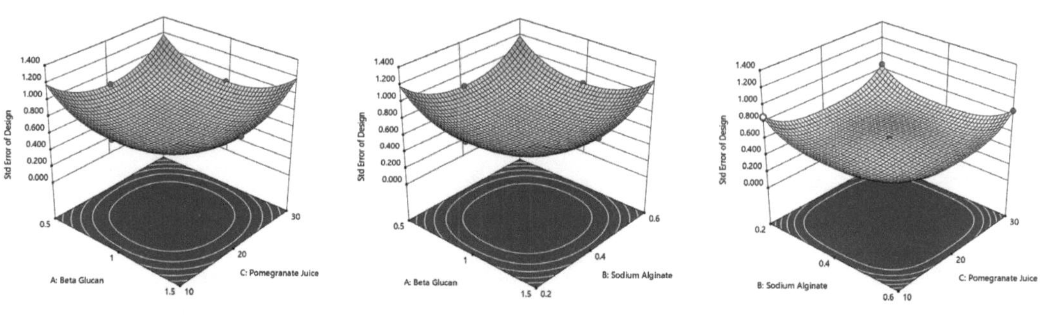

Figure 2. 3-D surface plot of the standard errors of the design of experiments.

Figure 2 shows the standard error of the design of the 15 experimental runs. The largest standard errors at the edges of the design space were found between the parameters β-glucan and pomegranate juice, respectively, and β-glucan and sodium alginate. Therefore, based on the figure, the measured root mean square error of the design space were relatively low, with values between 0.5 and 1.2 suggesting the model is expecting to provide reasonable prediction.

3.1. The Gelling Agent Characterization in Order to Obtain the Film-Forming Solution

Extremely important in making films is the sizing of the gelling agent so that the film-forming solution (FFS) is not too viscous. An instability of the polymer solution can occur by the deviation from the Cox–Merz rule which is attributed to high content of sodium alginate. Mancini et al., (1996) by investigating polymeric alginate solutions, concluded that, to fit into a pseudoplastic behavior, sodium alginate solutions should be between 0.125 and 1.5% (w/v) (on a temperature range of 5–35 °C) [43]. Thus, for the preparation of the films, the highest amount of sodium alginate used was 1 g in 150 mL of film-forming solution, an amount that would comply with the Cox–Merz rule and corresponding to a maximum of 0.66% (w/v). For data veracity, a separate solution with the highest amount of sodium alginate of 1 g in 150 mL of distilled water was subjected to rheological analysis (Haake Mars 40 rheometer, ThermoHaake, Germany, plate-plate geometry, 80/40 mm, 2 mm distance between plates). The results of the analysis are highlighted in Figure 3a,b by flow curves.

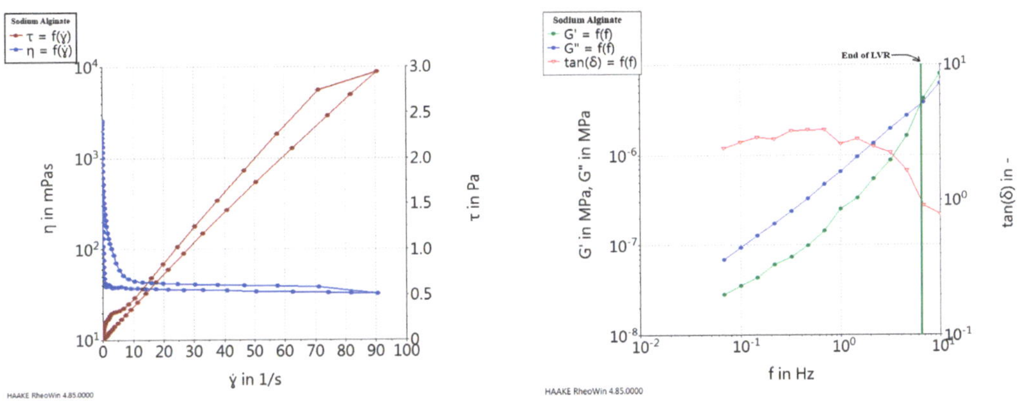

(a) (b)

Figure 3. (**a**) Rheological behavior of sodium alginate solution, η—viscosity function, mPas; r—shear stress, Pa; γ—shear rate, 1/s; (**b**) Elastic modulus (G′) and viscous modulus (G″) in scan tests as a function of frequency (Hz).

As can be seen in Figure 3a, the viscosity profile of the polymer solution (blue flow curve) is a descending one, as the applied shear stress increased (non-Newtonian behavior), in the first 0.1 s a steep decrease in the viscosity value was observed. On return, by decreasing the shear rate, the viscosity remained constant without any dilatancy. This pseudoplastic behavior leading to fluidization of the polymer system upon shear is particularly important in the dosing and dispersion of ingredients in a FFS when it is subjected to mixing, having an influence on film formation. From the rheological analysis we can conclude that the sodium alginate solution obeys the Cox–Merz rule, where the viscosity (η, mPas) is dependent on the shear rate (γ, 1/s) [44]. Studies of film-forming solutions with sodium alginate as gelling agent and pullulan (another type of glucan) it has been observed an overlapping of the pullulan chains with the formation of a transient network while the polymer chains of sodium alginate adopted a cross-linked network [45]. Further, Figure 3b gives details of the elastic modulus and viscous modulus of the sodium alginate solution. As can be seen, the viscous modulus (G″) exhibited values significantly higher than the elastic modulus (G′) until the end of the linear viscoelastic range (LVR), indicating that the polymer solution exhibits a viscous character. Both moduli (G′ and G″) are a function of frequency (Hz), and by gradually increasing both with frequency, information about the structure of the polymer gel is provided. Thus, at low frequencies (10^{-1} Hz) the polymer chains had more time to relax and form entangled chains but with increasing frequency towards 10 Hz the alginate chains unravel into random anchors [46].

3.2. Film Thickness

The thickness of the films was mostly influenced by the amount of dry matter in the film-forming solution. As can be seen in Figure 4, the optimization process of the pomegranate juice films, nor the addition of soybean oil to the bilberry or cranberry juice films, did not show a significant influence on the thickness, which was below 150 μm. Generally, edible films intended for packaging are within the thickness values between 31.2 μm [47] and 300 μm [48] with an average around 150 μm in order to achieve a uniform distribution of compounds from the FFS stage [49–51]. The maximum and minimum values are found in films with the highest amount of juice and solids with a peak at 147.3 μm in Sample 11 and, respectively, the lowest value at a low amount of solids at Sample 2 with 64.4 μm containing 0.5 g BG, 0.4 g SA, 1.24 g PJ.

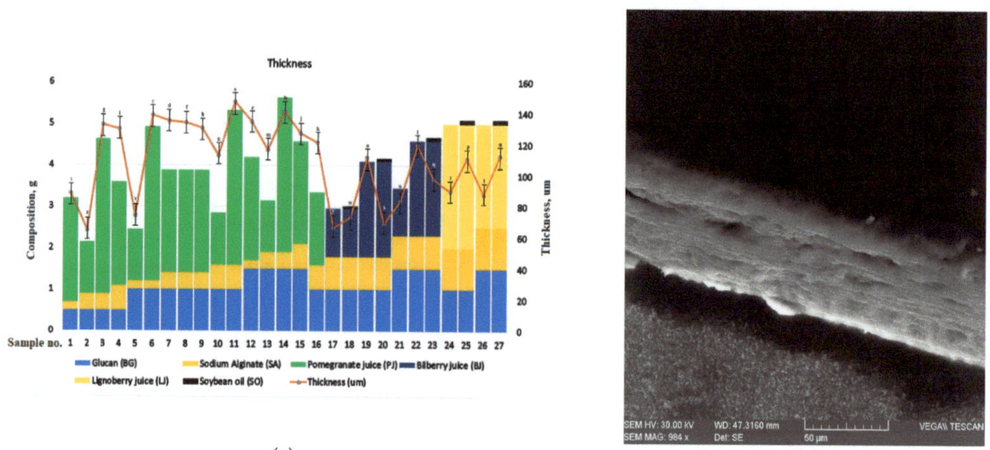

(a) (b)

Figure 4. Film thickness. (**a**) The thickness of the films related to the composition of the FFS; (**b**) 2-D SEM image of a cross-section film made from BG, SA and PJ. Values followed by different letters are significantly different at $p < 0.05$.

Figure 4b presents a cross-section image taken with the scanning electronic microscope (SEM). This technique is also often found in the scientific literature for film thickness measurement, although it is usually used for thin films [52,53]. Moreover, the uniform surface structure of the top of the film demonstrates good compatibility between β-glucan, sodium alginate and juice. In the cross-section, there are no pores or transverse microcracks, indicating a homogeneous interaction between the plasticizer and the polymers.

3.3. Water Vapor Transmission Rate (WVTR)

The barrier capacity of the film expressed by the amount of water vapor that can permeate per one unit of material area for certain time is particularly important in packaging (and especially for dry products that can absorb moisture from the external environment). The following figure shows the tendency of the rate of the water permeating through the film which occurs when water vapor diffuses from a higher partial pressure to a lower partial pressure [54]. From Figure 5, it can be observed with red line the WVTR trend of the film samples. In the case of the film optimization process of β-glucan (BG), sodium alginate, and pomegranate juice, high values can be observed. Of course, the highest value of all three types of films developed from different juices was found in this case at Sample 3 with 32.1896 g/h·m^2 having 0.5 g β-glucan content, 0.4 g sodium alginate (SA), and 30 mL (3.73 g dw.) pomegranate juice (PJ) [37]. From Sample 17, an abrupt decrease in the WVTR values was observed for the bilberry- and cranberry-based films. The lowest value of WVTR was found in the Sample 19 with 3.2562 g/h·m^2 containing 1 g BG, 0.8 SA, 20 g bilberry juice (BJ) [38]. In the case of films made of β-glucan/cranberry juice, with the highest amount of dry weight and the highest sodium alginate content, a minimum WVTR value of 7.6918 g/h·m^2 was identified. A drop in the WVTR values it could also be observed by the addition of soybean oil to the films based on bilberry and cranberry juices (Sample 18, 23, 25 and 27), compared to control samples (Sample 17, 22, 24 and 26). In all analyzed samples the amount of β-glucan no more than 1.5 g (dry weight) was added in the film-forming solution and varied at 0.5, 1 and 1.5 g. Sodium alginate did not exceed 1 g and the amount of plasticizer (glycerin) was 25% of total solids. However, such large differences between WVTR values (from 32.1896 to 3.2562 g/h·m^2) can only be explained by the change in juice in FFS.

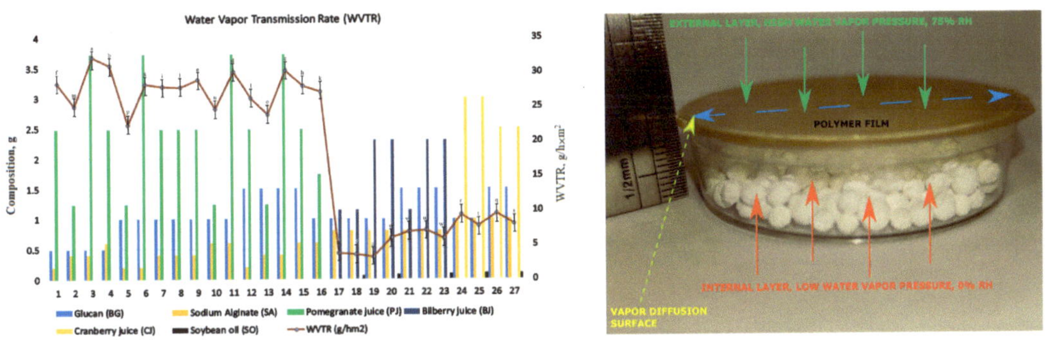

(a) (b)

Figure 5. (**a**) WVTR tendency of the β-glucan, sodium alginate and juices films; (**b**) Experimental model of dry cup with $CaCl_2$ as desiccant according to ASTM (4 mm air gap). Values followed by different letters are significantly different at $p < 0.05$.

3.4. Water Vapor Permeability (WVP)

The WVP results of the films (Figure 6a) showed values between 0.1057 and 1.8657 g·mm/kPa·h·m^2. A fairly strong positive relationship with a correlation coefficient of 0.8 was observed between WVP values and film thickness. The lowest value found in Sample 17 showed a thickness of 66.43 µm while the highest vapor flux through the film was found at Sample 11 with 147.3 µm thickness made of 30 mL PJ (3.73 g dw.), 1 g BG, 0.6 g SA [37]. The WVP results of the β-glucan/cranberry juice films (between 0.3523 and 0.3782 g·mm/kPa·h·m^2) are higher than β-glucan/bilberry juice films but in comparison with the scientific literature, Severo et al., (2021) after the development of films based on cranberry extract and chitosan for the antibacterial properties found values between 3.71 and 4 ($\times 10^{-12}$ mol·m/m^2·s·Pa) [51]. Owning that 1 mole of water weighs 18.01 g we can deduce that the values found for WVP by the authors were between 0.2405 and 0.2593 g·mm/kPa·h·m^2, close to our values. Azeredo et al., 2016 determined that a high pomegranate juice content in films will increase WVP values due to the plasticizing effect of sugars by adding significant amounts of juice and decreasing the concentration of the polymer matrix that weakens the films. The highest value found in that study was 10.91 g·mm/kPa·h·m^2 [55]. Of course, by all of three types of films made, it was found that the highest value of WVP was found in films based on pomegranate juice with 1.8657 g·mm/kPa·h·m^2 and 30 mL juice added. An innovative aspect that improved water vapor permeability was the addition of β-glucans to the films of various juices. It was observed in previous studies on films made of β-glucan/pullulan that the WVP values does not exceed 1.2 g·mm/kPa·d·m^2 [56]. In addition, films made of cranberry pomace extract, low methoxyl pectin, and glycerin showed increased values of 68.5 g·mm/kPa·d·m^2 (2.85 g·mm/kPa·h·m^2) compared to our samples that included β-glucans [57].

Based on the five stages for the gas transport described by Crank and Park (a diffusion surface layer of the side with higher concentration of penetrants, the absorption of the gas, diffusion the gas through the polymer of a certain thickness, desorption of the gas to the side with lower concentration of penetrant and diffusion of the gas [58]) in films based on pomegranates, bilberries, and cranberries juice, the composition and thickness of the layer are the only parameters that differs, whereas the rest of the parameters such as the surface of the film and the pressure on the two layers remained the same.

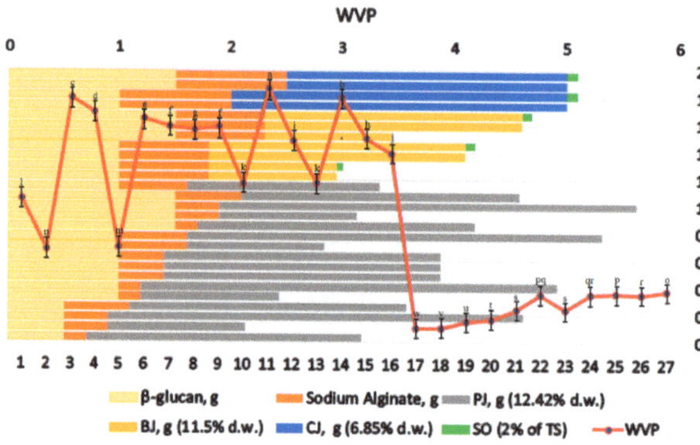

(a)

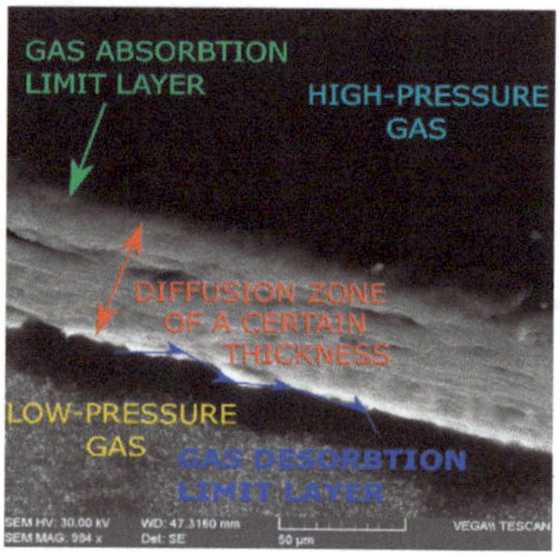

(b)

Figure 6. (**a**) Water vapor pressure tests of the films; (**b**) Cross-section SEM of a film obtained and the diffusion zones, adapted after [59]. Values followed by different letters are significantly different at $p < 0.05$.

However, another factors that affects the WVP values between samples are represented by the integrity of the film, hydrophilic-hydrophobic ratio, crystalline–amorphous ratio, or the polymeric chain mobility [60]. In the present study, in all 27 samples the FFS was a homogeneous mixture and the films showed uniformity, without coarse particles or air bubbles in the mass of the film. The plasticizer with the hydrophilic nature (glycerin) was kept at 25% from total solids while the addition of soybean oil did not show linear association with WVP (r = 0.09). The main use of glycerin was to impart flexibility and enhance toughness for films. In the absence of a plasticizer, the longitudinal microcracks visible in the cross-section of Figure 6b can initiate and over time it can lead to film breakage [61,62].

3.5. Dissolution Time

Depending on the destination of the packaged product, the films must fit within a certain time of solubilization to allow the release of the contents from the package. In the present study, films were intended for the packaging of dry powdered products that dissolves in the oral solution, thus the dissolution time must be as short as possible. As can be seen in Figure 7, the optimization process of β-glucan/pomegranate juice films managed to provide an optimal time of 4.5 min (Sample 16), taking into account a high content of juice compounds (1.74 g dw.), low WVTR, and WVP values.

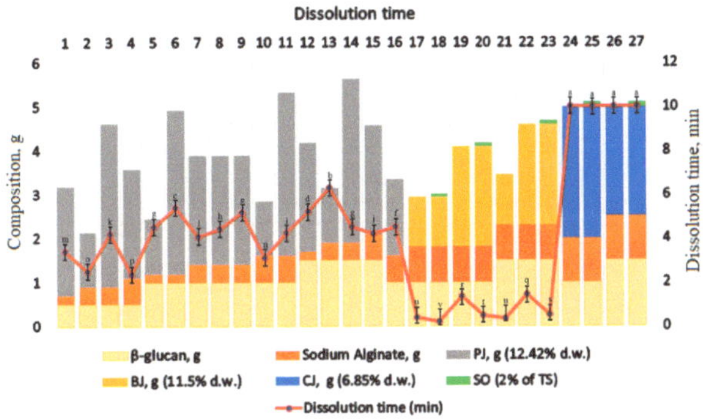

(a)

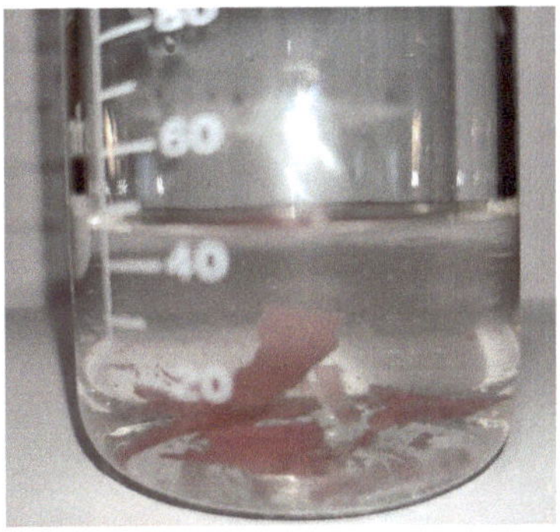

(b)

Figure 7. (**a**) Dissolution time of the films; (**b**) Image of suspended β-glucan/cranberry juice particles in solution after dissolution tests. Values followed by different letters are significantly different at $p < 0.05$.

On the other hand, in the case of β-glucan/bilberry juice films, if a surfactant was added to the film-forming solution, a halving in the dissolution time was found from the 1.45 min in Sample 22 to 0.55 min at Sample 23 which contains a surplus of only 0.092 g soybean oil (2% of total solids) [38]. Besides soybean oil, the amount of bilberry juice had a significant influence ($p < 0.05$) on the dissolution time. The blend compatibility of the natural compounds from the fast-dissolving film, the non-toxic, non-irritating and hydrophilic nature of juices make developed films an advantageous alternative to release the drugs [63]. Interestingly, in the case of β-glucan/cranberry, even after 10 min of shaking, a complete dissolution was not achieved. The use of soybean oil was also taken into account but without visible results. One possible explanation is that the gelling agent was increased to 1 g (SA) in all samples and the juice that was added at 36.49 and 43.79 g CJ (6.85% w/w) to reach 5 g of total solids in FFS. Another explanation could be the newly formed intermolecular bonds between polymer chains or the interaction between polyvalent cations from juice (such as calcium) that forms strong gels up to insoluble polymers [64,65].

3.6. Color and Optical Properties of the Film Samples

The color of the films is largely derived from the color of the juices introduced into the FFS. Only in the case of films made from bilberry juice there is an exception where the pH influences color changes because of the anthocyanins [66]. This property is widely used to the intelligent films for detection the food freshness [67,68].

As can be seen in Figure 8d–f, optical properties in the UV-VIS region were recorded for the representative samples. In the 200–400 nm region, the transmission was close to zero, sign that ultraviolet radiation was blocked by the film. With increasing the wavelength, transmittance increases and visible radiation penetrates through the film. It should be noted that, due to the intense red coloring given by the anthocyanin pigments such as cyanidins and peonidins from cranberry, a peak can be seen in the 500–600 nm region downward in the transmission of radiation in the visible range (arrow marked with 1 in Figure 8f). This shows us that films developed from β-glucan/cranberry juice absorb to some extent the green radiation of the visible spectrum.

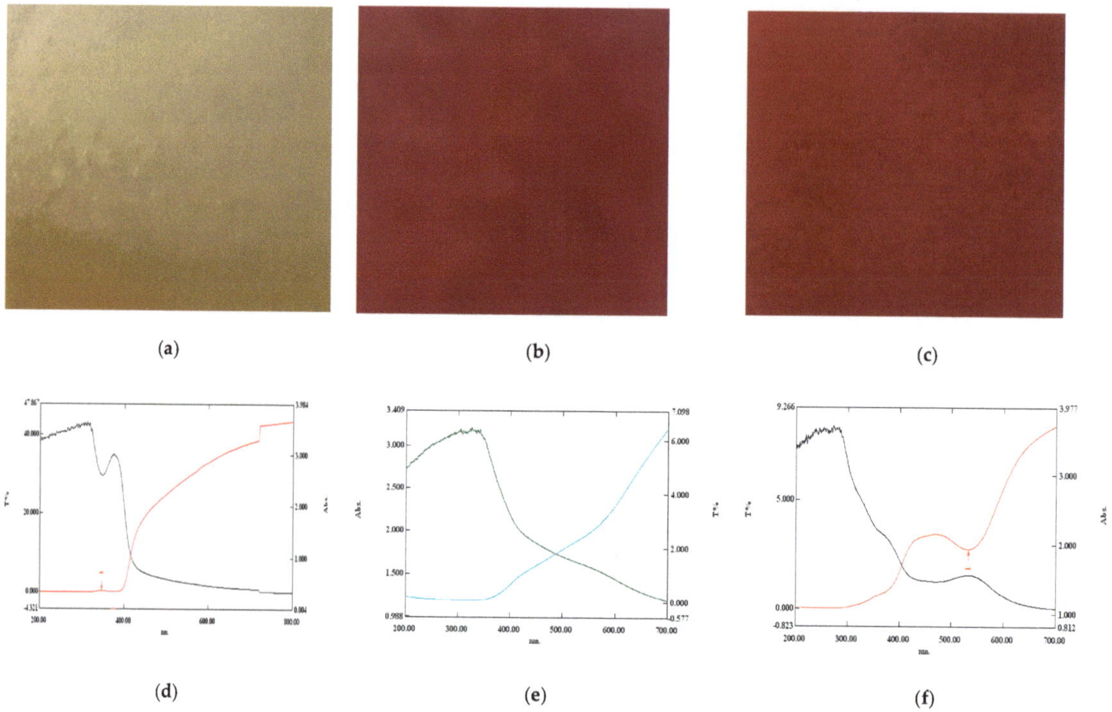

Figure 8. (**a**–**c**) Images of the films made of pomegranate, bilberry and cranberry juice; (**d**–**f**) Optical properties of the β–glucan/ juice films.

4. Conclusions

Bioactive films based on β-glucan and pomegranate, bilberry, or cranberry juice have been successfully developed. The results obtained in the order of film development (pomegranate, bilberry, and cranberry) showed differences in WVTR, WVP, solubilization time, and color.

The rheological study on the polymeric solution of sodium alginate (gelling agent) revealed a pseudoplastic behavior, leading to fluidization of the polymer system. This non-Newtonian behavior is particularly important in the dosing and dispersion of ingredients in a film-forming solution when subjected to mixing.

Film thickness was dependent on the amount of solids introduced into the film-forming solution as well as whether a constant layer thickness was maintained during casting. No more than 6 g of solids were added to the film-forming solution and the casting was conducted uniformly to obtain films below 200 μm thickness.

As the amount of sodium alginate increases, the WVTR decreases, but it increases when a higher concentration of juice was added. The hydrophobic nature of soybean oil decreases WVTR values to a small extent. In the case of the films with the highest amount of sodium alginate (1 g), a minimum WVTR of 7.6918 $g/h \cdot m^2$ was found, while the highest value of 32.1896 $g/h \cdot m^2$ was identified at a value of 0.4 g of SA and 30 mL of juice.

WVP was dependent on film thickness. The lowest WVP value was found in bilberry juice-based films (0.1057 $g \cdot mm/kPa \cdot h \cdot m^2$) followed by cranberry juice (0.3523 $g \cdot mm/kPa \cdot h \cdot m^2$) and finally in pomegranate juice-based films.

The addition of soybean oil to bilberry juice films reduces the dissolving time by half. The most noticeable effect of increasing the amount of sodium alginate and juice was to reduce solubility that has been found in films made from cranberry juice. Fast-dissolving films showed stability against water vapor, acceptable thickness, and no visible cracks in the film mass

Considering the results of the current study, the amount of juice and gelling agent are an important factor in attributing the physicochemical properties to bioactive films.

Author Contributions: Conceptualization, S.A. and I.A.; methodology, I.A. and S.A.; software, I.A.; validation, S.A. and I.A.; formal analysis, I.A.; investigation, I.A.; resources, S.A.; writing—original draft preparation, I.A.; writing—review and editing, I.A.; visualization, I.A.; supervision, S.A.; project administration, S.A.; funding acquisition, S.A. All authors have read and agreed to the published version of the manuscript.

Funding: This work was funded by Ministry of Research, Innovation and Digitalization within Program 1—Development of national research and development system, Subprogram 1.2—Institutional Performance—RDI excellence funding projects, under contract no. 10PFE/2021.

Institutional Review Board Statement: Not applicable.

Informed Consent Statement: Not applicable.

Data Availability Statement: The data presented in this study can be provided on request.

Conflicts of Interest: The authors declare no conflict of interest.

References

1. Cui, C.; Ji, N.; Wang, Y.; Xiong, L.; Sun, Q. Bioactive and intelligent starch-based films: A review. *Trends Food Sci. Technol.* **2021**, *116*, 854–869. [CrossRef]
2. Moeini, A.; Pedram, P.; Fattahi, E.; Cerruti, P.; Santagata, G. Edible Polymers and Secondary Bioactive Compounds for Food Packaging Applications: Antimicrobial, Mechanical, and Gas Barrier Properties. *Polymers* **2022**, *14*, 2395. [CrossRef]
3. Rezzani, G.D.; Choque, E.; Salvay, A.G.; Mathieu, F.; Peltzer, M.A. New Antioxidant Active Packaging Films Based on Yeast Cell Wall and Naphtho-γ-Pyrone Extract. *Polymers* **2022**, *14*, 2066. [CrossRef]
4. Chaudhary, V.; Thakur, N.; Kajla, P.; Thakur, S.; Punia, S. Application of encapsulation technology in edible films: Carrier of bioactive compounds. *Front. Sustain. Food Syst.* **2021**, *5*, 734921. [CrossRef]
5. La Storia, A.; Di Giuseppe, F.A.; Volpe, S.; Oliviero, V.; Villani, F.; Torrieri, E. Physical properties and antimicrobial activity of bioactive film based on whey protein and Lactobacillus curvatus 54M16 producer of bacteriocins. *Food Hydrocoll.* **2020**, *108*, 105959. [CrossRef]
6. Puscaselu, R.; Gutt, G.; Amariei, S. Rethinking the future of food packaging: Biobased edible films for powdered food and drinks. *Molecules* **2019**, *24*, 3136. [CrossRef]
7. Puscaselu, R.; Gutt, G.; Amariei, S. Biopolymer-Based Films Enriched with Stevia rebaudiana Used for the Development of Edible and Soluble Packaging. *Coatings* **2019**, *9*, 360. [CrossRef]
8. Hulda, N.M.C.; Bianca, S.d.C.; Wiliene, C.d.L.; Daniel, C.K.; Flvio, L.S. Fruit juices in polysaccharides edible films. *Afr. J. Food Sci.* **2020**, *14*, 53–62. [CrossRef]
9. Odjo, K.; Al-Maqtari, Q.A.; Yu, H.; Xie, Y.; Guo, Y.; Li, M.; Du, Y.; Liu, K.; Chen, Y.; Yao, W. Preparation and characterization of chitosan-based antimicrobial films containing encapsulated lemon essential oil by ionic gelation and cranberry juice. *Food Chem.* **2022**, *397*, 133781. [CrossRef]
10. Peltzer, M.A.; Salvay, A.G.; Delgado, J.F.; de la Osa, O.; Wagner, J.R. Use of Residual Yeast Cell Wall for New Biobased Materials Production: Effect of Plasticization on Film Properties. *Food Bioprocess Technol.* **2018**, *11*, 1995–2007. [CrossRef]
11. Topalović, A.; Knežević, M.; Ivanović, L.; Gačnik, S.; Mikulic-Petkovsek, M. Phytochemical screening of wild pomegranate (*Punica granatum* L.) juices from the market. *J. Food Compos. Anal.* **2021**, *100*, 103933. [CrossRef]
12. Kamau, C.M. Fruit Juices: Ellagic Acid Concentration and Sensory Appeal. Ph.D. Thesis, Bowling Green State University, Bowling Green, OH, USA, 2007.
13. Poyrazoğlu, E.; Gökmen, V.; Artık, N. Organic acids and phenolic compounds in pomegranates (*Punica granatum* L.) grown in Turkey. *J. Food Compos. Anal.* **2002**, *15*, 567–575. [CrossRef]
14. Adami, R.; Salvo, G.; Meneses, M.; Järvenpää, E.; Huopalahti, R.; Sesti Osséo, L.; Reverchon, E. Innovative treatment of bilberry by-products for a selective recovery of anthocyanin compounds. In Proceedings of the 10th Conference on Supercritical Fluids and Their Applications, Naples, Italy, 29 April–6 May 2013.
15. Laaksonen, O.; Sandell, M.; Kallio, H. Chemical factors contributing to orosensory profiles of bilberry (*Vaccinium myrtillus*) fractions. *Eur. Food Res. Technol.* **2010**, *231*, 271–285. [CrossRef]
16. Vilkickyte, G.; Petrikaite, V.; Pukalskas, A.; Sipailiene, A.; Raudone, L. Exploring *Vaccinium vitis-idaea* L. as a potential source of therapeutic agents: Antimicrobial, antioxidant, and anti-inflammatory activities of extracts and fractions. *J. Ethnopharmacol.* **2022**, *292*, 115207. [CrossRef]

17. Kowalska, K. Lingonberry (*Vaccinium vitis-idaea* L.) Fruit as a Source of Bioactive Compounds with Health-Promoting Effects—A Review. *Int. J. Mol. Sci.* **2021**, *22*, 5126. [CrossRef]
18. Fernandes, L.; Pereira, J.A.; Lopéz-Cortés, I.; Salazar, D.M.; González-Álvarez, J.; Ramalhosa, E. Physicochemical composition and antioxidant activity of several pomegranate (*Punica granatum* L.) cultivars grown in Spain. *Eur. Food Res. Technol.* **2017**, *243*, 1799–1814. [CrossRef]
19. Papanov, S.I.; Petkova, E.G.; Ivanov, I.G. Polyphenols Content and Antioxidant Activity of Bilberry Juice Obtained from Different Altitude Samples. *J. Pharm. Res. Int* **2021**, *33*, 218–223. [CrossRef]
20. Güder, A.; Gür, M.; Engin, M.S. Antidiabetic and antioxidant properties of bilberry (*Vaccinium myrtillus* Linn.) fruit and their chemical composition. *J. Agric. Sci. Technol.* **2015**, *17*, 401–414.
21. Rocha, D.M.U.P.; Caldas, A.P.S.; da Silva, B.P.; Hermsdorff, H.H.M.; Alfenas, R. de C.G. Effects of blueberry and cranberry consumption on type 2 diabetes glycemic control: A systematic review. *Crit. Rev. Food Sci. Nutr.* **2019**, *59*, 1816–1828. [CrossRef]
22. Vetvicka, V. Glucan-immunostimulant, adjuvant, potential drug. *World J. Clin. Oncol.* **2011**, *2*, 115. [CrossRef]
23. Dey, L.; Attele, A.S.; Yuan, C.-S. Alternative therapies for type 2 diabetes. *Altern. Med. Rev.* **2002**, *7*, 45–58.
24. Heiss, C.; Istas, G.; Feliciano, R.P.; Weber, T.; Wang, B.; Favari, C.; Mena, P.; Del Rio, D.; Rodriguez-Mateos, A. Daily consumption of cranberry improves endothelial function in healthy adults: A double blind randomized controlled trial. *Food Funct.* **2022**, *13*, 3812–3824. [CrossRef]
25. Carlos, O.S.; Enrique, F.I.H.; Eduardo, F.M.; Araceli, O.R.M.; Raquel, C.C.; Alberto, A.O.J.; Carlos, H.G.J.; Diana, O.; Carmen, V.V.; Helen, B.M.; et al. Potential Mechanisms of the Improvement of Glucose Homeostasis in Type 2 Diabetes by Pomegranate Juice. *Antioxidants* **2022**, *11*, 553. [CrossRef]
26. Kandylis, P.; Kokkinomagoulos, E. Food applications and potential health benefits of pomegranate and its derivatives. *Foods* **2020**, *9*, 122. [CrossRef]
27. Cui, Y.; Han, X.; Huang, X.; Xie, W.; Zhang, X.; Zhang, Z.; Yu, Q.; Tao, L.; Li, T.; Li, S. Effects of different sources of β-glucan on pasting, gelation, and digestive properties of pea starch. *Food Hydrocoll.* **2023**, *135*, 108172. [CrossRef]
28. Zhu, F.; Du, B.; Xu, B. A critical review on production and industrial applications of beta-glucans. *Food Hydrocoll.* **2016**, *52*, 275–288. [CrossRef]
29. Thomas, S.; Rezoagli, E.; Abidin, I.Z.; Major, I.; Murray, P.; Murphy, E.J. β-Glucans from Yeast—Immunomodulators from Novel Waste Resources. *Appl. Sci.* **2022**, *12*, 5208. [CrossRef]
30. Liu, Y.; Wu, Q.; Wu, X.; Algharib, S.A.; Gong, F.; Hu, J.; Luo, W.; Zhou, M.; Pan, Y.; Yan, Y. Structure, preparation, modification, and bioactivities of β-glucan and mannan from yeast cell wall: A review. *Int. J. Biol. Macromol.* **2021**, *173*, 445–456. [CrossRef]
31. Caruso, A.; Piermaria, J.A.; Abraham, A.G.; Micaela, M. β-glucans obtained from beer spent yeasts as functional food grade additive: Focus on biological activity. *Food Hydrocoll.* **2022**, *133*, 107963. [CrossRef]
32. Jaeger, A.; Arendt, E.K.; Zannini, E.; Sahin, A.W. Brewer's Spent Yeast (BSY), an Underutilized Brewing By-Product. *Fermentation* **2020**, *6*, 123. [CrossRef]
33. Zhang, L.; Yu, D.; Regenstein, J.M.; Xia, W.; Dong, J. A comprehensive review on natural bioactive films with controlled release characteristics and their applications in foods and pharmaceuticals. *Trends Food Sci. Technol.* **2021**, *112*, 690–707. [CrossRef]
34. Volman, J.J.; Ramakers, J.D.; Plat, J. Dietary modulation of immune function by β-glucans. *Physiol. Behav.* **2008**, *94*, 276–284. [CrossRef]
35. Baeghbali, V.; Niakousari, M.; Farahnaky, A. Refractance Window drying of pomegranate juice: Quality retention and energy efficiency. *LWT—Food Sci. Technol.* **2016**, *66*, 34–40. [CrossRef]
36. Bernaert, N.; Van Droogenbroeck, B.; Van Pamel, E.; De Ruyck, H. Innovative refractance window drying technology to keep nutrient value during processing. *Trends Food Sci. Technol.* **2019**, *84*, 22–24. [CrossRef]
37. Avramia, I.; Amariei, S. Formulation, Characterization and Optimization of β–Glucan and Pomegranate Juice Based Films for Its Potential in Diabetes. *Nutrients* **2022**, *14*, 2142. [CrossRef]
38. Avramia, I.; Amariei, S. Formulation of Fast Dissolving β-Glucan/Bilberry Juice Films for Packaging Dry Powdered Pharmaceuticals for Diabetes. *Plants* **2022**, *11*, 2040. [CrossRef]
39. *ASTM E96-00*; Standard Test Methods for Water Vapor Transmission of Materials 1. ASTM International: West Conshohocken, PA, USA, 2018; pp. 1–14.
40. Yıldırım-Yalçın, M.; Şeker, M.; Sadıkoğlu, H. Development and characterization of edible films based on modified corn starch and grape juice. *Food Chem.* **2019**, *292*, 6–13. [CrossRef]
41. Siddiqui, M.H.B.; Islam, M.S.; Razu, M.R.; Zaman, A.N.; Jadi, B.; Saha, T.; Pathan, M.S.I. Preparation and evaluation of sublingual film of ketorolac tromethamine. *Drug Dev. Ind. Pharm.* **2022**, *48*, 438–445. [CrossRef]
42. Kwon, S.; Orsuwan, A.; Bumbudsanpharoke, N.; Yoon, C.; Choi, J.; Ko, S. A short review of light barrier materials for food and beverage packaging. *Korean J. Packag. Sci. Technol.* **2018**, *24*, 141–148. [CrossRef]
43. Mancini, M.; Moresi, M.; Sappino, F. Rheological behaviour of aqueous dispersions of algal sodium alginates. *J. Food Eng.* **1996**, *28*, 283–295. [CrossRef]
44. Roland, C.M. Chapter 6—Rheological Behavior and Processing of Unvulcanized Rubber. In *The Science and Technology of Rubber*, 4th ed.; Mark, J., Erman, B., Roland, M., Eds.; Academic Press: Boston, MA, USA, 2013; pp. 285–336. ISBN 978-0-12-394584-6.
45. Xiao, Q.; Tong, Q.; Lim, L.T. Pullulan-sodium alginate based edible films: Rheological properties of film forming solutions. *Carbohydr. Polym.* **2012**, *87*, 1689–1695. [CrossRef]

46. Ma, J.; Lin, Y.; Chen, X.; Zhao, B.; Zhang, J. Flow behavior, thixotropy and dynamical viscoelasticity of sodium alginate aqueous solutions. *Food Hydrocoll.* **2014**, *38*, 119–128. [CrossRef]
47. Arham, R.; Mulyati, M.T.; Metusalach, M.; Salengke, S. Physical and mechanical properties of agar based edible film with glycerol plasticizer. *Int. Food Res. J.* **2016**, *23*, 1669–1675. [CrossRef]
48. Brasil, I.M.; Gomes, C.; Puerta-Gomez, A.; Castell-Perez, M.E.; Moreira, R.G. Polysaccharide-based multilayered antimicrobial edible coating enhances quality of fresh-cut papaya. *LWT* **2012**, *47*, 39–45. [CrossRef]
49. Nagendra, B.; Vignola, E.; Daniel, C.; Rizzo, P.; Guerra, G. Dependence on film thickness of guest-induced c perpendicular orientation in PPO films. *Polymers* **2021**, *13*, 4384. [CrossRef]
50. Riaz, A.; Lei, S.; Akhtar, H.M.S.; Wan, P.; Chen, D.; Jabbar, S.; Abid, M.; Hashim, M.M.; Zeng, X. Preparation and characterization of chitosan-based antimicrobial active food packaging film incorporated with apple peel polyphenols. *Int. J. Biol. Macromol.* **2018**, *114*, 547–555. [CrossRef]
51. Severo, C.; Anjos, I.; Souza, V.G.L.; Canejo, J.P.; Bronze, M.R.; Fernando, A.L.; Coelhoso, I.; Bettencourt, A.F.; Ribeiro, I.A.C. Development of cranberry extract films for the enhancement of food packaging antimicrobial properties. *Food Packag. Shelf Life* **2021**, *28*, 100646. [CrossRef]
52. Daskalova, A.; Trifonov, A.; Bliznakova, I.; Nathala, C.; Ajami, A.; Husinsky, W.; Declercq, H.; Buchvarov, I. Selective cell response on natural polymer bio-interfaces textured by femtosecond laser. *Appl. Phys. A* **2018**, *124*, 207. [CrossRef]
53. An, J.; Im, S.; Kang, H.; Kim, H.; Yoo, Y.; Noh, J.; Huh, S. A Study on the Wear Characteristics of Al7075 with Changes in Surface Roughness and Ti Thin Film Deposition Time. *Adv. Mater. Sci. Eng.* **2020**, *2020*, 7934842. [CrossRef]
54. Valovirta, I. Water Vapor Permeability and Thermal Conductivity as a Function of Temperature and Relative Humidity. 2004. Available online: https://trepo.tuni.fi/bitstream/handle/10024/128675/Valovirta_Water_Vapor_Permeability_and_Thermal_Conductivity_as_a_Function_of_Temperature_and_Relative_Humidity.pdf?sequence=1 (accessed on 16 February 2023).
55. Azeredo, H.M.C.; Morrugares-Carmona, R.; Wellner, N.; Cross, K.; Bajka, B.; Waldron, K.W. Development of pectin films with pomegranate juice and citric acid. *Food Chem.* **2016**, *198*, 101–106. [CrossRef]
56. Chang, J.; Li, W.; Liu, Q.; Zhou, Y.; Chen, X.; Lyu, Q.; Liu, G. Preparation, properties, and structural characterization of β-glucan/pullulan blend films. *Int. J. Biol. Macromol.* **2019**, *140*, 1269–1276. [CrossRef]
57. Park, S.I.; Zhao, Y. Development and characterization of edible films from cranberry pomace extracts. *J. Food Sci.* **2006**, *71*, E95–E101. [CrossRef]
58. Crank, J.; Park, G.S. Diffusion in high polymers: Some anomalies and their significance. *Trans. Faraday Soc.* **1951**, *47*, 1072–1084. [CrossRef]
59. Duan, Z. Water Vapour Permeability of Bio-Based Polymers. Ph.D. Thesis, Loughborough University, Loughborough, UK, 2013.
60. Ramakrishnan, R.K.; Miroslav, Č.; Padil, V.V.T. Biomacromolecule assembly based on gum kondagogu-sodium alginate composites and their expediency in flexible packaging films. *Int. J. Biol. Macromol.* **2021**, *177*, 526–534. [CrossRef]
61. Beigomi, M.; Mohsenzadeh, M.; Salari, A. Characterization of a novel biodegradable edible film obtained from Dracocephalum moldavica seed mucilage. *Int. J. Biol. Macromol.* **2018**, *108*, 874–883. [CrossRef]
62. Naushad Emmambux, M.; Stading, M. In situ tensile deformation of zein films with plasticizers and filler materials. *Food Hydrocoll.* **2007**, *21*, 1245–1255. [CrossRef]
63. Pacheco, M.S.; Barbieri, D.; da Silva, C.F.; de Moraes, M.A. A review on orally disintegrating films (ODFs) made from natural polymers such as pullulan, maltodextrin, starch, and others. *Int. J. Biol. Macromol.* **2021**, *178*, 504–513. [CrossRef]
64. Jeevahan, J.; Chandrasekaran, M.; Durairaj, R.; Mageshwaran, G.; Joseph, G.B. A brief review on edible food packing materials. *J. Glob. Eng. Probl. Solut.* **2017**, *1*, 9–19.
65. Tavassoli-Kafrani, E.; Shekarchizadeh, H.; Masoudpour-Behabadi, M. Development of edible films and coatings from alginates and carrageenans. *Carbohydr. Polym.* **2016**, *137*, 360–374. [CrossRef]
66. Luchese, C.L.; Abdalla, V.F.; Spada, J.C.; Tessaro, I.C. Evaluation of blueberry residue incorporated cassava starch film as pH indicator in different simulants and foodstuffs. *Food Hydrocoll.* **2018**, *82*, 209–218. [CrossRef]
67. Zhao, Y.; Du, J.; Zhou, H.; Zhou, S.; Lv, Y.; Cheng, Y.; Tao, Y.; Lu, J.; Wang, H. Biodegradable intelligent film for food preservation and real-time visual detection of food freshness. *Food Hydrocoll.* **2022**, *129*, 107665. [CrossRef]
68. Neves, D.; Andrade, P.B.; Videira, R.A.; de Freitas, V.; Cruz, L. Berry anthocyanin-based films in smart food packaging: A mini-review. *Food Hydrocoll.* **2022**, *133*, 107885. [CrossRef]

applied sciences

MDPI

Article

Nutritional Value, Mineral Composition, Fatty Acid Profile and Bioactive Compounds of Commercial Plant-Based Gluten-Free Flours

Ana M. Vivar-Quintana *, Yamina Absi, Miriam Hernández-Jiménez and Isabel Revilla

Food Technology, Escuela Politécnica Superior de Zamora, Universidad de Salamanca, 49022 Zamora, Spain
* Correspondence: avivar@usal.es; Tel.: +34-6-7069-9623

Abstract: Nowadays, a wide range of non-traditional gluten-free products can be purchased. Although consumers identify all these as being similar, their nutritional composition can differ depending on the raw materials used and the production process itself. The aim of the current study was to evaluate the nutritional quality of ten commercially available flours made from rice, pea, chickpea, soybean and hemp. The proximate composition (moisture, fat, fiber, protein, starch and carbohydrates), the mineral composition (Na, Mg, P, K, Ca, Cr, Ni, Se, Cu, Zn, Mn, Fe, Cd, Pb), the phenolic composition (total phenols, total flavonoids, flavanones and dihydrofavonols, and total antioxidant capacity) and the fatty acid profile were analyzed. Results showed an important influence of the vegetable species but also of the brand on the flour composition. Soybean flours, followed by hemp, showed the highest phenolic content and antioxidant capacity. Hemp flour showed a low n6/n3 ratio, while pea flour contained a small amount of fat with a large amount of oleic acid and a satisfactory n6/n3 ratio.

Keywords: chickpea; pea; hemp; soybean; rice; commercial flour; proximate composition; phenolic content; antioxidant activity; fatty acid profile

1. Introduction

Modern consumer requirements are reflecting more and more their concern about improving their diet, health and well-being [1]. In recent years, there has been a growing interest in the potential use of different sources of protein in human nutrition. Dairy and meat products are no longer the main sources of protein in diets, with an increase in the number of plant-based protein products [2]. In this context, non-traditional flours obtained from various raw products are attracting the interest of researchers and the food industry [3]. These kinds of products are suitable for use in vegan or vegetarian diets and for consumers who are aware of the need to reduce the consumption of animal products due to their relation to climate change. In addition, these flours can be used to meet the demand for gluten-free (GF) formulations [4] owing to the requirements of the growing population suffering from celiac disease [5]. Gluten-elimination studies performed on non-celiac individuals showed that a GF diet has beneficial effects in preventing not only gluten-related disorders (GRDs) but also other disorders such as irritable bowel syndrome (IBS) and diabetes [6].

On the other hand, these flours can be used to improve the textural properties of foods and/or improve their nutritional composition. Their ability to improve the rheological behavior of doughs has led to their incorporation into wheat for baking products [7,8]. They are also being used in fortification as a tool to balance the overall nutritional profile of diets, to restore nutrients lost in food processing, and to make products more attractive to consumers [9].

As mentioned above, growing interest has been observed in the use of gluten-free cereals and pseudocereals. These grains are mainly consumed in the form of flours and seeds, which can be easily added to different preparations. Among these cereals, rice is an important source of protein for humans. Like other cereals, rice seed proteins are deficient in some essential amino acids [10], which has led to intensive work on modifying the amino acid profile of rice grain to develop varieties of high nutritional value [11]. The components such as protein, available starch, and dietary fiber may be present in varying amounts in different rice cultivars [12]. Rice flour has been incorporated into gluten-free products such as cakes [13], yellow alkaline noodles [14], bread [15], and pasta [16], among other products.

Pulses have also attracted the interest of both the food industry and consumers. Pulse flours provide high levels of good-quality protein and dietary fiber [17]. Owing to the low level of sulfur amino acids, they are considered to be proteins of less biological value than animal proteins [18]. However, they are a rich source of other amino acids, such as lysine, leucine, isoleucine and phenylalanine and therefore provide a well-balanced essential amino acid profile when combined with other grains such as wheat or rice [19]. In addition, the high starch content of pulses makes their flour highly suitable for use in bakery products [20]. Legume flours have also been used to enhance the nutritional value of gluten-free foods such as pasta, breads and snacks [21,22]. Other minor components of pulse flour include proteins, such as lectins and enzyme inhibitors, and phytic acid. Although these have previously been considered anti-nutritional due to their ability to bind micronutrients and reduce the digestibility of macronutrients, some studies also show the potential health benefits of including low levels of these compounds in one's diet [23]. The correct ratio of anti-nutrients to nutrients can reduce the negative impact on digestibility and play a beneficial role in cellular processes, including antioxidant and anti-inflammatory activities [24]. Among the pulses, different flours have been studied, such as soy flour [25–27], chickpea and lentil flour [21], fava bean flour [28], carob flour [29], green and yellow pea flour [2] and flour of the *Prosopis* species [30], among others.

Other non-traditional seeds have been processed into flour, and their viability for incorporation into different products has been studied. Hemp is the non-pharmacological variety of *Cannabis sativa* subsp. *sativa*, the cultivation of which is legal in several countries [31]. Hemp seeds contain a high proportion of protein (20–25%) and a high fat content (25–35%), comprising mainly polyunsaturated fatty acids (PUFAs), and also interesting amounts of fiber and minerals [32]. Their flour has been used for the production of gluten-free bread [33] and bakery by-products [34]. Amaranth flour for the production of gluten-free breads and biscuits [35], quinoa flour for the production of dairy yogurt [36] or muffins [37] and sorghum flour for the preparation of pasta [38] and flatbread [39] have also been studied.

Plant-based foods are a growing trend among consumers, which has led to these flours quickly becoming popular [40]. At present, a wide range of these products can be purchased in supermarkets or online. Although consumers identify all these flours as similar products, their nutritional composition can be very different depending on the raw materials used and the production process itself. The aim of the current study was to evaluate the nutritional quality of ten commercially available flours made from rice, pea, chickpea, soybean and hemp in relation to current nutritional guidelines and dietary requirements.

2. Materials and Methods

2.1. Materials

Commercial flours from chickpea (*Cicer arietinum*), soybean (*Glycina maxima*), pea (*Pisum sativum*), hemp (*Cannabis sativa*) and rice (*Oryza sativa*) were purchased in local markets (Zamora, Spain) and by electronic commerce, each being of one type from two different food companies.

2.2. Proximate Composition of Flour

The samples were analyzed for nutritional composition (protein, fat, carbohydrates and starch) by using the AOAC procedures [41]. The crude protein content of the samples (N × 6.25) was estimated by the macro-Kjeldahl method; crude fat was determined using a Soxhlet apparatus by extracting a known weight of the sample with petroleum ether.

The moisture content of the flours was determined by using the AACC method 14–15 A [42]. Ash content was determined by incineration at 550 ± 10 °C using the AACC method 08-01.01 [43]. The total fiber was determined according to the AOAC Method 991.43 [44] by using the ANKOM dietary fiber analyzer (ANKOM technology, New York, NY, USA). Total carbohydrates were calculated by their difference, and the % energy provided was calculated considering that protein and carbohydrate provide 4 kcal/g and that fat provides of 9 kcal/g. All the determinations were performed in triplicate.

2.3. Mineral Analysis

Element concentrations were determined by ICP-MS (Na, Mg, P, K, Ca, Mn, Fe, Ni, Cu, Zn, Se, Cr, Cd and Pb). Amounts of approximately 0.2 g samples were placed in Teflon vessels with HNO3 and a Milestone digestion microwave system was used. Concentrations were determined using an Agilent 7800 ICP mass spectrometer (Agilent, Santa Clara, CA, USA) with the following operating conditions: Rf. Power 1550 W, Plasma Ar flow 15 L/min, Auxiliary Air flow 0.9 L/min, Nebulizer Air flow 0.99 L/min.

Quantification was carried out by using certified standard solutions (1 g/L) (Panreac, Castellar del Vallès, Spain) grouping the elements into two multi-elemental standards. The results were expressed as mg kg^{-1} of sample.

2.4. Fatty Acid Analysis

The extraction of lipids from 10 g of the flours was performed using a soxhlet extractor (Soxtherm Gerhardt Variostat; Soxtherm V7.5, Wiesbaden, Germany) using petroleum ether. The fatty acids were analyzed after methylation with methanolic KOH by gas chromatography using the method described by Lurueña et al. [45]. Two to five microliters of the sample were injected into the equipment (GC 6890 N, Agilent Technologies, Santa Clara, CA, USA) using a 100 m × 0.25 mm × 0.20 μm capillary column (SP-2560, Supelco, Inc., Bellefonte, PA, USA) equipped with a split/splitless injector and an FID detector. The fatty acids were identified according to their retention times by comparison with a mixture of commercial standards (47885-U Supelco, Sigma–Aldrich, Darmstadt, Germany). The quantification was expressed as g per 100 g of total fatty acid methyl esters. The determinations were performed in triplicate.

2.5. Determination of the Phenolic Composition and Antioxidant Capacity

2.5.1. Extraction of Flour Samples

The extracts of different flour samples were prepared as described by Betances Salcedo et al. [46]. The different flours (1 g each accurately weighed) were added to 20 mL of 70% ethanol aqueous solution. After keeping the mixture in an ultrasonic water bath for 8 min, it was centrifuged at 3000× *g* for 20 min at 20 °C. The supernatant was separated and then completed in a 25 mL volumetric flask with 70% ethanol. The extractions were performed in duplicate, and all the subsequent determinations were performed in triplicate.

2.5.2. Determination of the Total Phenolic Content

The total phenolic content of the extracts was measured using the Folin–Ciocalteu (FC) reagent using the method described by Millar et al. [2]. Approximately 0.5 mL of FC was added to 0.5 mL of the sample. Then, 10 mL of Na_2CO_3 10 H_2O (7.5%) was added, and the solution was mixed thoroughly. The volume of the mixture was made up to 25 mL with distilled water. The total phenolic content of the sample extract was determined at a 750 nm wavelength in a spectrophotometer (Shimadzu UV 1280, Shimadzu, Kyoto, Japan).

The results are expressed as mg gallic acid equivalent (GAE)/100g of fresh weight. The calibration equation used was y = 0.232x − 0.019; r^2 = 0.996.

2.5.3. Determination of the Total Flavonoid Content

The total flavonoid content (flavones and flavonols) of the sample extracts was determined using the colorimetric method based on the formation of the aluminum chloride complex as described by Valencia et al. [47], with slight modifications. Five hundred microliters (0.5 mL) of 5% AlCl3 was added to 2 mL of sample extracts and brought up to 25 mL with 96% ethanol. The mixture was kept for 30 min in the dark at room temperature. The absorbance was measured at 425 nm using a spectrophotometer (Shimadzu UV 1280, Shimadzu, Kyoto, Japan. The results are expressed as milligrams of rutin per 100 gram of sample. The calibration equation used was y = 2.839x − 0.004; r^2 = 0.997

2.5.4. Total Flavanone and Dihydroflavonol Content Determination

The content of flavanones and dihydroflavonols was determined spectrophotometrically, according to the method described by Popova et al. [48], with slight modifications. An aliquot (1 mL) of the flour extract and 2 mL of the DNP (2.4 dinitrophenylhydrazine) solution was heated to 50 °C for 50 min. After cooling to room temperature, 10% potassium hydroxide (KOH) in methanol (*w*/*v*) was added to the solution up to 10 mL. The absorbance at 486 nm of 1 mL of the resulting solution was measured using a spectrophotometer ((Shimadzu UV 1280, Shimadzu, Kyoto, Japan). The presence of pinocembrin has been described in the literature in rice and chickpea, so the authors decided to use it as a standard and to express the results as mg of pinocembrin/100 g of the flour sample extract The calibration equation used was y = 1.340x − 0.015; r^2 = 0.993.

2.5.5. Determination of the Total Antioxidant Capacity

The total equivalent antioxidant capacity (TEAC) was determined by the ABTS (2,2-azinobis (3-ethylenebenzothiazoline-6-sulfonic acid)) method according to the method described by Chen et al. [49] using the vitamin E analogue Trolox (6-hydroxy-2,5,7,8-tetramethylchorman-2-carboxylic acid) as a standard. An appropriate amount of the sample (20 µL) was added to the ABTS reagent, and the mixture was incubated at 25 °C. Absorbance was recorded every minute for 10 min using a Shimadzu spectrophotometer (Shimadzu UV 1280, Shimadzu, Kyoto, Japan). Appropriate solvent blanks were run in each assay. The percentage of inhibition of absorbance at 734 nm was calculated and plotted as a function of the concentration of Trolox to give the total equivalent antioxidant capacity (TEAC). The determinations were performed in triplicate and calculated from the calibration curve determined by linear regression (r^2: 0.9852): Trolox eq = (% decrease in Abs + 2.107)/1.1815.

2.6. Statistical Analysis

ANOVA and LSD Fisher Tests were used to identify significant differences between groups using the SPSS Package 25 (IBM, Chicago, IL, USA).

Unsupervised pattern recognition analysis through principal component analysis (PCA) was applied to group the samples. The data for PCA modeling were the analyzed physicochemical parameters, which showed values >0 in all the samples. These data were normalized, scaled, and mean-centered. Subsequently, the original variables were linearly transformed into a new set of variables (principal components PCs) which retain the information of the original set. The number of PCs for classifying purposes was determined by selecting those with an eigenvalue of >1. The calculations were performed using the SPSS Package 25 (IBM, Chicago, IL, USA).

3. Results and Discussion

3.1. Proximate Composition

The nutritional composition of the different flours analyzed can be seen in Table 1. The moisture content varied between 6.0% and 9.9%, with chickpea and pea flours having the highest moisture contents. There is a large variability in the moisture contents of these types of flours described in the bibliography, ranging from 0.2% described for hemp flour [50] to 10.11% described for pea flour [51]. The lowest protein content was found in rice flours (7.2%), while the highest was found in soybean flours (41.5%). In relation to fat content, soybean flours (23.0%) and hemp flours (11.5%) showed the highest concentrations, while the lowest concentrations were obtained in rice flours (0.6%). All the flours analyzed showed high amounts of carbohydrates (CHO) (>31.8%), with starch being the most important carbohydrate in the rice flour samples (76.6%), followed by pea flours (48.0%) and chickpea flours (41.1%). The highest fiber contents were found in hemp (50.2%) and soybean (28.6%) flours.

Chickpea and pea flours showed high carbohydrate contents and significant starch, fiber and protein contents. Similar values for chickpea flour (24.61%) and pea flour (22.96%) have been reported for protein content [52]. Similarly, the carbohydrate content reported for chickpea flour (57.78%) and pea flour (57.94%) agrees with the results obtained in this study [52]. Rice flours were characterized by a high carbohydrate content, with a high starch content and very low fat, fiber and ash contents, matching those described by Bolarinwa et al. [53]. Soybean and hemp flours showed high protein and fiber contents with higher amounts of fat and ash than the other flours tested, while their carbohydrate and starch contents were much lower. These results agree with those obtained by Uwem et al. [54] and Alphonce et al. [55] for soybean flours and [50] for hemp flours. As far as energy is concerned, the values are similar, and non-significant differences were observed between the products analyzed except for soybean, which showed significantly higher values owing to its higher fat content.

If we compare the different commercial brands analyzed, it can be observed that in all cases there are statistically significant differences in several of the parameters analyzed, with these being particularly striking in the fat, fiber and starch contents of soybean and hemp flours.

From the nutritional composition point of view, the content of fiber and fat in these flours is of particular interest. The current recommendation as outlined by the European Food Safety Authority (EFSA) is for consumption of >25 g fiber a day to maintain normal bowel function, prevent gastrointestinal disorders, aid in weight management and reduce the risks of coronary heart diseases and type 2 diabetes [56]. The role of fiber in weight management is due to its role in adding bulk to the diet to induce satiety and its lower energy density [57]. On the other hand, the high fat content can be a disadvantage for the use of these flours, owing to the lipoxygenase-catalyzed oxidation of unsaturated fatty acids to volatile compounds, which can affect both the taste and the shelf life of foods [58]. Among the tested flours, soybean and hemp flours differ from other flours in terms of their high fat and fiber, and pea and chickpea flours have a high concentration of fiber with a very low fat content; in contrast, rice flour has a low fiber and fat content.

The starch and protein content is another important aspect to take into account in the composition of these flours. The starch content affects human health through its contribution to the glycemic index and levels of resistant starch. Incomplete digestion–absorption of resistant starch in the small intestine results in non-digestible starch fractions with physiological functions similar to those of dietary fiber and therefore with important health benefits [59]. Starch contributes 50–70% of the energy in the human diet, as it provides a direct source of glucose, which is an essential substrate in brain and red blood cells for generating metabolic energy [60]. Considering the indications of the Food Safety Authority of Ireland, a product is considered to be high in protein when at least 20% of the total energy comes from protein. According to this approach, all the flours analyzed in this study are high-protein foods except rice flour. Soybean and hemp differ from other flours in terms of their low starch content and also in their high protein content. In contrast, rice flour is mainly composed of starch with a low protein content. Pea and chickpea flours have a high concentration of protein and starch, with a very low fat content, which makes them particularly interesting from a nutritional point of view.

3.2. Mineral Content

The concentrations of the 14 minerals analyzed (Na, Mg, P, K, Ca, Mn, Fe, Ni, Cu, Zn, Se, Cr, Cd and Pb) for each of the flours studied can be seen in Table 2. The minerals calcium, phosphorus, sodium, potassium and magnesium are called macroelements as their nutritional requirements are higher than 100 mg/dL day [61]. In all the flours analyzed, potassium is the major macroelement, followed by phosphorus. Sodium is the least abundant macroelement in chickpea, soybean and hemp flours, while in rice and pea flours, calcium is the minority macroelement. Soybean flours have the highest potassium content (18,223.35 mg/kg), while hemp flours have the highest concentrations of phosphorus (9908.3 mg/kg) and magnesium (4978.6 mg/kg). On the other hand, rice flours have the lowest macroelement content, with particularly low concentrations of calcium, potassium and magnesium compared to the other types of flour analyzed.

Those elements of which trace amounts are necessary are called microelements and include copper, iron, manganese, selenium, zinc, chromium and nickel. In all the flours analyzed, the microelement with the highest concentration is iron, followed by zinc and manganese. The highest concentrations of all of these elements were found in the hemp flour, whereas selenium was only detected in one sample of pea flour. The results obtained for macro- and microelements are similar to those described by other authors in chickpea flour [62], rice [63], soybean [64] and pea flour [2] but lower than those described for hemp flour [65]. In all the minerals analyzed, a large variability in their concentrations was found depending on the commercial brand analyzed. This is due to the fact that the mineral composition varies according to parameters such as phenotypic variation, climatic conditions, the fertilizers used, soil quality, processing and storage [66].

Minerals are essential elements in our diet for long and healthy life; some macro- and microelements are found in the structure of teeth (Ca, P and F) and bones (Ca, Mg, Mn, P, B and F) and in the control of blood pressure (Ca and K), and some act as catalysts for metabolic reactions (Zn, Cu, Se, Mg, Mn). They play a vital role as a structural part of many enzymes (Cu, Fe, Mn, Mg, Se and Zn) and are involved in the immune (Ca, Mg, Cu, Se and Zn) and brain (Cr and Mn) systems [67]. Plant foods are in themselves an important source of essential minerals, but they can also be used in the fortification of other foods as a strategy to reduce micronutrient malnutrition. Mineral deficiency is therefore more prevalent in developing countries, where access to fresh food is scarce, whereas in industrialized countries, Ca deficiency is a major problem [68]. However, it should be noted that the total amount of a mineral in a food does not reflect the amount available in the body through absorption, as only a certain amount of the mineral is bioavailable [67]. Dietary deficiency of some of these minerals may be due to low mineral intake, but in some cases, the cause is poor mineral absorption. Phytic acid present in cereals and some legumes has been described as one of the main inhibitors of Fe and Zn absorption and, to a lesser extent, of calcium, with oxalic acid being the main inhibitor of Ca absorption [69].

Table 1. The proximate composition (g/100 g fresh weight) of different flours. Data presented as means ± standard deviation from triplicate analysis.

	Chickpea		Pea		Soybean		Hemp		Rice	
	A *	B	A	B	A	B	A	B	A	B
Moisture	8.1 ± 0.3^{d}	9.0 ± 0.1^{e}	9.9 ± 0.5^{f}	8.9 ± 0.1^{e}	7.3 ± 0.1^{c}	6.0 ± 0.1^{a}	6.2 ± 0.1^{a}	6.6 ± 0.1^{b}	7.7 ± 0.2^{c}	9.8 ± 0.1^{f}
Ash	3.1 ± 0.0^{e}	2.7 ± 0.0^{c}	3.0 ± 0.0^{d}	3.2 ± 0.0^{f}	5.0 ± 0.0^{g}	5.3 ± 0.0^{h}	6.5 ± 0.0^{j}	6.0 ± 0.0^{i}	0.7 ± 0.1^{b}	0.5 ± 0.1^{a}
Fat	4.3 ± 0.9^{b}	5.6 ± 0.4^{c}	1.5 ± 0.5^{a}	1.1 ± 0.0^{a}	18.3 ± 0.6^{f}	23.0 ± 0.0^{g}	7.3 ± 0.0^{d}	11.5 ± 0.0^{e}	1.0 ± 0.1^{a}	0.6 ± 0.0^{a}
Fiber	16.7 ± 0.1^{d}	17.1 ± 0.0^{d}	8.6 ± 0.3^{b}	11.5 ± 0.2^{c}	28.6 ± 0.5^{f}	25.4 ± 0.4^{e}	50.2 ± 0.4^{h}	38.2 ± 0.9^{g}	1.0 ± 0.0^{a}	1.0 ± 0.0^{a}
Proteins	23.4 ± 0.0^{f}	19.3 ± 0.0^{c}	21.7 ± 0.1^{e}	21.3 ± 0.2^{d}	38.1 ± 0.0^{i}	41.5 ± 0.4^{j}	29.4 ± 0.1^{g}	32.1 ± 0.4^{h}	7.6 ± 0.0^{b}	7.2 ± 0.0^{a}
Starch	40.9 ± 0.0^{e}	41.1 ± 0.0^{e}	48.0 ± 0.0^{g}	45.6 ± 0.4^{f}	3.5 ± 0.1^{c}	4.1 ± 0.0^{d}	0.1 ± 0.0^{a}	0.8 ± 0.1^{b}	76.6 ± 0.2^{i}	72.4 ± 0.1^{h}
CHO	63.5 ± 4.5^{de}	65.2 ± 3.1^{de}	65.4 ± 2.5^{de}	65.8 ± 0.3^{e}	39.1 ± 13.5^{ab}	31.8 ± 13.4^{a}	53.0 ± 4.4^{cd}	47.6 ± 6.2^{bc}	83.3 ± 0.8^{f}	82.0 ± 0.3^{f}
Energy (Kcal/100 g)	373.6 ± 5.9^{ab}	371.7 ± 16.6^{ab}	357.4 ± 4.1^{a}	355.5 ± 3.5^{a}	418.3 ± 40.9^{bc}	431.2 ± 66.8^{c}	373.7 ± 20.8^{ab}	387.8 ± 33.3^{abc}	369.7 ± 2.1^{ab}	360.2 ± 2.0^{a}

* A and B are different commercial brands. a–j: values followed by different superscripts in the same row are significantly different ($p < 0.05$).

Table 2. The mineral composition (mg/kg fresh weight) of different flours. Data presented as means ± standard deviation from triplicate analysis.

	Chickpea		Pea		Soybean		Hemp		Rice	
	A *	B	A	B	A	B	A	B	A	B
					Macroelements					
Na	1196.8 ± 46.3^{e}	1028.8 ± 4.4^{d}	1057.9 ± 58.9^{d}	967.1 ± 41.2^{cd}	1009.9 ± 17.8^{d}	987.7 ± 12.6^{d}	641.4 ± 6.6^{a}	846.1 ± 5.6^{b}	991.2 ± 19.8^{d}	861.7 ± 4.9^{bc}
Mg	1430.7 ± 71.7^{c}	1316.9 ± 12.4^{c}	1224.1 ± 55.7^{bc}	958.4 ± 6.9^{b}	2390.9 ± 46.2^{d}	2144.2 ± 182.1^{d}	5255.2 ± 69.0^{f}	4702.0 ± 149.6^{e}	392.5 ± 7.1^{a}	374.7 ± 31.1^{a}
P	2268.9 ± 105.1^{b}	2580.0 ± 79.2^{b}	2530.2 ± 104.9^{b}	2524.9 ± 0.7^{b}	4147.8 ± 69.8^{c}	5347.0 ± 110.1^{d}	$10{,}717.6 \pm 24.5^{f}$	9099.0 ± 141.1^{e}	873.1 ± 19.3^{a}	1138.3 ± 60.3^{a}
K	$10{,}848.4 \pm 40.6^{c}$	9412.9 ± 56.0^{b}	$10{,}356.6 \pm 61.6^{c}$	$10{,}464.9 \pm 16.6^{c}$	$17{,}595.8 \pm 235.2^{d}$	$18{,}850.9 \pm 53.0^{e}$	$10{,}686.2 \pm 174.4^{c}$	$10{,}419.2 \pm 14.4^{c}$	1237.0 ± 12.7^{a}	1316.8 ± 12.5^{a}
Ca	1274.3 ± 52.9^{d}	1596.8 ± 37.3^{e}	677.6 ± 3.2^{c}	406.5 ± 8.4^{b}	1609.2 ± 22.7^{e}	2097.3 ± 136.6^{f}	1776.8 ± 17.9^{e}	1290.4 ± 0.7^{d}	69.7 ± 0.6^{a}	74.3 ± 0.4^{a}
					Microelements					
Cr	0.119 ± 0.07^{a}	$0.189.0 \pm 0.06^{b}$	0.119 ± 0.07^{a}	0.429 ± 0.01^{e}	0.950 ± 0.01^{g}	0.125 ± 0.09^{a}	0.331 ± 0.01^{d}	1.04 ± 0.02^{h}	0.485 ± 0.08^{f}	0.238 ± 0.01^{c}
Ni	4.150 ± 0.19^{d}	0.885 ± 0.09^{a}	3.596 ± 0.16^{c}	34.044 ± 0.18^{g}	8.436 ± 0.83^{e}	5.379 ± 0.15^{d}	3.443 ± 0.25^{c}	16.055 ± 0.07^{f}	1.152 ± 0.17^{b}	1.272 ± 0.23^{b}
Se	ND	ND	ND	0.322 ± 0.02	ND	ND	ND	ND	ND	ND
Cu	9.6 ± 0.4^{e}	4.9 ± 0.4^{bc}	7.6 ± 0.4^{d}	6.3 ± 0.1^{cd}	8.7 ± 0.0^{e}	14.1 ± 0.2^{f}	18.9 ± 1.0^{g}	14.9 ± 1.1^{f}	2.9 ± 0.1^{ab}	2.0 ± 0.1^{a}
Zn	32.7 ± 1.5^{cd}	21.7 ± 0.2^{ab}	26.1 ± 1.3^{bcd}	25.7 ± 0.1^{bc}	25.9 ± 0.1^{bc}	33.6 ± 5.6^{d}	77.0 ± 0.4^{f}	48.1 ± 0.8^{e}	14.7 ± 0.3^{a}	14.5 ± 0.1^{a}
Mn	22.4 ± 1.0^{b}	26.9 ± 1.1^{c}	11.0 ± 0.6^{a}	9.8 ± 0.3^{a}	27.2 ± 0.3^{c}	25.0 ± 0.2^{bc}	142.2 ± 0.6^{e}	102.6 ± 1.2^{d}	11.2 ± 0.2^{a}	11.4 ± 0.0^{a}
Fe	69.6 ± 3.6^{e}	49.2 ± 0.4^{c}	85.6 ± 4.2^{f}	35.1 ± 1.2^{b}	103.6 ± 1.3^{g}	59.3 ± 0.9^{d}	166.6 ± 0.9^{h}	204.0 ± 2.5^{i}	20.8 ± 0.3^{a}	68.1 ± 1.2^{e}
					Trace Toxic Elements					
Cd	ND	ND	0.014 ± 0.04^{a}	0.289 ± 0.08^{g}	0.034 ± 0.02^{b}	0.040 ± 0.05^{c}	0.057 ± 0.01^{e}	0.075 ± 0.06^{f}	0.053 ± 00.3^{d}	0.036 ± 0.07^{b}
Pb	0.108 ± 0.05^{a}	0.115 ± 00.7^{a}	0.327 ± 0.01^{d}	0.602 ± 0.01^{f}	0.492 ± 0.06^{e}	0.176 ± 0.01^{ab}	0.481 ± 0.09^{e}	2.945 ± 0.06^{g}	0.216 ± 0.04^{bc}	0.263 ± 0.09^{cd}

* A and B are different commercial brands. a–i Values followed by different superscripts in the same row are significantly different ($p < 0.05$). ND not detected.

In addition to these elements, some minerals such as Cd and Pb have no known biological activity but accumulate in living organisms and can be highly toxic. Cadmium has not been detected in any of the chickpea flours analyzed, and the concentrations found in the other flours are low except in one of the pea flour samples. Although there is no specific legislation regarding Cd and Pb content in plant-based flours, there is European legislation regulating the Cd and Pb content in cereals and legumes. As far as cadmium content is concerned, REGULATION (EU) 2021/1323 sets a limit of 0.04 mg/kg in dry legumes and 0.15 mg/kg in rice. The results obtained from the samples analyzed show that all of them are below the established limits, except for one sample of pea flour, which shows very high Cd contents (0.289 mg/kg). Ingestion of Cd-contaminated food has been linked to renal tubular damage, osteoporosis and cancer [70].

However, all the flours analyzed showed the presence of lead. In one of the samples of soybean flour and pea flour and in the two samples of hemp flour, high concentrations of this mineral were found, which were particularly high in one hemp flour (2.94 mg/kg). The levels found in the samples analyzed are higher than those recommended by REGULATION (EU) 2021/1317, which sets a maximum limit for cereals and pulses of 0.2 mg/kg of Pb. Only the two samples of chickpea flour and one sample of soybean flour were below this value. Prolonged exposure to Pb has been linked to kidney problems or high blood pressure in adults and delayed physical and mental development in children [71].

3.3. Phenolic Composition and Antioxidant Activity

All the studied parameters, the total phenolic content (TPC), the total flavonoid content (TFC), the flavanone and dihydroflavonol content, varied significantly in the different samples studied (Table 3), depending both on the plant species and on the commercial brand. Therefore, the total phenolic content (TPC) varied from 6.33 mg GAE/100 g in rice (B) flour to 139.31 mg GAE/100 g in soybean (B) flour; the total flavonoid content ranged from 1.47 for rice to 66.65 mg of rutin/100 g for hemp, and flavanones and dihydroflavonols showed values between 328.31 for rice and 638.45 mg of pinocembrin/100 g for pea. These results showed that soybean flour had the highest values of TPC, followed by hemp, which showed the highest values of TFC and flavanones and dihydroflavonols, while pea showed similar values to hemp for this group of phenolic compounds. On the other hand, rice showed the lowest values of all the bioactive compounds. Variations between brands were observed in the TPC for chickpea, pea and especially for soybean and in flavanones and dihydroflavonols for pea and soybean, but this variation was particularly noticeable in TFC also for chickpea, pea and soybean.

As can be observed in Table 3, all pulse flours showed significantly higher values than cereal flour (rice) for all the parameters determined. These results are in agreement with those previously reported by Millar et al. [2], which showed that pulses (fava, green pea and yellow pea) had higher phenolic contents than wheat flour, and with those of the Rocchetti et al. [72], which found lower TPC in white rice than in the chickpea. Regarding pulses, chickpeas showed a higher content of TPC than peas agreeing with the results of Xu et al. [73], although Magalhães et al. [74] showed the opposite for field peas. However, for TFC, the values found in this work for peas were similar to or higher than those found for chickpea, which was in agreement with the results found by Magalhães et al. [74] but not with those reported by other authors [73,75]. On the other hand, hemp showed higher contents of TPC than green pea [76], although this work pointed out that wheat (cereal) showed higher TPC content than green peas, differing from the results of Millar et al. [2] and those obtained in this study. Moreover, the research by Multari et al. [76] found similar contents of TFC for green pea and hemp, which were significantly higher than that observed for wheat, in agreement with the present study. Xu et al. [73] reported significantly higher values for TPC and TFC values in soybean than in pea and chickpea, as observed in this study, although the differences were lower.

Table 3. The phenolic composition and antioxidant capacity (per 100 g of fresh weight) of different flours. Data presented as means ± standard deviation from triplicate analysis.

	Chickpea		Pea		Soybean		Hemp		Rice	
	A *	B	A	B	A	B	A	B	A	B
TPC	41.85 ± 2.04^{c}	24.26 ± 1.63^{b}	21.91 ± 0.01^{b}	7.48 ± 1.64^{a}	95.45 ± 9.25^{f}	139.30 ± 1.74^{g}	79.53 ± 2.81^{e}	66.42 ± 0.72^{d}	6.90 ± 0.80^{a}	6.33 ± 0.81^{a}
TFC	39.73 ± 2.00^{cd}	17.08 ± 6.01^{b}	48.21 ± 2.04^{de}	32.65 ± 0.05^{c}	32.61 ± 0.05^{c}	56.66 ± 10.12^{ef}	58.24 ± 4.05^{fg}	66.65 ± 0.09^{g}	5.73 ± 2.02^{a}	1.48 ± 0.01^{a}
F/DF	471.59 ± 18.91^{c}	436.58 ± 50.17^{bc}	638.46 ± 10.89^{e}	570.56 ± 12.44^{d}	457.73 ± 40.93^{c}	390.75 ± 1.06^{ab}	620.31 ± 6.19^{de}	580.62 ± 37.71^{de}	331.33 ± 13.95^{a}	328.31 ± 35.28^{a}
TEAC	3.79 ± 0.79^{c}	2.82 ± 0.18^{bc}	1.86 ± 0.08^{ab}	1.08 ± 0.84^{a}	7.91 ± 0.64^{d}	8.69 ± 0.71^{d}	3.45 ± 0.09^{c}	3.75 ± 0.87^{c}	1.27 ± 0.04^{a}	1.66 ± 0.20^{ab}

* A and B are different commercial brands. TPC: total phenolic content (mg GAE/100 g), TFC: total flavonoid content (mg rutin/100 g), F/ DF: flavanones and dihydroflavonols (mg pinocembrin/100 g), TEAC: total antioxidant activity (nmol Trolox/100 g). a–g: values followed by different superscripts in the same row are significantly different ($p < 0.05$).

The TPC of chickpea was lower than that previously reported [73,77,78] but close to the value (54 mg/100 g) found by Fernandez-Orozco et al. [79] and higher than the values found by Magalhães et al. [74]. Previous studies also showed that the most important phenolic compounds in chickpeas are phenolic acids [74,80], with p-hidroxybenzoic acid being the most relevant in relation to the antioxidant properties [81]. Some studies also showed the presence of flavonoids, as was found in the present study, but this depended on the variety [80,81].

The TPC of peas was also lower than previously reported for both yellow and green pea [2,73]. Previous studies also showed that it is possible to find both phenolic acids (protocatechuic and p-hydroxybenzoic acids) and flavonoid compounds such as luteolin and apigenin [74,82]. Regarding soybean flour, the TPC was in the lower half of the range reported by Slavin et al. [83] and Kumar et al. [84] and slightly lower than that previously reported by Xu et al. [73] but slightly higher than the value found for soybean flour by Villalobos et al. [85]. The most relevant phenolic compounds in soybean are isoflavones daidzein and genistein, although it also contains phenolic acids, such as gallic and protocatechuic acids, and flavonoids such as rutin and quercetin [86]. As mentioned above, the rice flour showed the lowest content of TPC; indeed, the values observed were slightly lower than those found in white rice flour [87] and significantly lower than the values described for wild, black or violet rice flours [72] because pigmented varieties have higher TPCs related to reducing disease risk [88]. The main individual phenolic compounds present in rice flour are phenolic acids such as vanillic, chlorogenic and ferulic acids and flavonoids such as epicatequin [87,89]. Finally, hemp flour showed TPC values similar to or slightly lower than those previously reported, ranging from 74.4 to 98.7 [3], but lower than the values reported by Irakli et al. [90]. The most important phenols found in hemp flour are phenolic acids such as protocatechuic, p-hydroxybenzoic, cinnamic acid and p-coumaric [76,90].

The total equivalent antioxidant capacity (TEAC) also showed a significant variation among flours depending more on the plant species rather than the commercial brand, as no significant differences were detected regarding this factor. The TEAC values ranged from 8.69 nmols for soybean to 1.08 nmols for Trolox per 100 g of pea. Soybean showed the highest statistically significant values of TEAC, followed by hemp and chickpea which did not show significant differences between them, while rice and pea showed the lowest values without statistically significant differences between them. As antioxidant activity can be measured by different methods, the results of the ABTS methods in the bibliography are scarce, although in general, higher values were reported [85,90]. This can be related to the fact that the bioactivity depends on the extraction and assay conditions, and the lack of standardization in these methods can make comparisons difficult [91]. However, previous studies have pointed out that soybean and hemp have a higher antioxidant capacity than peas, as observed in these commercial flours [73,75]. Indeed, hemp showed higher DPPH values than wheat, which means that this species has a higher antioxidant capacity than cereals, as has been observed in this study in rice [3]. Other studies also found that chickpea has a higher antioxidant activity, measured by FRAP and ORAC methods, than rice [87], which is in agreement with the results obtained in this study for the ABTS method.

As previously mentioned, the highest antioxidant activity and the highest phenolic content (Table 3) were observed in soybean flour. Then, the Pearson correlations between phenolic groups and ABTS were subsequently calculated. A positive correlation between the ABTS and TPC was observed ($r^2 = 0.935$, $p < 0.01$), agreeing with previous reports by Millar et al. [2], and also between the TFC and ABTS ($r^2 = 0.652$, $p < 0.05$).

Phenolic compounds are known to interact with proteins to form complexes that, in turn, decrease the solubility of proteins and make protein complexes less susceptible to proteolytic attack than the same proteins in isolation [92]. Furthermore, they impair starch and disaccharide assimilation and interact with proteolytic enzymes to inhibit their activity. Then, the lower TPC content of some flours may result in greater protein bioavailability, because the total phenols include tannins and phenolic acids, which can affect the overall digestibility of pulse flour [93]. However, plant phenolics are attracting increased interest owing to their potential role as protective factors against free-radical-mediated pathologies, such as cancer and atherosclerosis in humans [94]. Indeed, phenolic compounds have a multitude of physiological functions such as anti-inflammatory and cardioprotective activity [95]. Owing to their intrinsic antioxidant activity, they are commonly used as food additives for the protection of food products against oxidation [96]. Knowledge of the biological activities of these plant chemicals is continually growing; they have the potential to play a preventative role in inflammatory conditions and the metabolic syndrome, as has been highlighted in several reviews [23,97,98].

The results presented in this study demonstrate the potential of incorporating such flours into cereal-based foods to increase the level of these beneficial plant chemicals in the diet. This was shown by Turco et al. [28], who used a wheat–pulse flour composite to increase the nutritive value of semolina pasta; a significant increase of 18% in antioxidant activity (ORAC) was observed. Similarly, [2] used wheat–pulse flour composites in the formulation of unleavened crackers. The authors reported an increase of up to 182% in antioxidant activity (DPPH).

3.4. Fatty Acid Profile

The results corresponding to the fatty acid composition of the various flours analyzed are shown in Table 4. Significant differences due to the vegetal species were observed for all the fatty acids but also between brands for pea, soybean and hemp, where these differences are very important for the first two.

The most abundant fatty acid in pulses and hemp was C18:2 n6, ranging from 40.7 to 56.5%, except for the B brand of soybean (28.1%), whereas rice flour showed a very low value of this fatty acid (4.8% on average). For these flours, the most abundant fatty acid was C16:0, with average values of 51.5% on average. The chickpea and hemp flours showed the highest values for C18:2 n6 acid without significant differences between them, followed by pea and soybean, with similar values. In fact, the significant differences observed for pea and soybean were between brands, with the difference being very important for this soybean flour.

The second most abundant fatty acid in pulses and rice flour was C18:1 (24.8–33.6%), except for the B brand of soybean (33.6%) for which C18:1 was the most abundant. Then, the highest values of this fatty acid were found for soybean and pea flours, followed by chickpea and rice, while hemp flours showed the lowest values.

However, for hemp flour and the B brand of pea, the second most abundant fatty acid was C18:3 n3, with C18:1 being the third most abundant fatty acid. In fact, marked differences among samples were observed for C18:3 n3, with hemp flours and the A brand of pea showing the highest levels, followed by soybean (A brand) and chickpea, while rice flour showed very low values.

On the other hand, chickpea, the A brand of pea and soybean flours showed higher values of the C16:0 (9.9–11.1%) than those observed for C18:3 n3, with the content in the B brand of soybean sample (21.7%) being especially high. As mentioned above, rice flour showed the highest values of this fatty acid, and it was the most abundant and different from all other flours. The pulses showed intermediate values, while hemp flours had the lowest contents of this fatty acid.

As far as C18:0 is concerned, significant differences were observed between vegetable species but also between brands. Therefore, the highest content of this fatty acid was therefore observed for the B brand of soybean, followed by the A brand of soybean, rice and pea without significant differences being found between the samples.

Table 4. The fatty acid composition of different flours (g per 100 g of total fatty acid methyl esters). Data presented as means ± standard deviation from triplicate analysis.

	Chickpea		Pea		Soyabean		Hemp		Rice	
	A *	B	A	B	A	B	A	B	A	B
C14: 0	0.14 ± 0.00 ^{b}	0.14 ± 0.00 ^{b}	0.51 ± 0.01 ^{bc}	0.81 ± 0.10 ^{c}	0.07 ± 0.0 ^{a}	0.16 ± 0.00 ^{b}	ND	ND	3.01 ± 0.35 ^{e}	1.79 ± 0.29 ^{d}
C15:0	0.06 ± 0.01 ^{a}	0.06 ± 0.00 ^{a}	ND	ND	0.02 ± 0.0 ^{a}	0.045 ± 0.01 ^{a}	ND	ND	0.26 ± 0.07 ^{b}	0.52 ± 0.32 ^{b}
C16:0	9.89 ± 0.01 ^{c}	10.76 ± 0.90 ^{c}	10.65 ± 0.51 ^{c}	10.83 ± 1.23 ^{c}	11.11 ± 1.17 ^{c}	21.69 ± 0.31 ^{d}	6.98 ± 0.04 ^{a}	8.67 ± 0.37 ^{b}	51.08 ± 0.81 ^{e}	52.50 ± 0.59 ^{e}
C16:1	0.24 ± 0.00 ^{cd}	0.26 ± 0.01 ^{de}	0.08 ± 0.01 ^{a}	0.18 ± 0.01 ^{b}	0.08 ± 0.01 ^{a}	0.11 ± 0.00 ^{a}	0.11 ± 0.00 ^{a}	0.21 ± 0.02 ^{bc}	0.29 ± 0.04 ^{e}	0.20 ± 0.03 ^{bc}
C17:0	0.07 ± 0.00 ^{a}	0.07 ± 0.01 ^{a}	ND	ND	0.13 ± 0.00 ^{a}	0.22 ± 0.02 ^{a}	ND	ND	1.08 ± 0.40 ^{b}	0.42 ± 0.43 ^{a}
C17:1	0.07 ± 0.00 ^{a}	0.18 ± 0.14 ^{a}	0.05 ± 0.02 ^{a}	0.94 ± 0.58 ^{a}	ND	ND	ND	ND	0.53 ± 0.60 ^{a}	0.15 ± 0.07 ^{a}
C18:0	1.56 ± 0.02 ^{a}	1.57 ± 0.04 ^{a}	4.50 ± 0.37 ^{d}	2.83 ± 0.70 ^{b}	4.61 ± 0.18 ^{d}	9.62 ± 0.25 ^{e}	3.16 ± 0.01 ^{bc}	2.17 ± 0.47 ^{ab}	4.82 ± 1.15 ^{d}	4.04 ± 0.04 ^{cd}
C18:1 n9	0.05 ± 0.01 ^{ab}	0.02 ± 0.01 ^{a}	0.10 ± 0.01 ^{ab}	0.49 ± 0.08 ^{cd}	0.07 ± 0.03 ^{ab}	0.38 ± 0.03 ^{bc}	0.02 ± 0.00 ^{a}	0.84 ± 0.42 ^{e}	1.23 ± 0.09 ^{f}	0.78 ± 0.05 ^{de}
C18:1	26.51 ± 0.24 ^{c}	26.63 ± 0.67 ^{c}	31.51 ± 0.52 ^{d}	15.15 ± 1.64 ^{b}	24.81 ± 1.80 ^{c}	33.65 ± 0.10 ^{d}	13.55 ± 0.06 ^{b}	10.81 ± 1.33 ^{a}	25.80 ± 0.21 ^{c}	25.25 ± 1.56 ^{c}
C18:1n7	1.36 ± 0.00 ^{c}	0.85 ± 0.11 ^{b}	ND	ND	1.03 ± 0.11 ^{b}	1.67 ± 0.06 ^{a}	0.63 ± 0.02 ^{a}	0.94 ± 0.13 ^{b}	ND	ND
C18:2 n6t	0.01 ± 0.00 ^{a}	0.08 ± 0.01 ^{a}	ND	ND	ND	ND	ND	ND	2.83 ± 0.82 ^{b}	2.58 ± 1.37 ^{b}
C18:2 n6	54.51 ± 0.18 ^{ef}	53.81 ± 0.50 ^{e}	40.72 ± 0.36 ^{c}	47.82 ± 1.03 ^{d}	48.09 ± 1.90 ^{d}	28.14 ± 0.78 ^{b}	56.58 ± 0.05 ^{f}	52.85 ± 1.61 ^{e}	4.25 ± 1.80 ^{a}	5.37 ± 0.74 ^{a}
C20:0	0.93 ± 0.02 ^{a}	0.97 ± 0.01 ^{a}	0.94 ± 0.23 ^{a}	0.92 ± 0.93 ^{a}	0.46 ± 0.00 ^{a}	0.88 ± 0.01 ^{a}	0.89 ± 0.01 ^{a}	0.71 ± 0.21 ^{a}	0.83 ± 0.09 ^{a}	0.67 ± 0.19 ^{a}
C20:1 n9	0.02 ± 0.00 ^{a}	0.01 ± 0.00 ^{a}	ND	ND	0.03 ± 0.00 ^{b}	0.06 ± 0.00 ^{c}	ND	ND	ND	ND
C18:3 n3	3.45 ± 0.03 ^{a}	3.43 ± 0.01 ^{a}	9.72 ± 1.48 ^{b}	17.33 ± 3.15 ^{c}	8.36 ± 1.22 ^{b}	1.50 ± 0.07 ^{a}	16.75 ± 0.06 ^{c}	15.19 ± 1.35 ^{c}	0.96 ± 1.11 ^{a}	1.75 ± 0.11 ^{a}
C21:0	0.10 ± 0.00 ^{b}	0.09 ± 0.00 ^{b}	ND	ND	0.05 ± 0.00 ^{a}	0.08 ± 0.01 ^{b}	ND	ND	ND	ND
C20:2 n6	0.09 ± 0.00 ^{a}	0.10 ± 0.00 ^{a}	ND	ND	0.04 ± 0.00 ^{a}	0.06 ± 0.00 ^{a}	ND	ND	0.48 ± 0.20 ^{b}	0.39 ± 0.15 ^{b}
C22:0	0.68 ± 0.02 ^{de}	0.64 ± 0.02 ^{cde}	0.28 ± 0.05 ^{ab}	0.19 ± 0.10 ^{a}	0.62 ± 0.04 ^{bcde}	0.89 ± 0.03 ^{e}	0.44 ± 0.01 ^{abcd}	0.56 ± 0.43 ^{bcde}	0.30 ± 0.04 ^{abc}	0.40 ± 0.02 ^{abcd}
C22:1 n9	0.04 ± 0.00 ^{a}	0.04 ± 0.01 ^{a}	0.08 ± 0.02 ^{a}	0.43 ± 0.10 ^{a}	0.01 ± 0.00 ^{a}	0.06 ± 0.00 ^{a}	ND	ND	0.66 ± 0.15 ^{a}	2.19 ± 1.77 ^{b}
C20:3 n3	ND	ND	0.01 ± 0.01 ^{a}	0.09 ± 0.07 ^{a}	ND	ND	ND	ND	0.54 ± 0.05 ^{b}	0.40 ± 0.15 ^{b}
C20:4 n6	0.11 ± 0.01 ^{b}	0.11 ± 0.00 ^{b}	ND	ND	0.06 ± 0.01 ^{a}	0.11 ± 0.01 ^{b}	ND	ND	ND	ND
C22:2 n6	0.04 ± 0.04 ^{a}	0.10 ± 0.05 ^{a}	0.05 ± 0.04 ^{a}	0.33 ± 0.25 ^{a}	ND	ND	ND	ND	ND	ND
C24:0	ND	ND	0.69 ± 0.06 ^{b}	0.29 ± 0.05 ^{a}	ND	ND	ND	ND	0.80 ± 0.09 ^{b}	0.46 ± 0.03 a
C20:5 n3	ND	ND	ND	ND	0.20 ± 0.01 a	0.36 ± 0.02 ^{b}	0.19 ± 0.00 ^{a}	0.26 ± 0.10 a^{b}	ND	ND
C24:1 n9	0.02 ± 0.01 ^{a}	0.01 ± 0.01 ^{a}	0.06 ± 0.03 ^{a}	1.33 ± 0.44 ^{b}	0.08 ± 0.04 ^{a}	0.18 ± 0.04 ^{a}	ND	ND	ND	ND
SFA	13.45 ± 0.02 ^{ab}	14.34 ± 0.83 ^{abc}	17.58 ± 1.20 ^{c}	15.88 ± 2.61 ^{bc}	17.08 ± 1.33 ^{c}	33.60 ± 0.56 ^{d}	11.47 ± 0.07 ^{a}	12.13 ± 1.48 ^{a}	62.20 ± 2.60 ^{e}	60.83 ± 0.75 ^{e}
MUFA	28.31 ± 0.21 ^{c}	28.03 ± 0.38 ^{c}	31.90 ± 0.59 ^{d}	18.52 ± 1.75 ^{b}	26.11 ± 1.78 ^{c}	36.11 ± 0.23 ^{e}	14.32 ± 0.04 ^{a}	12.81 ± 1.90 ^{a}	28.52 ± 0.72 ^{c}	28.59 ± 0.21 ^{c}
PUFA	58.23 ± 0.23 ^{d}	57.63 ± 0.45 ^{d}	50.52 ± 1.80 ^{c}	65.59 ± 4.36 ^{e}	56.81 ± 3.11 ^{d}	30.28 ± 0.78 ^{b}	74.21 ± 0.11 ^{f}	75.05 ± 3.39 ^{f}	9.27 ± 1.87 ^{a}	10.58 ± 0.54 ^{a}
n3	3.45 ± 0.03 ^{a}	3.43 ± 0.01 ^{a}	9.75 ± 1.48 ^{b}	17.43 ± 3.08 ^{c}	8.61 ± 1.21 ^{b}	1.97 ± 0.01 ^{a}	16.95 ± 0.06 ^{c}	15.45 ± 1.24 ^{c}	1.51 ± 1.16 ^{a}	2.15 ± 0.05 ^{a}
n6	54.78 ± 0.21 ^{e}	54.20 ± 0.44 ^{e}	40.77 ± 0.32 ^{c}	48.15 ± 1.28 ^{d}	48.19 ± 1.90 ^{d}	28.31 ± 0.77 ^{b}	57.26 ± 0.06 ^{f}	59.60 ± 2.13 ^{f}	7.77 ± 0.72 ^{a}	8.42 ± 0.49 ^{a}
n6/n3	15.89 ± 0.08 ^{c}	15.79 ± 0.06 ^{c}	4.23 ± 0.61 ^{a}	2.80 ± 0.42 ^{a}	5.63 ± 0.57 ^{b}	14.34 ± 0.27 ^{c}	3.38 ± 0.01 ^{a}	3.86 ± 0.17 ^{a}	5.0 ± 2.96 ^{b}	3.90 ± 0.14 ^{a}

* A and B are different commercial brands. a–f: values followed by different superscripts in the same row are significantly different ($p < 0.05$). ND not detected.

Some fatty acids were found in certain vegetable species but not in others such as C15:0, C17:0 or C18:1n9t. Other fatty acids were determined only in some samples such as linolenic-γ, which was found in hemp; the amount of the two brands analyzed was different, as that of the B brand was 6 times higher than that of the A bran (6.75 vs. 0.68%). Eicosatrienoic was found only in rice, and DHA was determined only in soybean flours; in both cases, the levels were very low (<1%) and there were no significant differences between brands. It is noteworthy that hemp had the lowest variety of fatty acids, while chickpeas showed the highest number of fatty acids.

According to these values, pulses showed very high contents of polyunsaturated fatty acids (PUFA) (50–65%), especially n6 PUFA (48–54%), followed by monounsaturated fatty acids (MUFA) (18–31%) and with low values of saturated fatty acids (SFA) (13–17%), except for the B brand of soybean, which showed a different behavior, and the levels of the three FA groups were similar. Moreover, pea, followed by soybean, was characterized by the high contents of n3 PUFA, while chickpea had the lowest values. These results are in agreement with those previously reported by Jahreis et al. [99].

The findings of this study regarding the fatty acid profile of chickpeas are similar to those reported by other authors who found the SFA to be between 12.6 and 16.5% (13.4%), the MUFA between 19 and 39% (28.2%) and the PUFA between 44 and 63% (58%) [99–103]. Among the PUFAs, the percentage of n3 and the n6/n3 ratio were within the range previously reported (15–21%) [99,102]. These studies also found that the most dominant fatty acid in chickpea is linoleic acid (42.1–61.4%), followed by oleic acid (20.1–38.5%), palmitic acid (9.1–12.1%), linolenic acid (2.1–6.6%) and stearic acid (1.3–2.4%). The variations reported can be attributed to environmental factors and the cultivar [103].

The results from pea indicate that the values found for the A brand for SFA, MUFA and PUFA are within the range described by previous studies, i.e., 11–23%, 26–34% and 41–56%, respectively [99,104,105]; however the B brand showed slightly higher values for PUFA and lower values for MUFA. The n6/n3 ratios in the analyzed A and B flours were slightly lower than the values previously found [99,105], and the values were significantly lower than those found for chickpeas. For chickpeas and other pulses, the main fatty acid was linoleic acid, followed by oleic acid, and the values were similar to those reported in other studies (34.7–47.7% and 25.0–33.3%, respectively) [99,103,105,106]. In general, the amount of linoleic acid and oleic acid was similar to that of other pulses, such as beans and lentils, which have 46.7 and 44.4% linoleic acid and 28.1 and 20.9% oleic acid, respectively [103]. These fatty acids were followed by linolenic acid, in contrast to chickpeas, and showed slightly higher values than those previously reported (6.5–8.3%), while palmitic acid showed lower values (15.5–12.6%) [99,105].

Soybean flour A showed a similar percentage of SFA, MUFA and PUFA to pea and chickpea. This result and the percentages observed are in agreement with the data reported by Jahreis et al. [99] and Dahmer et al. [107], who also found that the n6/n3 ratio varied between 5.8 and 8.2, as was observed for the A flour. Then, the main fatty acids of the A brand flour were linoleic, oleic, palmitic and α-linolenic acid, as previously observed [99,108–111], with the percentages observed for these four fatty acids within the range described by these authors being 40.4–54.8%, 17.5–21.4%, 10.8–13.1% and 5.3–9.9%, respectively. However, the B flour showed a lower amount of PUFA and a higher amount of the two other fatty acid groups and a very low amount of n3. The differences in the technological process, which led to a variation in the fat content (18 vs. 23%), may be the origin of the difference in the level of fatty acids, together with differences in environmental factors and the cultivar [112].

Hemp flour showed the highest PUFA contents among the flours analyzed, with a value of 74–75%, while the MUFA and SFA were about 11–14%. These results are in agreement with the values described in both flours and hemp seeds [90,113–115]. The high proportion of n3 fatty acids, which makes the ratio of n6/n3 one of the lowest among the flours analyzed, is noteworthy. These studies also highlight the high percentage of linoleic acid (51.6–56.4%) followed by α-linolenic (10.5–17.3%) and oleic acid (10.7–17.9%), with significant levels of γ-linolenic acid (3–18%), which is in accordance with the findings of this study.

Finally, rice flour showed a very different fatty acid profile, with high amounts of SFA, low amounts of PUFA compared with the remainder of the flours and with the data reported in the literature pointing to levels of SFA between 25 and 39.3%, and low amounts of MUFA between 34.9 and 50.9% and of PUFA between 9.9 and 37.2% for milled rice flour [116–118]. The closest values were found for parboiled milled rice, with an SFA content of 55.5% [116]. These results highlight the strong influence of the rice variety and the technological process on the fatty acid profile of rice flour [116]. The main fatty acids in the flours analyzed were palmitic acid, followed by oleic, linoleic and stearic acids, in contrast to the findings of previous studies, which reported that the main fatty acids were oleic, palmitic and linoleic acid for white rice flours and oleic, linoleic and palmitic acids for brown rice flours [118,119].

Fatty acids of specific chain length and saturation are required by humans for their structural and metabolic needs. Linoleic acid is responsible for the production of prostaglandins, which prevent the concentration of heavy cholesterol on the blood vessel walls [120], but is also correlated with an improved blood lipid profile, lower incidence of type 2 diabetes, improved insulin sensitivity and anti-arrhythmic effects. Oleic acid is known to have beneficial effects on autoimmune and inflammatory diseases, as well as cancer and the healing of wounds [121]. α-Linolenic acid is an essential fatty acid correlated to inflammatory and immunosuppressive activities by means of the formation of several intermediates such as leukotrienes, thromboxane and prostaglandins [122]. However, unsaturated fatty acids are susceptible to lipid oxidation, which results in undesirable changes in organoleptic properties owing to the presence of off flavors [123]. In this sense, oleic and linoleic acids show less propensity towards oxidation compared to linolenic acid during processing [124]. Pulses and hemp flours were characterized by high levels of all these fatty acids, which are correlated with the above-mentioned health benefits, with pea and hemp flours showing the highest levels of linolenic acid.

However, not only PUFA levels but also the n-6/n-3 PUFA ratio should be taken into account. According to the World Health Organization, this ratio should be below 4:1, as is shown by both brands of both pea and hemp flours. In this context, if any flour with a high n6/n3 ratio is to be partially replaced, it would be desirable to use hemp flour, owing to its low n6/n3 ratio or pea flour, as this contains only a small amount of fat with a large amount of oleic acid and an adequate n6/n3 ratio.

3.5. PCA Method

Principal component analysis yielded six PCs with an eigenvalue >1, which explained 96.43% of the variance. Figure 1 shows the projection plot of the samples and the original variables in the space defined by the first two PCs, which explain 70.36% of the variance.

Soybean samples were found in the upper-right quadrant, with positive values of both PCs, due to the strong contribution of protein, ash, TPC, Ca and K to PC1, together with the correlation of C22:0 and C18:1 to the PC2. The hemp flour samples presented positive PC1 values due to the correlation of this PC1 with ash and fiber, as previously mentioned, but also with TFC and Cu contents. These samples also showed negative PC2 values, which were correlated with C18:2, Mg and Fe variables, but mainly with C18:3 levels. In the opposite quadrant were chickpea samples showing positive PC2 values, which correlated with the observed levels of C18:1 and C18:0 (as for soybean and one pea sample, which have positive PC1 values). In fact, chickpea samples had negative PC1 values due to the strong correlation with Na.

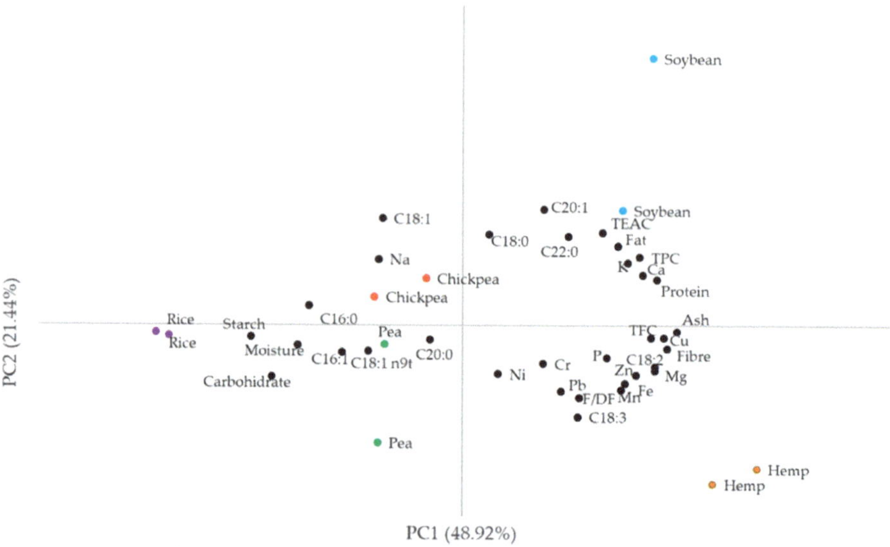

Figure 1. Projection plot of the flour samples in the space defined by the two firsts PCs. TPC: total phenolic content, TFC: total flavonoid content, F/DF: flavanones and dihydroflavonols, TEAC: total equivalent antioxidant activity (nmol Trolox/100 g).

Pea and rice samples were in the lower-left quadrant and showed negative PC1 and PC2 values. Thus, rice samples had the highest levels of C16:0 and C18:1n9t but also of carbohydrates and starch, together with the lowest values of TFC, TPC, C18:2 and most minerals. The pea samples had a high moisture content but also low values of Mg, P and Ca. However, they showed the highest values of Pb, which was associated with the negative values of PC2. As in the case of rice, the pea samples showed relatively low TPC and TEAC values. One sample showed very low amounts of C18:1, which justifies the low PC2 value observed.

The figure also shows that the samples of the two brands of rice, hemp and chickpea were close, while the two samples of pea and especially soybean were separated due to the observed compositional differences. Indeed, the soybean, rice and hemp flours were very different from each other, while the chickpea flour and one of the pea samples were quite close.

4. Conclusions

Although plant-based flours are referred to as a generic term, not all products under this name have a similar composition and therefore do not make a similar contribution from a nutritional point of view. All the flours analyzed in this study (chickpea, pea, soybean and hemp) have carbohydrates as their main component and provide a similar amount of energy (Kcal/100 g). However, pea and chickpea flours contain large amounts of protein and fiber, while soybean and hemp flours provide significant amounts of fat. Regarding the mineral composition, potassium and phosphorus as macro-elements and iron as a microelement are the major components in all flours. The phenolic composition, antioxidant activity and fatty acid profile of these meals vary greatly depending on the plant species. In addition to these differences, plant-based flours have been shown to have highly variable compositions between different commercial brands, which makes it necessary to educate consumers to check the nutritional labels of this type of food in order to ensure healthy diets. Surprisingly, the presence of high amounts of lead has also been detected in some of these flours, so health authorities should be aware of this fact in order to implement control mechanisms for plant-based flours.

Author Contributions: Conceptualization and methodology, A.M.V.-Q. and I.R.; formal analysis, Y.A.; data curation, M.H.-J.; writing—original draft preparation, M.H.-J. and Y.A.; writing—review and editing, A.M.V.-Q. and I.R. All authors have read and agreed to the published version of the manuscript.

Funding: This research received no external funding.

Data Availability Statement: The data presented in this study are available on request from the corresponding author.

Acknowledgments: Y. Absi acknowledges the Algerian government for the grant long-term residential doctoral program abroad and M. Hernández-Jimenez acknowledges the Own Program III: Grants for Pre-doctoral Contracts of the University of Salamanca co-funded by Banco Santander.

Conflicts of Interest: The authors declare no conflict of interest.

References

1. Bigliardi, B.; Galati, F. Innovation trends in the food industry: The case of functional foods. *Trends Food Sci. Technol.* **2013**, *31*, 118–129. [CrossRef]
2. Millar, K.A.; Gallagher, E.; Burke, R.; McCarthy, S.; Barry-Ryan, C. Proximate composition and anti-nutritional factors of fava-bean (*Vicia faba*), green-pea and yellow-pea (*Pisum sativum*) flour. *J. Food Compos. Anal.* **2019**, *82*, 103233. [CrossRef]
3. Mikulec, A.; Kowalski, S.; Sabat, R.; Skoczylas, L.; Tabaszewska, M.; Wywrocka-Gurgul, A. Hemp flour as a valuable component for enriching physicochemical and antioxidant properties of wheat bread. *LWT* **2019**, *102*, 164–172. [CrossRef]
4. Gambus, H.; Gambus, F.; Pastuszka, D.; Wrona, P.; Ziobro, R.; Sabat, R.; Mickowska, B.; Nowotna, A.; Sikora, M. Quality of glutenfree supplemented cakes and biscuits. International. *J. Food Sci. Nutr.* **2009**, *60*, 31–50. [CrossRef]
5. Niro, S.; D'Agostino, A.; Fratianni, A.; Cinquanta, L.; Panfili, G. Gluten-Free Alternative Grains: Nutritional Evaluation and Bioactive Compounds. *Foods* **2009**, *8*, 208. [CrossRef]
6. Di Liberto, D.; Carlisi, D.; D'Anneo, A.; Emanuele, S.; Giuliano, M.; De Blasio, A.; Calvaruso, G.; Lauricella, M. Gluten Free Diet for the Management of Non Celiac Diseases: The Two Sides of the Coin. *Healthcare* **2020**, *8*, 400. [CrossRef]
7. Doxastakis, G.; Zafiriadis, I.; Irakli, M.; Marlani, H.; Tananaki, C. Lupin, soya and triticale addition to wheat flour doughs and their effect on rheological properties. *Food Chem.* **2002**, *77*, 219–227. [CrossRef]
8. Hefnawy, T.M.H.; El-Shourbagy, G.A.; Ramadan, M.F. Impact of adding chickpea (*Cicer arietinum*, L.) flout to wheat flour on the rheological properties of toast bread. *Int. Food Res. J.* **2012**, *19*, 521–525.
9. Dwyer, J.T.; Wiemer, K.L.; Dary, O.; Keen, C.L.; King, J.C.; Miller, K.B.; Philbert, M.A.; Tarasuk, V.; Taylor, C.L.; Gaine, P.C.; et al. Fortification and Health: Challenges and Opportunities. *Adv. Nutr.* **2015**, *6*, 124–131. [CrossRef]
10. Xu, Q.C. Studies of the differences of amino acid contents in rice varieties and the relationship between the amino acid content and genetic code numbers. *J. Hunan Agric. Coll.* **1990**, *16*, 221–225.
11. Wu, J.G.; Shi, C.; Zhang, X. Estimating the amino acid composition in milled rice by near-infrared reflectance spectroscopy. *Field Crops Res.* **1990**, *75*, 1–7. [CrossRef]
12. Storck, C.R.; Picolli da Silva, L.; Alves Fagundes, C.A. Categorizing rice cultivars based on differences in chemical composition. *J. Food Compos. Anal.* **2005**, *18*, 333–341. [CrossRef]
13. Das, A.B.; Bhattacharya, S. Characterization of the batter and gluten-free cake from extruded red rice flour. *LWT* **2019**, *102*, 197–204. [CrossRef]
14. Seetapan, S.; Limparyoon, N.; Yooberg, R.; Leelawat, B.; Charunuch, C. Influence of addition of extruded rice flour on preparation and quality of fresh gluten-free yellow alkaline noodles. *J. Cereal Sci.* **2019**, *90*, 102828. [CrossRef]
15. Charoenthaikij, P.; Jangchud, K.; Jangchud, A.; Prinyawiwatkul, W.; Tungtrakul, P. Germination conditions affect selected quality of composite wheat-germinated brown rice flour and bread formulations. *J. Food Sci.* **2010**, *75*, S312–S318. [CrossRef] [PubMed]
16. Marti, A.; Caramanico, R.; Bottega, G.; Pagani, M.A. Cooking behavior of rice pasta: Effect of thermal treatments and extrusion conditions. *LWT* **2013**, *54*, 229–235. [CrossRef]
17. Messina, V. Nutritional and health benefits of dried beans. *Am. J. Clin. Nutr.* **2014**, *100*, 437S–442S. [CrossRef] [PubMed]
18. Hoffman, J.R.; Falvo, M.J. Protein—Which is Best? *J. Sport. Sci. Med.* **2004**, *3*, 118–130.
19. Boye, J.; Zare, F.; Pletch, A. Pulse proteins: Processing, characterization, functional properties and applications in food and feed. *Food Res. Int.* **2010**, *43*, 414–431. [CrossRef]
20. Herranz, B.; Canet, W.; Jiménez, M.J.; Fuentes, R.; Alvarez, M.D. Characterisation of chickpea flour-based gluten-free batters and muffins with added biopolymers: Rheological, physical and sensory properties. *Int. J. Food Sci. Technol.* **2016**, *51*, 1087–1098. [CrossRef]
21. Han, J.; Janz, J.A.M.; Gerlat, M. Development of gluten-free cracker snacks using pulse flours and fractions. *Food Res. Int.* **2010**, *43*, 627–633. [CrossRef]
22. Laleg, K.; Barron, C.; Santé-Lhoutellier, V.; Walrand, S.; Micard, V. Protein enriched pasta: Structure and digestibility of its protein network. *Food Funct.* **2016**, *7*, 1196–1207. [CrossRef] [PubMed]

23. Campos-Vega, G.; Loarca-Piña, B.; Oomah, D. Minor components of pulses and their potential impact on human health. *Food Res. Int.* **2010**, *43*, 461–482. [CrossRef]
24. Amarakoon, A.; Thavarajah, D.; McPhee, K.; Thavarajah, P. Iron-, zinc-, and magnesium-rich field peas (*Pisum sativum* L.) with naturally low phytic acid: A potential food-based solution to global micronutrient malnutrition. *J. Food Compos. Anal.* **2012**, *27*, 8–13. [CrossRef]
25. Sanchez, H.D.; Osella, C.A.; de la Torre, M.A. Use of Response Surface Methodology to Optimize Gluten-Free Bread Fortified with Soy Flour and Dry Milk. *Food Sci. Technol. Int.* **2004**, *10*, 5–9. [CrossRef]
26. Shogren, R.L.; Hareland, G.A.; Wu, Y.V. Sensory evaluation and composition of Spaghetti fortified with soy flour. *J. Food Sci.* **2006**, *71*, 428–432. [CrossRef]
27. Shin, D.J.; Kim, W.; Kim, Y. Physicochemical and sensory properties of soy bread made with germinated, steamed and roasted soy flour. *Food Chem.* **2013**, *141*, 517–523. [CrossRef] [PubMed]
28. Turco, I.; Ferretti, G.; Bacchetti, T. Review of the health benefits of Faba bean (*Vicia faba* L.) polyphenols. *J. Food Nutr. Res.* **2016**, *55*, 283–293. [CrossRef]
29. Durazzo, A.; Turfani, V.; Narducci, V.; Azzini, E.; Maiani, G.; Carcea, M. Nutritional characterisation and bioactive components of commercial carobs flours. *Food Chem.* **2014**, *153*, 109–113. [CrossRef]
30. González-Montemayor, A.M.; Flores-Gallegos, A.C.; Contreras-Esquivel, J.C.; Solanilla-Duque, J.F.; Rodríguez-Herrera, R. *Prosopis* spp. functional activities and its applications in bakery products. *Trends Food Sci. Technol.* **2019**, *94*, 12–19. [CrossRef]
31. Petrović, M.; Debeljak, Ž.; Kezić, N.; Džidara, P. Relationship between cannabinoids content and composition of fatty acids in hempseed oils. *Food Chem.* **2015**, *170*, 218–225. [CrossRef] [PubMed]
32. Hofmanová, T.; Švec, I.; Hrušková, M. Nutritional properties of non-traditional seeds. *J. Life Med.* **2014**, *2*, 10–14.
33. Korus, J.; Witczak, M.; Ziobro, R.; Juszczak, L. Hemp (*Cannabis sativa* subsp. *sativa*) flour and protein preparation as natural nutrients and structure forming agents in starch based gluten-free bread. *LWT* **2017**, *84*, 43–150. [CrossRef]
34. Pojić, M.; Dapčević Hadnađev, T.; Hadnađev, M.; Rakita, S.; Brlek, T. Bread supplementation with hemp seed cake: A by-product of hemp oil processing. *J. Food Qual.* **2015**, *38*, 431–440. [CrossRef]
35. de la Barca, A.M.C.; Rojas-Martínez, M.E.; Islas-Rubio, A.R.; Cabrera-Chávez, F. Gluten-free breads and cookies of raw and popped amaranth flours with attractive technological and nutritional qualities. *Plant Foods Hum. Nutr.* **2010**, *65*, 241–246. [CrossRef]
36. Codină, G.G.; Franciuc, S.G.; Mironeasa, S. Rheological Characteristics and Microstructure of Milk Yogurt as Influenced by Quinoa Flour Addition. *J. Food Qual.* **2016**, *39*, 559–566. [CrossRef]
37. Miranda-Villa, P.P.; Mufari, J.R.; Bergesse, A.E.; Calandri, E.L. Effects of whole and malted quinoa flour addition on gluten-free muffins quality. *J. Food Sci.* **2018**, *84*, 147–153. [CrossRef] [PubMed]
38. Khan, I.; Yousif, A.; Johnson, S.K.; Gamlath, S. Effect of sorghum flour addition on resistant starch content, phenolic profile and antioxidant capacity of durum wheat pasta. *Food Res. Int.* **2013**, *54*, 578–586. [CrossRef]
39. Yousif, A.; Nhepera, D.; Johnson, S. Influence of sorghum flour addition on flat bread in vitro starch digestibility, antioxidant capacity and consumer acceptability. *Food Chem.* **2012**, *134*, 880–887. [CrossRef]
40. Aschemann-Witzel, J.; Futtrup-Gantriis, R.; Fraga, P.; Perz-cueto, F.J.A. Plant-based food and protein trend from a businness perpective: Markets, consumers and the challenges and opportunities in the future. *Crit. Rev. Food Sci. Nutr.* **2020**, *61*, 3119–3128. [CrossRef]
41. AOAC International. *Official Methods of Analysis*, 20th ed.; Association of Official Analytical Chemists: Gaithersbg, MD, USA, 2016.
42. AACC. American Association of Cereal Chemists Method 44-15.02., Moisture: Air Oven Methods. In *Approved Methods of Analysis*, 11th ed.; AACC International: St. Paul, MN, USA, 2001.
43. AACC. American Association of Cereal Chemists Method 08-01.01. Ash—Basic method. In *Approved Methods of Analysis*, 11th ed.; AACC International: St. Paul, MN, USA, 1999.
44. AOAC International. *Official Methods of Analysis*, 16th ed.; Association of Official Analytical Chemists: Washington, DC, USA, 1995.
45. Lurueña-Martínez, M.A.; Palacios, C.; Vivar-Quintana, A.M.; Revilla, I. Effect of the addition of calcium soap to ewes' diet on fatty acid composition of ewe milk and subcutaneous fat of suckling lambs reared on ewe milk. *Meat Sci.* **2010**, *84*, 677–683. [CrossRef] [PubMed]
46. Betances-Salcedo, E.; Revilla, I.; Vivar-Quintana, A.M.; González-Martín, M.I. Flavonoid and antioxidant capacity of propolis prediction using near infrared spectroscopy. *Sensors* **2017**, *17*, 1647. [CrossRef] [PubMed]
47. Valencia, D.; Alday, E.; Robles-Zepeda, R.; Garibay-Escobar, A.; Galvez-Ruiz, J.C.; Salas-Reyes, M.; Jiménez-Estrada, M.; Velazquez-Contreras, E.; Hernandez, J.; Velazquez, C. Seasonal effect on chemical composition and biological activities of Sonoran propolis. *Food Chem.* **2012**, *131*, 645–651. [CrossRef]
48. Popova, M.; Bankova, V.; Butovska, D.; Petkov, V.; Nikolova-Damyanova, B.; Sabatini, A.G.; Marcazzan, G.L.; Bogdanov, S. Validated methods for the quantification of biologically active constituents of poplar-type propolis. *Phytochem. Anal. Int. J. Plant Chem. Biochem. Tech.* **2004**, *15*, 235–240. [CrossRef]
49. Chen, J.; Lindmark-Månsson, H.; Gorton, L.; Åkesson, B. Antioxidant capacity of bovine milk as assayed by spectrophotometric and amperometric methods. *Int. Dairy J.* **2003**, *13*, 927–935. [CrossRef]

50. Lan, Y.; Zha, F.; Peckrul, A.; Hanson, B.; Johnson, B.; Rao, J.; Chen, B. Genotype x environmental effects on yielding ability and seed chemical composition of industrial hemp (*Cannabis sativa* L.) varieties grown in North Dakota, USA. *J. Am. Oil Chem. Soc.* **2019**, *96*, 1417–1425. [CrossRef]
51. Vatansever, S.; Rao, J.; Hall, C. Effects of ethanol modified supercritical carbon dioxide extraction and particle size on the physical, chemical, and functional properties of yellow pea flour. *Cereal Chem.* **2020**, *97*, 1133–1147. [CrossRef]
52. Wani, S.A.; Kumar, P. Comparative study of chickpea and green pea flour based on chemical composition, functional and pasting properties. *J. Food Res. Technol.* **2014**, *2*, 124–129.
53. Bolarinwa, I.F.; Lim, P.T.; Muhammad, K. Quality of gluten-free cookies from germinated brown rice flour. *Food Res.* **2019**, *3*, 199–207. [CrossRef]
54. Uwem, U.M.; Babafemi, A.A.; Sunday, D.M. Proximate composition, phytoconstituents and mineral contents of soybean (*Glycine max*) flour grown and processed in Northern Nigeria. *Adv. Appl. Sci.* **2017**, *2*, 48. [CrossRef]
55. Alphonce, S.; Kaale, L.D.; Rweyemamu, L.M.P. Proximate composition of fermented cassava meal "mchuchume" fortified with soya bean flour and Moringa oleifera leaves powder. *J. Food Sci. Technol.* **2019**, *56*, 3660–3667. [CrossRef]
56. EFSA Panel on Dietetic Products, Nutrition, and Allergies (NDA). Scientific opinion on dietary reference values for fats, including saturated fatty acids, polyunsaturated fatty acids, monounsaturated fatty acids, trans fatty acids, and cholesterol. *EFSA J.* **2010**, *8*, 1461. [CrossRef]
57. Lattimer, J.M.; Haub, M.D. Effects of dietary fiber and its components on metabolic health. *Nutrients* **2010**, *2*, 1266–1289. [CrossRef] [PubMed]
58. Guichard, E. Interactions between flavor compounds and food ingredients and their influence on flavor perception. *Food Rev. Int.* **2002**, *18*, 49–70. [CrossRef]
59. Kharabian-Masouleh, A.; Waters, D.; Reinke, R.F.; Reinke, R.; Rachelle Ward, R.J.; Henry, R. SNP in starch biosynthesis genes associated with nutritional and functional properties of rice. *Sci. Rep.* **2012**, *2*, 557. [CrossRef] [PubMed]
60. Copeland, L.; Blazek, J.; Salman, H.; Tang, M.C. Form and functionality of starch. *Food Hydrocoll.* **2009**, *23*, 1527–1534. [CrossRef]
61. Soetan, K.O.; Olaiya, C.O.; Oyewole, O.E. The importance of mineral elements for humans, domestic animals and plants-A review. *Afr. J. Food Sci.* **2010**, *4*, 200–222. [CrossRef]
62. Mittal, R.; Nagi, H.P.S.; Sharma, P.; Sharma, S. Effect of processing on chemical composition and antinutritional factors in chickpea flour. *J. Food Sci. Eng.* **2012**, *2*, 180.
63. Pérez-Rodríguez, M.; Dirchwolf, P.M.; Rodríguez-Negrín, Z.; Pellerano, R.G. Assessing mineral profiles for rice flour fraud detection by principal component analysis-based data fusion. *Food Chem.* **2021**, *339*, 128125. [CrossRef]
64. Anjum, F.M.; Hussain, S.; Tariq, M.T. Effect of soy flour supplementation on mineral and phytate contents of unleavened flat bread (chapatis). *Nutr. Food Sci.* **2005**, *35*, 163–168. [CrossRef]
65. Rusu, I.-E.; Marc, R.A.; Mureşan, C.C.; Mureşan, A.E.; Filip, M.R.; Onica, B.-M.; Csaba, K.B.; Alexa, E.; Szanto, L.; Muste, S. Advanced Characterization of Hemp Flour (*Cannabis sativa* L.) from Dacia Secuieni and Zenit Varieties, Compared to Wheat Flour. *Plants* **2021**, *10*, 1237. [CrossRef] [PubMed]
66. Sujatha, S.J.; Ahmad, R.; Bhat, P.R. Physicochemical properties and cooking qualities of two varieties of raw and parboiled rice cultivated in the costal region of Dakshina Kannada, India. *Food Chem.* **2004**, *86*, 211–216. [CrossRef]
67. Gharibzahedi, S.M.T.; Jafari, S.M. The importance of minerals in human nutrition: Bioavailability, food fortification, processing effects and nanoencapsulation. *Trends Food Sci. Technol.* **2017**, *62*, 119–132. [CrossRef]
68. Gomez-Galera, S.; Rojas, E.; Sudhakar, D.; Zhu, C.M.; Pelacho, A.; Capell, T.; Christou, P. Critical evaluation of strategies for mineral fortification of staple food crops. *Transgenic Res.* **2010**, *19*, 165–180. [CrossRef] [PubMed]
69. Frossard, E.; Bucher, M.; Mächler, F.; Mozafar, A.; Hurrell, R. Potential for increasing the content and bioavailability of Fe, Zn and Ca in plants for human nutrition. *J. Sci. Food Agric.* **2000**, *80*, 861–879. [CrossRef]
70. Liu, X.; Song, Q.; Tang, Y.; Li, W.; Xu, J.; Wu, J.; Wang, F.; Brookes, P.C. Human health risk assessment of heavy metals in soil-vegetable system: A multi-medium analysis. *Sci. Total Environ.* **2013**, *463–464*, 530–540. [CrossRef]
71. Rahman, M.M.; Asaduzzaman, M.; Naidu, R. Consumption of arsenic and other elements from vegetables and drinking water from an arsenic-contaminated area of Bangladesh. *J. Hazard. Mater.* **2013**, *262*, 1056–1063. [CrossRef]
72. Rocchetti, G.; Lucini, L.; Rodriguez, J.M.L.; Barba, F.J.; Giuberti, G. Gluten-free flours from cereals, pseudocereals and legumes: Phenolic fingerprints and in vitro antioxidant properties. *Food Chem.* **2019**, *271*, 157–164. [CrossRef] [PubMed]
73. Xu, B.J.; Yuan, S.H.; Chang, S.K.C. Comparative Analyses of Phenolic Composition, Antioxidant Capacity, and Color of Cool Season Legumes and Other Selected Food Legumes. *J. Food Sci.* **2007**, *72*, S167–S177. [CrossRef]
74. Magalhães, S.C.Q.; Taveira, M.; Cabrita, A.R.J.; Fonseca, A.J.M.; Valentão, P.; Andrade, P.B. European marketable grain legume seeds: Further insight into phenolic compounds profiles. *Food Chem.* **2017**, *215*, 177–184. [CrossRef]
75. Nithiyanantham, S.; Selvakumar, S.; Siddhuraju, P. Total phenolic content and antioxidant activity of two different solvent extracts from raw and processed legumes, *Cicer arietinum* L. and *Pisum sativum* L. *J. Food Compos. Anal.* **2012**, *27*, 52–60. [CrossRef]
76. Multari, S.; Neacsu, M.; Scobbie, L.; Cantlay, L.; Duncan, G.; Vaughan, N.; Derek Stewart, D.R.; Russell, W. Nutritional and Phytochemical Content of High-Protein Crops. *J. Agricultural. Food Chem.* **2016**, *64*, 7800–7811. [CrossRef] [PubMed]
77. Sreerama, Y.N.; Sashikala, V.B.; Pratape, V.M. Variability in the Distribution of Phenolic Compounds in Milled Fractions of Chickpea and Horse Gram: Evaluation of Their Antioxidant Properties. *J. Agric. Food Chem.* **2010**, *58*, 8322–8330. [CrossRef] [PubMed]

78. Sreerama, Y.N.; Sashikala, V.B.; Pratape, V.M. Phenolic compounds in cowpea and horse gram flours in comparison to chickpea flour: Evaluation of their antioxidant and enzyme inhibitory properties associated with hyperglycemia and hypertension. *Food Chem.* **2012**, *133*, 156–162. [CrossRef]
79. Fernandez-Orozco, R.; Frias, J.; Zielinski, H.; Muñoz, R.; Piskula, M.K.; Kozlowska, H.; Vidal-Valverde, C. Evaluation of bioprocesses to improve the antioxidant properties of chickpeas. *LWT* **2009**, *42*, 885–892. [CrossRef]
80. Kalogeropoulos, N.; Chiou, A.; Ioannou, M.T.; Karathanos, V.; Hassapidou, M.; Andrikopoulos, N. Nutritional evaluation and bioactive microconstituents (phytosterols, tocopherols, polyphenols, triterpenic acids) in cooked dry legumes usually consumed in the Mediterranean countries. *Food Chem.* **2010**, *121*, 682–690. [CrossRef]
81. Aguilera, O.; Estrella, I.; Benitez, V.; Esteban, R.M.; Martín-Cabrejas, M. Bioactive phenolic compounds and functional properties of dehydrated bean flours. *Food Res. Int.* **2011**, *44*, 774–780. [CrossRef]
82. Dueñas, M.; Estrella, I.; Hernández, T. Occurrence of phenolic compounds in the seed coat and the cotyledon of peas (*Pisum sativum* L. *Eur. Food Res. Technol.* **2004**, *219*, 116–123. [CrossRef]
83. Slavin, M.; Kenworthy, W.; Yu, L.L. Antioxidant Properties, Phytochemical Composition, and Antiproliferative Activity of Maryland-Grown Soybeans with Colored Seed Coats. *J. Agric. Food Chem.* **2009**, *57*, 11174–11185. [CrossRef]
84. Kumar, V.; Rani, A.; Dixit, A.K.; Pratap, D.; Bhatnagar, D. A comparative assessment of total phenolic content, ferric reducing-anti-oxidative power, free radical-scavenging activity, vitamin C and isoflavones content in soybean with varying seed coat colour. *Food Res. Int.* **2010**, *43*, 323–328. [CrossRef]
85. Villalobos, M.D.C.; Serradilla, M.J.; Martín, A.; Ordiales, E.; Ruiz-Moyano, S.; Córdoba, M.D.G. Antioxidant and antimicrobial activity of natural phenolic extract from defatted soybean flour by-product for stone fruit postharvest application. *J. Sci. Food Agric.* **2016**, *96*, 2116–2124. [CrossRef] [PubMed]
86. Tangkhawanit, E.; Siriamornpun, S. Bioactive compounds, biological activity, and starch digestibility of dried soy residues from the soybean oil industry and the effects of hot-air drying. *J. Sci. Food Agric.* **2021**, *102*, 1719–1728. [CrossRef]
87. Rocchetti, G.; Chiodelli, G.; Giuberti, G.; Masoero, F.; Trevisan, M.; Lucini, L. Evaluation of phenolic profile and antioxidant capacity in gluten-free flours. *Food Chem.* **2017**, *228*, 367–373. [CrossRef]
88. Deng, G.-F.; Xu, X.-R.; Zhang, Y.; Li, D.; Gan, R.-Y.; Li, H.-B. Phenolic compounds and bioactivities of pigmented rice. *Crit. Rev. Food Sci. Nutr.* **2013**, *53*, 296–306. [CrossRef] [PubMed]
89. Aalim, H.; Luo, Z. Insight into rice (*Oryza sativa L.*) cooking: Phenolic composition, inhibition of α-amylase and α-glucosidase, and starch physicochemical and functional properties. *Food Biosci.* **2021**, *40*, 100917. [CrossRef]
90. Irakli, M.; Tsaliki, E.; Kalivas, A.; Kleisiaris, F.; Sarrou, E.; Cook, M.C. Effect of Genotype and Growing Year on the Nutritional, Phytochemical, and Antioxidant Properties of Industrial Hemp (*Cannabis sativa* L.) Seeds. *Antioxidants* **2019**, *8*, 491. [CrossRef]
91. Šibul, F.S.; Orčić, D.Z.; Svirčev, E.; Mimica-Dukić, N.M. Optimization of extraction conditions for secondary biomolecules from various plant species. *Hem. Ind.* **2016**, *70*, 473–483. [CrossRef]
92. Reddy, N.R.; Pierson, M.D. Dry Bean Tannins: A Review of Nutritional Implications. *J. Am. Oil Chem. Soc.* **1985**, *62*, 541–549. [CrossRef]
93. Gilani, G.S.; Xiao, C.W.A.; Cockell, K. Impact of Antinutritional Factors in Food Proteins on the Digestibility of Protein and the Bioavailability of Amino Acids and on Protein Quality. *Br. J. Nutr.* **2012**, *108*, S315–S332. [CrossRef] [PubMed]
94. Kehrere, J. Free Radicals as Mediators of Tissue Injury and Disease. *Crit. Rev. Toxicol.* **1993**, *23*, 21–48. [CrossRef]
95. Balasundram, N.; Sundram, K.; Samman, S. Phenolic compounds in plants and agri-industrial by-products: Antioxidant activity, occurrence, and potential uses. *Food Chem.* **2006**, *99*, 191–203. [CrossRef]
96. Vonapartis, E.; Aubin, M.P.; Seguin, P.; Mustafa, A.F.; Charron, J.B. Seed composition of ten industrial hemp cultivars approved for production in Canada. *J. Food Compos. Anal.* **2015**, *39*, 8–12. [CrossRef]
97. Babu, P.V.A.; Liu, D.R.; Gilbert, E. Recent advances in understanding the anti-diabetic actions of dietary flavonoids. *J. Nutr. Biochem.* **2013**, *24*, 1777–1789. [CrossRef] [PubMed]
98. Wang, T.; He, F.; Chen, G. Improving bioaccessibility and bioavailability of phenolic compounds in cereal grains through processing technologies: A concise review. *J. Funct. Foods* **2014**, *7*, 101–111. [CrossRef]
99. Jahreis, G.; Brese, M.; Leiterer, M.; Schaefer, U.; Boehm, V. Legume flours: Nutritionally important sources of protein and dietary fiber. *Ernahr. Umsch.* **2016**, *63*, 36–42. [CrossRef]
100. Nikolić, N.C.; Todorović, Z.B.; Stojanović, J.S.; Veličković, D.T.; Lazić, M.L. *The Fatty Acids and Acylglycerols in Chickpea and Lentil Flour*; Agro FOOD Industry Hi Tech: Milano, Italy, 2013; p. 20.
101. Shariati-Ievari, S.; Ryland, D.; Edel, A.; Nicholson, T.; Suh, M.; Aliani, M. Sensory and physicochemical studies of thermally micronized chickpea (*Cicer arietinum*) and green lentil (*Lens culinaris*) flours as binders in low-fat beef burgers. *J. Food Sci.* **2016**, *81*, S1230–S1242. [CrossRef]
102. Summo, C.; De Angelis, D.; Rochette, I.; Mouquet-Rivier, C.; Pasqualone, A. Influence of the preparation process on the chemical composition and nutritional value of canned purée of kabuli and Apulian black chickpeas. *Heliyon* **2019**, *5*, e01361. [CrossRef] [PubMed]
103. Kaur, R.; Prasad, K. Technological, processing and nutritional aspects of chickpea (*Cicer arietinum*)—A review. *Trends Food Sci. Technol.* **2021**, *109*, 448–463. [CrossRef]
104. Kosson, R.; Czuchajowska, Z.; Pomeranz, Y. Smooth and wrinkled peas. 2. Distribution of protein, lipid, and fatty acids in seed and milling fractions. *J. Agric. Food Chem.* **1994**, *42*, 96–99.

105. Fahmi, R.; Ryland, D.; Sopiwnyk, E.; Malcolmson, L.; Shariati-Ievari, S.; McElrea, A.; Barthet, V.; Blewett, H.; Aliani, M. Effect of Revtech thermal processing on volatile organic compounds and chemical characteristics of split yellow pea (*Pisum sativum* L.) flour. *J. Food Sci.* **2021**, *86*, 4330–4353. [CrossRef]
106. Dandachy, S.; Mawlawi, H.; Obeid, O. Effect of processed chickpea flour incorporation on sensory properties of mankoushe zaatar. *Foods* **2019**, *8*, 151. [CrossRef]
107. Dahmer, A.M.; Rigo, A.A.; Steffens, J.; Steffens, C.; Carrão-Panizzi, M.C. Thermal treatment for soybean flour processing with high-quality color and reduced Kunitz trypsin inhibitor. *J. Food Process Eng.* **2018**, *41*, e12925. [CrossRef]
108. Osuna, M.B.; Judis, M.A.; Romero, A.M.; Avallone, C.M.; Bertola, N.C. Improvement of fatty acid profile and studio of rheological and technological characteristics in breads supplemented with flaxseed, soybean, and wheat bran flours. *Sci. World J.* **2014**, *2014*, 401981. [CrossRef] [PubMed]
109. Prabakaran, M.; Lee, K.-J.; An, Y.; Kwon, C.; Kim, S.; Yang, Y.; Ahmad, A.; Kim, S.-H.; Chung, I.-M. Changes in soybean (*Glycine max* L.) flour fatty-acid content based on storage temperature and duration. *Molecules* **2018**, *23*, 2713. [CrossRef] [PubMed]
110. Dahmer, A.M.; Rigo, A.A.; Valduga, E.; Steffens, J.; Steffens, C.; Carrão-Panizzi, M.C. Quality characteristics of rotative-type biscuits free of gluten prepared with soya flour and cassava starch. *Curr. Nutr. Food Sci.* **2020**, *16*, 176–184. [CrossRef]
111. Li, S.; Jin, Z.; Hu, D.; Yang, W.; Yan, Y.; Nie, X.; Lin, J.; Zhang, Q.; Gai, D.; Ji, Y.; et al. Effect of solid-state fermentation with Lactobacillus casei on the nutritional value, isoflavones, phenolic acids and antioxidant activity of whole soybean flour. *LWT* **2020**, *125*, 109264. [CrossRef]
112. Gül, M.K.; Egesel, C.Ö.; Turhan, H. The effects of planting time on fatty acids and tocopherols in chickpea. *Eur. Food Res. Technol.* **2008**, *226*, 517–522. [CrossRef]
113. Werz, O.; Seegers, J.; Schaible, A.M.; Weinigel, C.; Barz, D.; Koeberle, A.; Allegrone, G.; Pollastro, F.; Zampieri, L.; Grassi, G.; et al. Cannflavins from hemp sprouts, a novel cannabinoid-free hemp food product, target microsomal prostaglandin E2 synthase-1 and 5-lipoxygenase. *PharmaNutrition* **2014**, *2*, 53–60. [CrossRef]
114. Siano, F.; Moccia, S.; Picariello, G.; Russo, G.L.; Sorrentino, G.; Di Stasio, M.; La Cara, F.; Volpe, M.G. Comparative Study of Chemical, Biochemical Characteristic and ATR-FTIR Analysis of Seeds, Oil and Flour of the Edible Fedora Cultivar Hemp (*Cannabis sativa* L.). *Molecules* **2019**, *24*, 83. [CrossRef]
115. Lančaričová, A.; Kuzmiaková, B.; Porvaz, P.; Havrlentová, M.; Nemeček, P.; Kraic, J. Nutritional quality of hemp seeds (*Cannabis sativa* L.) in different environments. *J. Cent. Eur. Agric.* **2021**, *22*, 748–761. [CrossRef]
116. Khatoon, S.; Gopalakrishna, A.G. Fat-soluble nutraceuticals and fatty acid composition of selected Indian rice varieties. *J. Am. Oil Chem. Soc.* **2004**, *81*, 939–943. [CrossRef]
117. Monks, J.L.F.; Vanier, N.L.; Casaril, J.; Berto, R.M.; de Oliveira, M.; Gomes, C.B.; de Carvalho, M.P.; Dias, A.R.G.; Elias, M.C. Effects of milling on proximate composition, folic acid, fatty acids and technological properties of rice. *J. Food Compos. Anal.* **2013**, *30*, 73–79. [CrossRef]
118. Aleman, R.S.; Morris, A.; Prinyawiwatkul, W.; Moncada, M.; King, J.M. Physicochemical properties of Frontière rice flour and its application in a gluten-free cupcake. *Cereal Chem.* **2022**, *99*, 303–315. [CrossRef]
119. Zubair, M.; Anwar, F.; Ashraf, M.; Uddin, M.K. Characterization of high-value bioactives in some selected varieties of Pakistani rice (*Oryza sativa* L.). *Int. J. Mol. Sci.* **2012**, *13*, 4608–4622. [CrossRef] [PubMed]
120. Sadigova, M.K.; Buhovets, V.A.; Belova, M.V.; Rysmuhambetova, G.E. Technology solutions in case of using chickpea flour in industrial bakery. *Sci. Study Res. Chem. Chem. Eng. Biotechnol. Food Ind.* **2018**, *19*, 169–180.
121. Kris-Etherton, P.M.; Hecker, K.D.; Binkoski, A.E. Polyunsaturated fatty acids and cardiovascular health. *Nutr. Rev.* **2004**, *62*, 414–426. [CrossRef] [PubMed]
122. Gogus, U.; Smith, C. n-3 Omega fatty acids: A review of current knowledge. *Int. J. Food Sci. Technol.* **2010**, *45*, 417–436. [CrossRef]
123. Feng, D.; Xia, A.; Liao, Q.; Nizami, A.-S.; Sun, C.; Huang, Y.; Zhu, X.; Zhu, X. Carbon cloth facilitates semi-continuous anaerobic digestion of organic wastewater rich in volatile fatty acids from dark fermentation. *Environ. Pollut.* **2021**, *272*, 116030. [CrossRef] [PubMed]
124. Yu, L.; Li, G.; Li, M.; Xu, F.; Beta, T.; Bao, J. Genotypic variation in phenolic acids, vitamin E and fatty acids in wholegrain rice. *Food Chem.* **2016**, *197*, 776–782. [CrossRef]

applied sciences

MDPI

Article

Carcass Yield and Meat Quality of Broiler Chicks Supplemented with Yeasts Bioproducts

Daniela-Mihaela Grigore [1], Silvia Mironeasa [2,*], Georgeta Ciurescu [3], Mădălina Ungureanu-Iuga [2,*], Ana Batariuc [2] and Narcisa Elena Babeanu [1]

1 Faculty of Biotechnology, University of Agronomic Sciences and Veterinary Medicine of Bucharest, 59 Marasti Blvd., District 1, 011464 Bucharest, Romania
2 Faculty of Food Engineering, Ştefan cel Mare University, 720229 Suceava, Romania
3 National Research-Development Institute for Biology and Animal Nutrition, Calea Bucuresti No. 1, 077015 Balotesti, Romania
* Correspondence: silviam@fia.usv.ro (S.M.); madalina.iuga@usm.ro (M.U.-I.)

Abstract: The utilization of synthetic additives (vitamins, proteins, and pigments) in broiler chicks' feeds may cause problems in the future, such as competitive availability, extra productive induced costs, and human health risks relayed on sole ingredients commonly used on human and animal food and feeds. A total of 320-day-old broiler chicks were randomly assigned to eight experimental groups (a four × two factorial design), receiving two dietary yeasts, lyophilizate lysates yeasts:brewer's spent yeasts: *Saccharomyces cerevisiae* (SC: 0, 0.6, 1 and 1.3 g/kg) and *Rhodotorula mucilaginosa* (RM: with 0.3 kg/t or without), during a 42-day trial. The broilers' 24 h post–slaughtering performance, meat quality (pH, color, proximate chemical composition, texture profile analysis), and meat sensorial evaluation were performed at the end of the trial. Dietary SC supplementation had a significant effect on fat deposits ($p < 0.05$), thigh meat protein content ($p < 0.05$), and breast meat lightness ($p < 0.05$). Meanwhile, RM dietary supplementation positively affected all meat color parameters ($p < 0.05$). Considerable interactions (SC × RM) were obtained on the broilers' breast and thigh meat moisture and yellowing index. Dietary lyophilizes lysates yeasts supplementation had an effect on meat color and proximate chemical composition. Furthermore, investigation is needed to elucidate the effects of microbial pigment supplements on broiler meat biochemistry and its pigment metabolism.

Keywords: broiler meat; meat quality; meat color; pigment yeast; sensorial evaluation; texture profile

1. Introduction

The future use of synthetic additives (vitamins, proteins, and pigments) in broiler chicks' feeds poses a number of problems [1] including competitive availability, increased productive induced costs, and health risks associated with sole ingredients commonly used in human and animal feeds and food. Moreover, the projection concerning the increased demand for broiler meat and meat products is directly affected by the consumer's choice. Recently, consumers' preferences are directed towards safe and high-quality broiler meat and meat products, often associated with slow-growing broiler breeds [2], reared in organic systems [3], obtained using an economic circular approach by introducing organic valuable waste and by-products [4].

After twelve–fifteen successive fermentation batches, the brewing biological catalyst *Saccharomyces cerevisiae* loses its vitality and viability and might be biologically compromised, being no longer appropriate for beer production. Brewers' spent yeast (SC) was previously described as a nutritive complex ingredient, an excellent source of protein (more than 40%) [5], presenting a high content of essential limitative amino acids, with high levels of biotin, folic acid, microelements, and intracellular flavoring compounds [6,7]. It is known that *Saccharomyces cerevisiae* (SC) consumption as a protein source for animal and human nutrition is restricted, due to the high level of nucleic acids (up to 15%), mainly causing seric and tissue uric acid accumulation [8]. Despite these facts, SC was mainly subjected to debittering, efficiently valuing all biologic active compounds [9], that can conventionally be achieved by physical, chemical, and biological processes [10], while avoiding toxicity [11].

Poultry cannot synthesize carotenoids; synthetic pigment feed additives are often employed in order to adjust the retinoic acid ratio via carotenoid dietary supplementation. Most commonly used food and feed colorant additives in poultry nutrition are xanthophylls (canthaxanthin, astaxanthin, and zeaxanthin) that originate from almost 90% mainly synthetic resources. Nowadays, poultry meat consumers' preference is directing the meat producing sector towards organic and less intensive poultry rearing techniques, including the use of natural and organic feed coloring additives. Recent scientific evidence points out the advantages of employing microbial pigment sources, lowering the costs covering the geo-climate variability and time spent on the vegetal pigment producing industry, obtaining a short-time valuable pigment product, mostly based on waste fermentation, and low-cost materials [12–14]. Yeast such as *Rhodotorula mucilaginosa* (RM) represents a perfect choice as a natural resource of secondary metabolites: carotenoids, lipids, and extracellular enzymes [13,15,16].

Previous studies indicate an active/inactive SC dietary supplementation improves broiler growth performance [11,17,18], immunity response [13], carcass yield [17], and meat quality [19,20]. Scientific evidence for using RM on broiler chick feed is very limited [21], but recent studies suggest that it could have growth promoting, health promoting, and production quality effects on laying hens [22], fish [23], and swine [24].

The study was performed to probate the hypothesis that yeasts lyophilizate lysates dietary supplements might impair the carcass yield and the meat quality attributes. Therefore, the main goals were to evaluate the singular and the interaction effects of the dietary yeasts lyophilizate lysates supplements on the broiler carcass yield and breast and thigh meat quality attributes (pH, color, proximate chemical composition, texture profile analysis, and sensorial evaluation).

2. Materials and Methods

The Animal Ethics Committee approved the study protocol of the National Research and Development Institute for Animal Biology and Nutrition (INCDBNA-IBNA) Baloteşti, Romania, under the EU Directive 2010/63/EU and Romanian Law on Animal Protection (no. 199/2018). The bird's slaughtering was carried out following the applicable rules on handling animals at the time of slaughter, including humane treatment. We confirm that all engaged methods on chickens were carried out in accordance with relevant guidelines and regulations.

A total of 320-day-old unsexed Cobb 500 broiler chicks (body weight = 43.11 ± 0.7 g/bird) were randomly assessed in eight experimental groups (Table 1) to receive a corn-soybean meal based diet, supplemented with yeasts dietary lyophilizate lysates l: SC (0, 0.6, 1 and 1.3 g/kg) and RM (with 0.3 kg/t or without) as previously described [18], and listed in Table 2.

Table 1. Basal diets of broiler chicks, per each growth period [1].

Groups	Dietary Treatments
1	Negative control group – fed with basal diet (Corn-soybean meal);
2	Basal diet + SC 0.6 g/kg feed;
3	Basal diet + SC 1.0 g/kg feed;
4	Basal diet + SC 1.3 g/kg feed;
5	Positive control group – fed with basal diet (Corn-soybean meal) + RM (0.3 kg/t feed);
6	Basal diet + SC 0.6 g/kg feed + RM (0.3 kg/t feed);
7	Basal diet + SC 1.0 g/kg feed + RM (0.3 kg/t of feed);
8	Basal diet + SC 1.3 g/kg feed + RM (0.3 kg/t feed);

[1] after [18].

Table 2. Basal diets of broiler chicks, per each growth period [1].

	Starter	Grower	Finisher
Ingredients (g/kg)			
Corn	557.9	567.3	629.6
Soybean meal	331.0	311.0	255.0
Corn gluten	43.0	43.0	35.0
Soybean oil	14.6	29.8	34.0
Monocalcium phosphate	16.9	16.6	14.5
Calcium carbonate	16.9	14.6	13.2
Salt	2.8	2.8	2.8
L-lysine HCl	3.3	1.8	2.7
Dl-methionine	2.8	2.3	2.5
Choline-chloride 50%	0.8	0.8	0.7
Vitamin − mineral mixture *	10.0	10.0	10.0
RM [2]	+/−	+/−	+/−
SC [3]	0/0.6/1 or 1.3	0/0.6/1 or 1.3	0/0.6/1 or 1.3
Calculated composition			
ME (MJ/kg)	12.55	13.02	13.40
CP (%)	23.0	22.0	20.0
Lysine, total (%)	1.41	1.24	1.05
Lysine, digestible (%)	1.28	1.16	0.98
Methionine + cysteine, total (%)	1.02	0.95	0.86
Methionine + cysteine, digestible (%)	0.94	0.87	0.75
Ca (%)	1.00	0.90	0.80
Available P (%)	0.45	0.45	0.45
Crude fat (%)	4.34	5.85	6.23
Crude fiber (%)	2.85	2.77	2.56

* Supplied per kg diet: 12,000 IU vitamin A, 5000 IU vitamin D3, 75 mg vitamin E, 3 mg vitamin K3, 3 mg vitamin B1, 8 mg vitamin B2, 5 mg vitamin B6, 0.016 mg vitamin B12, 13 mg pantothenic acid, 55 mg nicotinic acid, 2 mg folic acid, 0.2 mg biotin, 120 mg Mn, 100 mg Zn, 40 mg Fe, 16 mg Cu, 1.25 mg I and 0.3 mg Se, 70 mg Monteban G100. [1] after [18]; [2] RM—*Rhodotorula mucilaginosa*— lyophilized lysed yeast 0.3 kg/T feed − = not supplemented in the broilers' diet; + = supplemented in the broilers' diet; [3] SC—*Saccharomyces cerevisiae*—not included or supplemented as 0.6, 1, or 1.3 in the broilers' diet.

2.1. Slaughtering and Post–Slaughtering Performance

At the end of the experimental trial (day 42), all birds were individually weighed using a commercial scale. Five birds from each pen were electrically stunned and slaughtered by carotid amputation. Carcasses were plucked, eviscerated, and individually weighed. For the carcass yield, slaughter performance and the rates of retailed meat cuts (carcass yield, feet and hock joint removed carcass, front part of carcass, back part of carcass, breast fillets, breast tenders, leg quarters, thighs, drumsticks, and whole wings) and internal organs (heart, liver, and gizzard) were weighed and reported as a percentage from the commercial carcass weight.

The deboned breast ($n = 40$) and thigh ($n = 40$) were collected [25], vacuum sealed, and refrigerated (2 ± 2 °C) for further analysis.

2.2. pH

After 24 h of refrigeration, the pH of the breast and thigh meat samples ($n = 5$) were recorded (SR ISO 2917:2007) using a calibrated (4.01, 7.00 and 9.21 buffer solutions, at 20 °C) pH-meter (WTW 315i, Weilheim, Germany).

2.3. Meat Color

Breast and thigh meat colors ($n = 5$) were surface measured using the portable, calibrated (white ceramic tile) CR 410 color meter (Konica Minolta INC, Osaka, Japan) on middle area of the left-side breast–pectoralis major and left thigh (homogeneous, and fat–free area–8 mm). The broiler meat colors were reported in the CIE Lab system (L^*, a^*, b^*) [26] and calculated for yellowness index (YI), browning index (BI), chroma (C), and hue (h) [27].

2.4. Proximate Chemical Composition

For the proximate chemical composition, the breast and thigh meat samples were minced and individually analyzed (n = 5) using the DA6200 automatic analyzer (PerkinElmer, Inc., Waltham, MA, USA) equipped with a magnetic tray (14 mm height and 170 mL volume) and diode array detector (λ = 850–1050 nm). To minimize the sampling error, the replicates (n = 5) were analyzed in triplicate. Optical calibration was performed before meat sample analysis, using the polystyrene check sample.

2.5. Texture Profile Analysis

The texture profile analysis (TPA) of raw breast and thigh meat samples (n = 5) was individually analyzed, employing the texture analyzer (Brookfield CT3, Massachusetts, USA). The double cycle compression method was adopted on the cylinder-catted meat samples (20 mm diameter and 5 $\pm$ 0.2 mm height). The analyzer was supplied with a 50 kg loading cell, a cylinder probe (76.2 $\times$ 10 mm), and a fixture bare table. The experimental physical conditions were set as previously described [28,29], pre-test (speed = 2.0 mm s^{-1}), test (speed = 1.0 mm s^{-1}), and post-test (speed = 2.0 mm s^{-1}).

2.6. Sensorial Evaluation

Sensorial raw meat analysis was performed on the breast and thigh meat samples (n = 3) by employing ten naive assessors (ISO 5492: 2008, gender-balanced, aged between 20–25 years), chosen for their ability to perform a sensory test, acquainted with meat and meat product quality traits. The sensorial evaluation was by the environmental guidelines set up for this purpose (ISO 8589:2007). The descriptive analysis was applied, using a scoring scale (ISO 6658:2017; sored between 1–5; 1 for indicating the lowest score and 5 for indicating the highest score). The scale parameters included meat appearance (MFA), fat appearance (FTA), odor (ODO), consistency (CON), juiciness (JCY), and meat tenderness (TEN); the scoring scale included the quality sensorial attributes description as standing from 1–unaccepted, unrecommended, up to 5—very desirable, recommended [30].

2.7. Statistics

The experimental data RM was analyzed using the General Linear Model procedure (SPSS v25, IBM) and reported as means and standard error of the means (SEM). The one-way ANOVA for the differences among the experimental groups and Tukey's multiple range of tests for differencing the significant means at $p < 0.05$ level were employed. The relationships between characteristics were evaluated trough Pearson correlations and Principal Component Analysis (PCA), the significance level being $p < 0.05$.

The factorial design equation:

$$X_{ijk} = \mu + \alpha_i + \beta_j + \alpha\beta_{ij} + e_{ijk} \quad (1)$$

μ = general mean, α_i = the effect of SC, having the levels i = 1–4; β_j = the effect of RM, having the levels j = 1–2; $\alpha\beta_{ij}$–the interaction of SC and RH, having 2 $\times$ 4 levels; and e_{ijk} = the experimental unit associated error, having the levels k = 1–8.

3. Results

3.1. Post-Slaughtering Performance

The post-slaughtering performance of dietary supplemented broiler chicks are detailed in Table 3. The RM dietary supplement had no significant effect ($p > 0.05$) on the broiler's carcass yield, slaughter performance, nor the rates of retailed meat cuts. Although, supplementing SC on the broiler chick's diet had significant effect on gizzard weight (p = 0.04) and abdominal fat accumulation (p = 0.001). Considerably higher differences were observed in groups four and eight, when compared with all experimental groups. Through dietary supplementation, different levels of SC linearly decreased the abdominal fat deposits, having the lowest values on the 1.3 g/kg SC supplemented groups (four and eight), when compared with all experimental groups.

Table 3. Carcass yield and retailed body parts of broilers' [1] diet supplemented with yeast.

Group	SC	RM	LW [2]	CW [2]	CY [3]	Breast [3]	Thigh [3]	Drumsticks [3]	Liver [3]	Gizzard [3]	Heart [3]	Abd. Fat [3]
1	0	No	2761.80	2070.40	74.98	41.59	12.51	10.35	2.83	1.90	0.620	2.61
2	0.6	No	2789.60	2064.00	74.06	41.76	12.08	10.39	2.73	2.04	0.589	2.11
3	1	No	2723.80	2049.40	75.24	41.94	12.37	10.57	2.81	2.06	0.628	1.64
4	1.3	No	2525.80	1902.40	75.38	41.88	12.51	10.95	3.08	2.20	0.685	1.44
5	0	Yes	2744.60	2046.20	74.57	41.32	12.10	10.84	2.61	1.58	0.529	2.12
6	0.6	Yes	2659.40	1975.00	73.61	41.65	12.13	10.97	2.78	1.71	0.643	2.16
7	1	Yes	2555.80	1925.20	75.32	41.71	12.53	10.79	2.74	1.85	0.638	1.96
8	1.3	Yes	2463.00	1872.60	75.93	41.97	12.81	10.35	2.96	2.89	0.644	1.47
SEM			75.102	62.267	1.018	1.153	0.507	0.363	0.167	0.195	0.041	0.205
p value			0.751	0.798	0.956	0.999	0.906	0.366	0.865	0.060	0.390	0.263
						Main effect—SC levels						
0			2753.20 [a]	2058.30 [a]	74.77	41.46	12.31	10.60	2.74	1.88 [b]	0.56	2.37 [a]
0.6			2724.50 [a]	2010.30 [ab]	73.84	41.71	12.11	10.68	2.76	1.88 [b]	0.62	2.14 [ab]
1			2639.80 [ab]	1987.30 [ab]	75.28	41.88	12.45	10.68	2.78	1.96 [b]	0.63	1.80 [bc]
1.3			2494.40 [b]	1888.30 [b]	75.66	41.93	12.66	10.65	3.03	2.54 [a]	0.66	1.45 [c]
SEM			53.105	44.029	0.720	0.815	0.359	0.257	0.118	0.138	0.029	0.145
p value			0.007	0.066	0.327	0.977	0.739	0.995	0.288	0.004	0.201	0.001
						Main effect—RM addition						
No			2700.25	2022.05	74.92	41.79	12.37	10.56	2.86	2.05	0.633	1.95
Yes			2605.70	1950.25	74.86	41.69	12.39	10.74	2.77	2.08	0.614	1.92
SEM			37.551	31.133	0.509	0.576	0.254	0.182	0.084	0.098	0.021	0.102
p value			0.085	0.113	0.933	0.895	0.942	0.507	0.427	0.841	0.516	0.884

SC—*Saccharomyces cerevisiae*, g/kg feed; RM—*Rhodotorula mucilaginosa* 0/0.3 kg/t of feed; [1] LW—final live weight; [2] CW—eviscerated carcass weight; [3] CY—carcass yield; Abd. Fat—abdominal fat; SEM—standard error mean; SC × RM—interaction between the dietary supplements; [1] $n = 5$/replicate pen; [2] expressed in gram; [3] expressed as percentage of CW; [abcd] Columns means with different superscripts differ significantly at $p < 0.05$.

3.2. Meat pH and the Chemical Proximate Analysis

The broiler raw breast and thigh meat pH and the proximate chemical composition (PCC) results are displayed in Table 4. There was no significant difference ($p > 0.05$) on the interaction of the yeasts (SC × RM) dietary supplemented broiler meat samples after 24 h post-mortem refrigerated breast and thigh meat samples. No significant effects ($p > 0.05$) were observed on the broiler breast and thigh meat obtained with SC dietary supplementation, nor on the broiler breast and thigh meat that received RM dietary supplementation.

Data concerning dietary yeasts supplementation on the broiler breast and thigh meat samples are indicating significant differences among the PCC on all the experimental groups. There were significant linear differences in the SC main effects groups, resulting in lower protein and fat percent on the breast and thigh meat, indicating that SC dietary linear increasement (0, 0.6, 1 and 1.3 g/kg feed) decreases the meat protein and fat percentage. Conversely, the broiler breast and thigh meat were significantly influenced ($p < 0.05$) by the dietary linear supplementation of SC, indicating a direct proportionality between moisture and collagen percentage and the SC dietary increasement (0, 0.6, 1 and 1.3 g/kg feed).

Table 4. Yeasts supplemented broiler's breast [1] and thigh meat pH and PCC.

			Broiler Breast Meat					Broiler Thigh Meat				
Group	**SC**	**RM**	**pH**	**Protein ***	**Fat ***	**Moisture ***	**Collagen ***	**pH**	**Protein ***	**Fat ***	**Moisture ***	**Collagen ***
1	**0**	**No**	6.02	21.77	2.23	76.17	0.84	5.91	19.31	10.44	70.10	0.98
2	**0.6**	**No**	5.99	21.63	2.27	75.76	0.87	5.96	18.59	8.83	71.26	1.09
3	**1**	**No**	6.02	21.17	1.87	76.99	0.89	6.00	18.85	8.26	71.64	1.11
4	**1.3**	**No**	6.01	20.87	1.70	76.89	0.93	6.01	17.68	8.99	71.03	1.25
5	**0**	**Yes**	5.99	21.54	2.67	76.95	0.83	5.97	19.07	10.25	69.75	0.98
6	**0.6**	**Yes**	5.99	21.42	2.61	78.55	0.85	5.99	18.41	9.48	71.87	1.02
7	**1**	**Yes**	5.97	21.03	2.23	76.74	0.88	6.02	18.90	8.77	72.92	1.11
8	**1.3**	**Yes**	6.00	20.82	1.99	76.88	0,90	5.99	17.48	8.42	72.47	1.24
SEM			0.032	0.151	0.058	0.072	0.833	0.033	0.077	0.276	0.205	0.021
***p* value**			0.889	0.938	0.900	<0.000	0.600	0.663	0.232	0.109	<0.000	0.342
						Main effect—SC levels						
0			6.00	21.66 [a]	2.48 [a]	76.76 [c]	0.83 [d]	5.94	19.19 [a]	10.35 [a]	69.93 [c]	0.98 [c]
0.6			5.99	21.53 [a]	2.44 [a]	77.16 [a]	0.86 [c]	5.98	18.45 [c]	9.15 [b]	71.56 [b]	1.05 [b]
1			5.99	21.10 [b]	2.05 [b]	76.87 [b]	0.89 [b]	6.01	18.87 [b]	8.51 [b]	72.28 [a]	1.11 [b]
1.3			6.01	20.86 [c]	1.85 [c]	76.89 [b]	0.91 [a]	6.00	17.58 [d]	8.70 [b]	71.75 [ab]	1.25 [a]
SEM			0.22	0.107	0.041	0.051	0.006	0.023	0.054	0.195	0.145	0.015
***p* value**			0.867	<0.000	<0.000	<0.000	<0.000	0.133	<0.000	<0.000	<0.000	<0.000
						Main effect—RM addition						
No			6.01	21.36	2.03 [b]	76.45 [b]	0.88	5.97	18.61 [a]	9.13	71.01 [b]	1.11
Yes			5.99	21.21	2.38 [a]	77.23 [a]	0.87	5.99	18.47 [b]	9.23	71.75 [a]	1.09
SEM			0.022	0.107	0.041	0.051	0.006	0.16	0.038	0.138	0.103	0.011
***p* value**			0.300	0.159	<0.000	<0.000	0.110	0.380	0.013	0.606	<0.000	0.282

[1], * $n = 5$/replicate pen; SC—*Saccharomyces cerevisiae*, g/kg feed; RM—*Rhodotorula mucilaginosa*, 0/0.3 kg/t of feed; SEM—standard error mean; SC × RM—interaction between the dietary supplements; [abcd] Columns means with different superscripts differ significantly at $p < 0.05$.

3.3. Broiler Breast and Thigh Color Profile

Manipulating broiler meat color by dietary supplementation of different inactive yeast sources resulted in significant changes in the breast and thigh meat color parameters (Table 5). Significant interactions ($p < 0.05$) between both of the main dietary supplements (SC × RM) were shown on broiler breast (YI) and thigh meat (a^*, b^*, C, BI, YI and ΔE). The results show a bifactorial positive interaction on the broiler meat color parameters values, which linearly increased within the dietary supplementation.

As expected, the dietary RM supplement had highly significant effects ($p < 0.005$) on all analyzed color parameters of the broiler breast meat. Our results shows that broiler dietary RM supplementation significantly lowers the breast meat color parameters: a^* ($p < 0.000$), b^* ($p < 0.000$), C ($p = 0.001$), BI ($p < 0.000$), YI ($p < 0.000$), and ΔE ($p < 0.000$) of the experimental groups (five, six, seven, eight), while the L^* ($p < 0.000$) and h ($p < 0.000$) increases, when compared with the control group (1).

The SC dietary supplementation was found to have significant influence ($p < 0.05$) on the broiler breast meat L^*, YI, BI, and ΔE. The SC dietary supplementation showed that L^* values were lowered in the broiler breast meat experimental groups (two, three, and four) than the control (one), while higher values were obtained for the total color difference. Although conversely, the broiler thigh meat brightness (L^*) values were higher in the control group than in the experimental groups (two, three, and four), indicating that SC treatment might influence the carotenoid muscular accumulation, thus carotenoid abundance. The SC × RM interaction had an effect ($p < 0.05$) on the broiler thigh meat L^*, a^*, b^*, C, BI, YI, and ΔE.

Table 5. Yeasts supplemented broilers' breast [1] and thigh meat color.

			Broiler Breast Meat								Broiler Thigh Meat							
Group	**SC**	**RM**	L^*	a^*	b^*	h	C	**BI**	**YI**	**ΔE**	L^*	a^*	b^*	h	C	**BI**	**YI**	**ΔE**
1	**0**	**No**	54.14	10.05	13.18	0.92	16.58	41.00	34.75	43.80	57.95	12.62	11.50	0.74	17.12	37.12	28.11	40.20
2	**0.6**	**No**	56.16	10.77	12.29	0.85	16.35	38.26	31.27	41.89	57.47	14.61	12.25	0.70	19.07	41.09	30.45	41.98
3	**1**	**No**	55.51	10.36	13.80	0.93	17.26	41.78	35.51	42.73	54.73	13.38	12.87	0.77	18.57	41.43	31.73	41.23
4	**1.3**	**No**	56.85	10.68	12.83	0.88	16.72	38.82	32.21	41.35	57.81	11.91	11.50	0.77	16.55	36.72	28.41	40.54
5	**0**	**Yes**	48.52	12.16	14.16	0.86	18.71	52.21	34.75	49.86	56.99	13.52	10.85	0.68	17.35	39.55	28.35	43.82
6	**0.6**	**Yes**	51.35	12.83	14.27	0.84	19.20	50.25	39.70	47.41	56.67	12.12	10.23	0.70	15.86	34.98	25.78	41.44
7	**1**	**Yes**	51.20	12.72	14.13	0.84	19.01	49.92	39.45	47.48	58.99	12.58	11.23	0.73	16.87	37.53	28.16	41.48
8	**1.3**	**Yes**	50.69	12.97	14.66	0.85	19.58	52.25	41.32	48.15	55.58	13.84	11.77	0.70	18.19	41.15	30.07	43.07
SEM			0.540	0.295	0.360	0.018	0.334	0.736	0.709	0.456	0.303	0.357	0.309	0.021	0.310	0.779	0.794	0.271
***p* value**			0.336	0.954	0.096	0.188	0.274	0.009	0.005	0.167	<0.000	<0.000	0.003	0.303	<0.000	<0.000	0.001	<0.000
									Main effect—SC levels									
0			51.33 [b]	11.11	13.67	0.89	17.64	46.6	38.17	48.82 [a]	57.47 [a]	13.07	11.18 [b]	0.71	17.23	38.34	28.23	42.00
0.6			53.75 [a]	11.80	13.28	0.85	17.78	44.26	35.48	44.65 [b]	57.07 [ab]	13.37	11.24 [ab]	0.69	17.47	38.44	28.12	41.71
1			53.35 [a]	11.83	13.97	0.88	18.14	45.85	37.48	45.10 [b]	56.61 [ab]	12.98	12.05 [a]	0.75	17.73	39.48	29.94	43.36
1.3			53.77 [a]	11.80	13.74	0.86	18.15	45.54	36.76	44.75 [b]	56.84 [ab]	12.87	11.63 [ab]	0.74	17.38	38.94	29.24	41.81
SEM			0.382	0.209	0.255	0.013	0.236	0.521	0.502	0.322	0.214	0.252	0.219	0.015	0.219	0.551	0.562	0.191
***p* value**			<0.000	0.072	0.306	0.081	0.335	0.072	0.081	<0.000	0.048	0.560	0.029	0.089	0.456	0.449	0.087	0.128
									Main effect—RM addition									
No			55.66 [a]	10.47 [b]	13.02 [b]	0.89 [a]	16.73 [b]	39.96 [b]	33.43 [b]	42.44 [b]	57.93 [a]	13.13	12.03 [a]	0.744 [a]	17.83 [a]	30.30	29.68 [a]	40.99 [b]
Yes			50.44 [b]	12.67 [a]	14.31 [a]	0.85 [b]	19.13 [a]	51.16 [a]	40.52 [a]	48.23 [a]	56.06 [b]	13.02	11.02 [b]	0.702 [b]	17.07 [b]	38.31	28.09 [b]	42.45 [a]
SEM			0.270	0.148	0.180	0.009	0.167	0.368	0.355	0.228	0.152	0.179	0.155	0.010	0.155	0.389	0.397	0.135
***p* value**			<0.000	<0.000	<0.000	0.001	<0.000	<0.000	<0.000	<0.000	<0.000	0.657	<0.000	0.008	0.002	0.082	0.008	<0.000

SC—*Saccharomyces cerevisiae*, g/kg feed; RM—*Rhodotorula mucilaginosa* 0/0.3 kg/t of feed; L^*—meat brightness, a^*—meat redness, b^*—meat yellowness, h—meat hue; C—meat chroma; BI—meat browning index; YI—meat yellowing index; ΔE—meat total color difference; SEM—standard error mean; SC × RM—interaction between the dietary supplements; [ab] Columns means with different superscripts differ significantly at $p < 0.05$.

3.4. Texture Profile Analysis

The broiler breast and thigh meat texture profiles analysis are displayed in the Tables 6 and 7. In this study, the broiler breast and thigh meat texture profiles were not affected by the SC ($p > 0.05$) nor the RM dietary supplementation ($p > 0.05$). Similar results for hardness, adhesiveness, chewiness, cohesiveness, springiness, gumminess, and resilience were found on all experimental groups (1–8).

Table 6. Yeasts supplemented broilers' breast meat texture profile.

Group	SC	RM	Hard	Adhe	Cohe	Spri	Gumm	Chew	Resi
1	0	No	2806.00	0.266	0.260	2.238	729.69	13.27	0.204
2	0.6	No	2789.20	0.264	0.238	2.280	665.52	12.10	0.186
3	1	No	2721.00	0.266	0.266	2.392	724.54	13.17	0.203
4	1.3	No	2720.60	0.278	0.262	2.216	711.07	12.93	0.199
5	0	Yes	2779.40	0.264	0.240	2.390	667.35	12.34	0.187
6	0.6	Yes	2758.80	0.266	0.256	2.194	709.41	12.90	0.198
7	1	Yes	2709.40	0.270	0.252	2.268	683.92	12.44	0.191
8	1.3	Yes	2709.60	0.280	0.236	2.316	635.26	11.55	0.178
SEM			82.146	0.010	0.012	0.074	39.473	0.718	0.011
p value			0.999	0.992	0.279	0.186	0.439	0.439	0.439
				Main effect—SC levels					
0			2792.70	0.265	0.250	2.314	698.52	12.70	0.195
0.6			2774.00	0.270	0.247	2.237	687.47	12.50	0.192
1			2715.20	0.268	0.259	2.330	704.23	12.80	0.197
1.3			2714.90	0.279	0.249	2.266	673.16	12.24	0.188
SEM			58.086	0.007	0.008	0.052	27.911	0.507	0.008
p value			0.700	0.467	0.759	0.571	0.867	0.867	0.867
				Main effect—RM addition					
No			2759.10	0.269	0.257	2.28	707.71	12.87	0.20
Yes			2739.30	0.270	0.246	2.29	673.98	12.25	0.19
SEM			41.073	0.005	0.006	0.037	19.98	0.359	0.006
p value			0.735	0.835	0.225	0.841	0.236	0.236	0.236

SC—*Saccharomyces cerevisiae*, g/kg feed; RM—*Rhodotorula mucilaginosa*, 0/0.3 kg/t of feed; SEM—standard error mean; SC × RM—interaction between the dietary supplements; Adhe—adhesiveness; Cohe—cohesiveness; Spri—springiness; Gumm—gumminess; Chew—chewiness; Resi—resilience;

Table 7. Yeasts supplemented broilers' thigh meat texture profile.

Group	SC	RM	Hard	Adhe	Cohe	Spri	Gumm	Chew	Resi
1	0	No	4124.82	0.458	0.335	2.563	1732.19	30.36	0.209
2	0.6	No	4189.36	0.444	0.307	2.611	1620.02	28.40	0.196
3	1	No	3953.39	0.434	0.343	2.739	1713.82	30.04	0.207
4	1.3	No	3915.09	0.492	0.338	2.537	1668.12	29.24	0.202
5	0	Yes	4022.42	0.463	0.310	2.737	1573.22	27.58	0.190
6	0.6	Yes	4264.88	0.414	0.330	2.512	1775.92	31.13	0.215
7	1	Yes	4108.10	0.438	0.325	2.597	1677.89	29.41	0.203
8	1.3	Yes	4119.53	0.468	0.304	2.652	1565.94	27.45	0.189
SEM			246.213	0.038	0.015	0.084	99.064	1.736	0.012
p value			0.929	0.945	0.279	0.186	0.425	0.425	0.425

Table 7. *Cont.*

Group	SC	RM	Hard	Adhe	Cohe	Spri	Gumm	Chew	Resi
				Main effect—SC levels					
0			4073.62	0.460	0.323	2.650	1652.71	28.97	0.200
0.6			4227.12	0.429	0.319	2.561	1697.97	29.76	0.205
1			4030.75	0.436	0.334	2.668	1695.86	29.73	0.205
1.3			4017.31	0.480	0.321	2.595	1617.03	28.34	0.195
SEM			174.099	0.027	0.011	0.060	70.049	1.228	0.008
***p* value**			0.821	0.533	0.759	0.571	0.822	0.822	0.822
				Main effect—RM addition					
No			4045.66	0.457	0.331	2.612	1683.54	29.51	0.204
Yes			4128.73	0.446	0.317	2.624	1648.23	28.89	0.199
SEM			123.107	0.019	0.008	0.042	49.532	0.868	0.006
***p* value**			0.637	0.685	0.255	0.841	0.618	0.618	0.618

SC—*Saccharomyces cerevisiae*, g/kg feed; RM—*Rhodotorula mucilaginosa*, 0/0.3 kg/t of feed; SEM—standard error mean; SC × RM—interaction between the dietary supplements; Adhe—adhesiveness; Cohe—cohesiveness; Spri—springiness; Gumm—gumminess; Chew—chewiness; Resi—resilience;

No considerable interaction (SC × RM) was observed on the broiler breast or thigh meat texture profile.

3.5. Sensorial Evaluation

The dietary yeast supplemented broiler breast and thigh raw meat sensorial evaluation is presented in Figure 1.

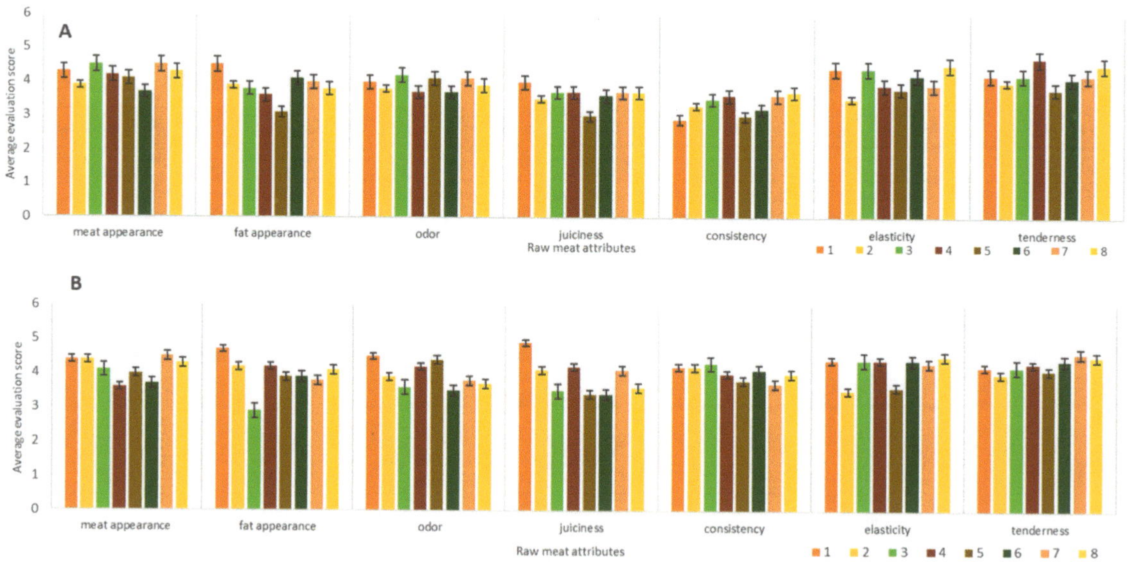

Figure 1. Yeasts supplemented broiler's breast (**A**) and thigh (**B**) meat sensorial attributes. 1—Negative control group-fed with basal diet (Corn-soybean meal); 2—Basal diet + SC 0.6 g/kg feed; 3—Basal diet + SC 1.0 g/kg feed; 4—Basal diet + SC 1.3 g/kg feed; 5—Positive control group-fed with basal diet (Corn-soybean meal) + RM (0.3 kg/t feed); 6—Basal diet + SC 0.6 g/kg feed + RM (0.3 kg/t feed); 7—Basal diet + SC 1.0 g/kg feed + RM (0.3 kg/t of feed); 8—Basal diet + SC 1.3 g/kg feed + RM (0.3 kg/t feed).

Compared with the control, similar values for the broiler breast meat appearance and odor were obtained. Significant differences ($p < 0.05$) were shown for the broilers' raw breast meat fat appearance, consistency, juiciness, elasticity, and tenderness. The highest evaluation fat appearance score was found in the control group, and the lowest in the experimental group five (RM positive control). The same trend was observed for the meat juiciness attribute. Meat consistency was found to be influenced by the dietary SC linear increasement (0, 0.6, 1, and 1.3 g/kg feed), showing a direct influence on the evaluation average score, having the lowest values in the control groups (one and five), and the highest in groups four and eight. Similar score evaluation was found in the broiler thigh meat appearance, consistency, elasticity, and meat tenderness. Although, fat appearance, odor, and juiciness were significantly influenced by the dietary SC supplementation.

3.6. Correlations and Principal Component Analysis

The Principal Component Analysis bi-plot (Figure 2) revealed the relationships between the studied variables and the samples. A total of 85.06% of the data variance was explained by the two principal components: 67.87% being attributed to the first component (PC1) and 17.19% to the second one (PC2). The first component, PC1, was associated with the chemical composition parameters (moisture, protein, fat, collagen content), texture parameters (springiness, adhesiveness, hardness, gumminess, chewiness, and cohesiveness) and some color parameters (b^*, YI, h). The second component, PC2, was associated with pH, resilience, a^* and C. As it can be seen in Figure 2, the breast samples were clustered on the left quadrants of the graphic, while the thigh samples were grouped on the right side. Significant correlations at $p < 0.05$ can be observed between some of the characteristics (Table 8, Figure 2).

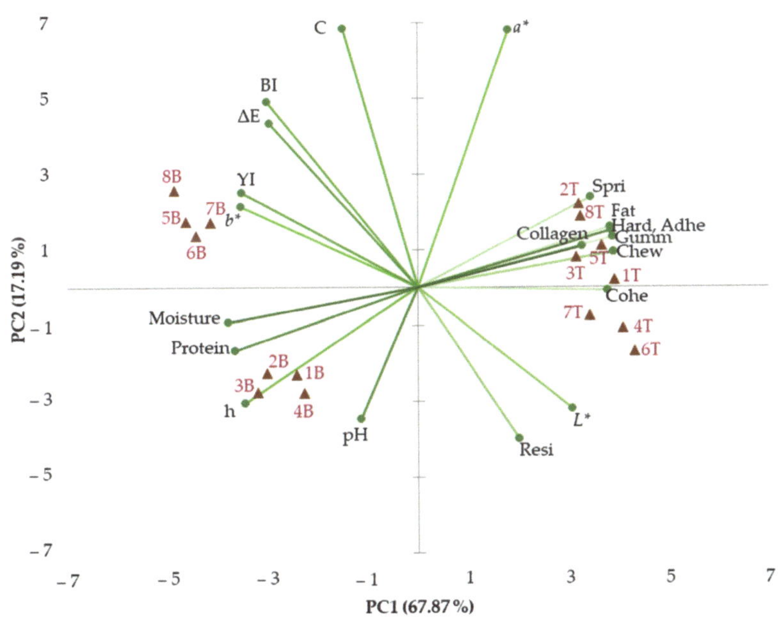

Figure 2. Principal Component Analysis bi-plot: red triangles represent the chicken groups (B—breast (1–8 groups), T—thigh (1–8 groups)) and green dots represent the variables; Adhe—adhesiveness, Cohe—cohesiveness, Spri—springiness, Gumm—gumminess, Chew—chewiness, Resi—resilience, L^*, a^*, b^*—color parameters.

Table 8. Pearson correlations coefficients.

Variables	Hard	Adhe	Cohe	Spri	Gumm	Chew	Resi	L^*	a^*	b^*	h	C	BI	YI	ΔE	pH	Prot	Fat	Moist	Coll
Hard	**1.00**																			
Adhe	**0.96**	**1.00**																		
Cohe	**0.92**	**0.92**	**1.00**																	
Spri	**0.89**	**0.90**	**0.83**	**1.00**																
Gumm	**0.99**	**0.97**	**0.97**	**0.88**	**1.00**															
Chew	**0.99**	**0.97**	**0.97**	**0.88**	**0.99**	**1.00**														
Resi	0.44	0.35	**0.66**	0.23	**0.52**	**0.52**	**1.00**													
L^*	**0.65**	**0.66**	**0.70**	**0.60**	**0.67**	**0.67**	0.46	**1.00**												
a^*	**0.59**	**0.59**	0.41	**0.63**	**0.54**	**0.54**	−0.20	0.06	**1.00**											
B^*	**−0.84**	**−0.79**	**−0.77**	**−0.65**	**−0.82**	**−0.82**	−0.47	**−0.76**	−0.18	**1.00**										
h	**−0.93**	**−0.89**	**−0.76**	**−0.82**	**−0.89**	**−0.89**	−0.18	**−0.51**	**−0.76**	**0.77**	**1.00**									
C	−0.23	−0.19	−0.32	−0.04	−0.25	−0.26	**−0.54**	**−0.56**	**0.61**	**0.67**	0.05	**1.00**								
BI	**−0.63**	**−0.60**	**−0.68**	−0.45	**−0.65**	**−0.65**	**−0.63**	**−0.88**	0.18	**0.87**	0.44	**0.83**	**1.00**							
YI	**−0.82**	**−0.77**	**−0.78**	**−0.68**	**−0.81**	**−0.82**	**−0.52**	**−0.76**	−0.11	**0.95**	**0.69**	**0.68**	**0.88**	**1.00**						
ΔE	**−0.61**	**−0.61**	**−0.71**	−0.46	**−0.65**	**−0.65**	**−0.64**	**−0.89**	0.11	**0.74**	0.39	**0.68**	**0.95**	**0.78**	**1.00**					
pH	−0.35	−0.32	−0.23	−0.27	−0.32	−0.32	−0.01	0.07	−0.49	0.21	0.46	−0.20	0.03	0.19	0.06	**1.00**				
Prot	**−0.92**	**−0.95**	**−0.87**	**−0.84**	**−0.92**	**−0.92**	−0.32	**−0.63**	**−0.59**	**0.74**	**0.87**	0.15	**0.55**	**0.71**	**0.57**	0.19	**1.00**			
Fat	**0.98**	**0.97**	**0.93**	**0.89**	**0.98**	**0.98**	0.43	**0.62**	**0.59**	**−0.83**	**−0.92**	−0.23	**−0.61**	**−0.81**	**−0.58**	−0.42	**−0.90**	**1.00**		
Moist	**−0.94**	**−0.95**	**−0.89**	**−0.89**	**−0.94**	**−0.94**	−0.37	**−0.70**	**−0.51**	**0.83**	**0.86**	0.29	**0.66**	**0.82**	**0.65**	0.44	**0.86**	**−0.96**	**1.00**	
Coll	**0.80**	**0.88**	**0.79**	**0.74**	**0.81**	**0.81**	0.25	**0.63**	0.49	**−0.61**	**−0.71**	−0.13	**−0.52**	**−0.62**	**−0.58**	0.01	**−0.94**	**0.76**	**−0.73**	**1.00**

Values in bold are significant at $p < 0.05$, Adhe—adhesiveness, Cohe—cohesiveness, Spri—springiness, Gumm-gumminess, Chew—chewiness, Resi—resilience, Prot—protein, Mois—moisture, Coll—collagen content, L^*—meat brightness, a^*—meat redness, b^*—meat yellowness, h-meat hue; C—meat chroma; BI—meat browning index; YI—meat yellowing index; ΔE—meat total color difference.

The color parameters L^*, a^*, b^*, h, BI, YI, and ΔE exhibited some significant correlations ($p < 0.05$) with the texture parameters such as adhesiveness, cohesiveness, springiness, gumminess, and chewiness (Table 8).

The chemical composition showed significant correlations ($p < 0.05$) with the texture parameters, except for resilience, and with the color parameters, except for C and a^* which were not correlated with collagen content. In addition, some significant relationships ($p < 0.05$) can be observed between the chemical components of the groups (Table 8).

4. Discussion

No research appears to have been reported on the effect of inactive *Rhodotorula mucilaginosa* biomass supplementation on broilers' carcass yield and meat quality. Furthermore, our study's aim could be considered as a novel applied investigation subject, providing new insights on valuable microbial pigment additives usage in poultry nutrition.

In this study, the yeasts dietary supplementation affected the gizzard weight and the abdominal fat deposits of broiler chicks. Gizzard weight was significantly influenced by the highest level of SC supplementation (1.3 g/kg feed), when compared with all experimental groups. Lower abdominal fat deposits were observed by increasing the level of SC dietary supplementation (up to 1.3 g/kg diet). Similar to our findings, Paryad's results [17] suggest that supplementing different levels of SC dietary supplements significantly affects the broiler chicks' gizzard weights and abdominal fat accumulations. Relying on our previous results [18], a possible explanation could be attributed to the SC growth promotive effect, having a direct correlation between the final live body weight and the abdominal fat deposits, rather than lipogenic modulative effects, having no correlation with the broilers' serum triglycerides nor with their serum cholesterol.

In our study, pH was not influenced by the inactive yeasts dietary supplementation. The raw broiler meat pH mean values ranged for raw breast (5.97–6.02) and thigh (5.91–6.02) and fitted in the 24 h raw refrigerated broiler's meat normal range of values [31,32] without abnormality encounters such as pale soft exudative and wooden-like meat characteristics. Generally, the meat pH is an important quality indicator, firmly associated with slaughtering, meat processing, and storage management. In addition, meat pH is strongly correlated with color and appearance attributes [19], indicating the main meat quality characteristics and abnormalities. It is known that darker breast meat is related to high pH values [20,32].

Color is an important meat attribute since it is directly influencing the consumer's perception, often associated with meat freshness [33] and rearing nutritional conditions [34]. Results obtained on this work suggest that dietary supplementation of SC increases the meat brightness and total color differences ($p < 0.05$), causing the meat's visual appearance to shift to a paler nuance, when compared with the control group. In accordance with our findings, Hou et al. [35] found that yeast dietary supplementation increases the broiler meat lightness (L^*). However, a recent study [36] reports that live *Saccharomyces sp.* supplementation decreases effects on the lightness (L^*) and yellowness (b^*) attributes, and increases the broiler chickens meat redness (a^*). Moreover, red yeast (RM) dietary supplementation had a major influence on the broiler meat color parameters. Previous studies confirm our results [37,38], indicating significant effects of red yeast administration on broiler's meat brightness (L^*) and pronounced redness nuance (a^*). A brief explanation for the opposite effect and the color variations in this study concerning the RM dietary supplement on the broiler breast and thigh meat brightness could be attributed to the RM carotenoid complex, mainly constituted by the xanthophylls (torularhodin) and less carotenoids (β-carotene) [39], which has a greater pigmentation capacity [40], thus, directing the broiler carotenoid metabolism towards retinol accumulation via β-carotene conversion and extending the flesh and skin xanthophylls deposits [35]. The broiler meat color was significantly influenced by both SC and RM supplementation, presenting an interaction ($p < 0.05$) with the thigh meat L^*, a^*, b^*, C, BI, YI, and ΔE. The broiler thigh meat lightness, redness, chroma, and total color values were increased, while the yellowness, yellow index, and browning index were decreased, when comparing all experimental groups.

Furthermore, the raw broiler meat is strongly correlated with physical-chemical characteristics and meat functional properties, depending on the post-slaughtering treatments [9] and sarcoplasmic myoglobin content [40,41].

The chemical proximate analysis (total protein, fat, moisture, and collagen) was significantly influenced by the dietary supplementation. The broiler breast and thigh meat moisture were significantly higher on the experimental groups supplemented with SC (0, 0.6, and 1 g/kg feed). The highest values of moisture and fat content could be attributed to the growth promoting effect of SC and correlated with the final live body weight. The lowest fat content was observed in the 1.3 g/kg diet supplemented group, indicating that a higher supplementation dosage of SC might not only influence the final body weight but also the meat fat content. It is postulated that higher levels of SC administration could negatively influence the growth performance and the meat quality, due to the high levels of nucleic acids, disturbing the protein metabolism.

Supplementing the experimental broiler diets with high protein sources (SC) significantly decreased the meat protein content, when compared with the negative control group ($p < 0.05$). Opposite to our findings, Hou et al.'s [35] results indicate that the dietary supplementation of live SC could positively modulate the meat protein content. A possible explanation could be connected with the SC amino acids content, which has higher levels of lysine, leucine, and methionine [42], and lower cystine levels [43], when compared with the soybean meal. In broiler diets, the most important and expensive component is the protein source, and it is known that the ideal vegetal resource is soybean meal, having all essential amino acids, especially those that birds need the most: sulphuric amino acids lysine, methionine and cystine [44].

Meat collagen is an important attribute for broiler meat quality and it is the main abundant protein in the broiler chicks' connecting tissues and body [45], and directly contributes to the meat tenderness and gumminess [46]. In our study, the results show that dietary SC supplementation (0.6, 1, and 1.3 g/kg diet) on broiler chicks' diets significantly increases the meat collagen, when compared with the negative control group. Lysine, glycine, and proline are the major amino acid constituents in the collagen [47]. We strongly believe that the SC dietary supplementation influenced the meat collagen synthesis, due to the supplementary levels of lysine and proline, enhancing the hydroxyproline and hydroxylysine accumulation, thus the collagen formation.

The RM dietary supplementation had significantly increased the broiler breast meat moisture and fat content, when compared with the negative and positive control groups ($p < 0.05$). There is a deficit of information regarding the chemical proximate composition of the broiler chicks' diet supplemented with inactive RM, and indeed, with lyophilizate lysates pigment yeasts. To the best of our knowledge, this is the first study about the effects of dietary RM supplementation on broiler chicks' meat quality attributes and carcass yields. Therefore, future studies are needed to determine the effects of inactive yeasts dietary supplementation on broiler metabolism, thus the meat proximate chemical composition.

Meat textural profile analysis is generally associated with meat processing proprieties [48] and consumer eating satisfaction [26]. This current study revealed similar textural properties ($p > 0.05$) for all broilers' breast and thigh meat samples. All the texture parameters were highly correlated with each other ($0.83 < r < 0.99$, $p < 0.05$), except for resilience which was correlated significantly only with cohesiveness, gumminess, and chewiness ($0.52 < r < 0.66$, $p < 0.05$). Moreover, meat textural profile analysis and sensorial correlation could not be appropriate for evaluation, considering the fact that the mechanical force applied differs and could not be an exact replica of the consumer's compressive experience.

Significant differences were recorded for broilers' raw breast meat fat appearance, consistency, juiciness, elasticity, and tenderness. The opposite trend was observed for fat appearance and juiciness meat attributes, that were significantly decreased when compared with the control groups. In discordance with our results, recent investigations [27,28] indicate that live yeast culture (*SC*) supplementation significantly increased the sensorial broiler tenderness and juiciness and odor meat attributes. The meat fat appearance score was decreased, when compared with the negative control group. The meat fat appearance was influenced by the RM dietary supplement, with the positive control group in thigh meat samples having the lowest score. On combining these results with the RM significant effect on the color parameter a^* (responsible for red–green spectra), it could be concluded that the meat's fat appearance was influenced, inducing a darker nuance. It is very likely that xanthophylls present in the RM dietary supplement were very quickly metabolized, and deposited in the body fat, due to the fact that carotenoids mainly conjugate with fatty acids and esters, forming chylomicrons [49,50].

The meat tenderness, elasticity, and consistency are directly related to the meat collagen content. The dietary yeasts supplemented groups were more appreciated in the consumer's preference score than in the control groups, indicating a positive modulative effect of yeasts dietary supplementation on the consumer's perception. Similar to our findings, Ma et al. [51] show that supplementing SC could modulate the broiler meat elasticity, gumminess, and tenderness. In meat sensorial attributes, juiciness is generally correlated with the meat moisture content, but also could be influenced by the meat cutting processes. In our study, the juiciest meat score was obtained from the negative control group (1), with the interpretation as least juicy from the experimental group who received inactive yeast supplements (2–8). In discordance with our results, there could be a positive effect of administrating inactive SC on broiler cooked meat quality, due to the fact that SC possesses 5′-guanosine monophosphate and 5′-inosine monophosphate enzymatic nucleases [52], that might be appropriate as flavor enhancers, featuring the supposed umami taste.

5. Conclusions

This current paper indicated that inactive yeasts SC and RM are valuable nutritional supplements for broilers, in physical and mechanical meat attributes. In addition, bivalent yeast supplementation could synergically enhance meat quality attributes and might positively modulate the consumer's preference, increasing meat moisture, lightness, redness, and decreasing the browning index. Furthermore, based on the sensorial meat texture, elasticity, consistency attributes, and collagen content we validate our dietary yeasts supplements as possible alternatives for feed additives, having both growth promoting and product quality enhancing benefits. Further investigations are required in order to elucidate the effects of microbial pigment supplements on broiler meat biochemistry and pigment metabolism.

Author Contributions: Conceptualization, D.-M.G., M.U.-I. and S.M.; methodology, D.-M.G.; software, M.U.-I., A.B. and D.-M.G.; validation, G.C., S.M. and N.E.B.; formal analysis, D.-M.G. and A.B.; investigation, D.-M.G.; resources, M.U.-I.; data curation, M.U.-I. and D.-M.G.; writing—original draft preparation, D.-M.G. and M.U.-I.; writing—review and editing, D.-M.G. and M.U.-I.; visualization, G.C., S.M. and N.E.B.; supervision, N.E.B., S.M. and G.C.; project administration, funding acquisition, N.E.B. and S.M. All authors have read and agreed to the published version of the manuscript.

Funding: This work was funded by the Ministry of Research, Innovation and Digitalization within Program 1—Development of national research and development system, Subprogram 1.2—Institutional Performance—RDI excellence funding projects, under contract no. 10PFE/2021.

Institutional Review Board Statement: The Animal Ethics Committee approved the study protocol of the National Re-search and Development Institute for Animal Biology and Nutrition (INCDBNA-IBNA) Baloteşti, Romania, under the EU Directive 2010/63/EU and Roma-nian Law on Animal Protection (no. 199/2018).

Informed Consent Statement: Not applicable.

Data Availability Statement: No new data were created or analyzed in this study. Data sharing is not applicable to this article.

Acknowledgments: This work was funded by the Ministry of Research, Innovation and Digitalization within Program 1—Development of national research and development system, Subprogram 1.2—Institutional Performance—RDI excellence funding projects, under contract no. 10PFE/2021.

Conflicts of Interest: The authors declare no conflict of interest.

References

1. Rodriguez-Concepcion, M.; Avalos, J.; Bonet, M.L.; Boronat, A.; Gomez-Gomez, L.; Hornero-Mendez, D.; Limon, M.C.; Meléndez-Martínez, A.J.; Olmedilla-Alonso, B.; Palou, A.; et al. A Global Perspective on Carotenoids: Metabolism, Biotechnology, and Benefits for Nutrition and Health. *Prog. Lipid Res.* **2018**, *70*, 62–93. [CrossRef] [PubMed]
2. Dal Bosco, A.; Mattioli, S.; Cartoni Mancinelli, A.; Cotozzolo, E.; Castellini, C. Extensive Rearing Systems in Poultry Production: The Right Chicken for the Right Farming System. A Review of Twenty Years of Scientific Research in Perugia University, Italy. *Animals* **2021**, *11*, 1281. [CrossRef]
3. Guarino Amato, M.; Castellini, C. Adaptability Challenges for Organic Broiler Chickens: A Commentary. *Animals* **2022**, *12*, 1354. [CrossRef]
4. Van Horne, P.L.M. *Competitiveness of the EU Poultry Meat Sector, Base Year 2017: International Comparison of Production Costs*; Wageningen Economic Research: Den Haag, The Netherlands, 2018; ISBN 978-94-6343-387-7.
5. Jacob, F.F.; Hutzler, M.; Methner, F.-J. Comparison of Various Industrially Applicable Disruption Methods to Produce Yeast Extract Using Spent Yeast from Top-Fermenting Beer Production: Influence on Amino Acid and Protein Content. *Eur. Food Res. Technol.* **2019**, *245*, 95–109. [CrossRef]
6. Vieira, E.F.; Carvalho, J.; Pinto, E.; Cunha, S.; Almeida, A.A.; Ferreira, I.M.P.L.V.O. Nutritive Value, Antioxidant Activity and Phenolic Compounds Profile of Brewer's Spent Yeast Extract. *J. Food Compos. Anal.* **2016**, *52*, 44–51. [CrossRef]
7. Fărcaş, A.C.; Socaci, S.A.; Mudura, E.; Dulf, F.V.; Vodnar, D.C.; Tofană, M.; Salanţă, L.C. *Exploitation of Brewing Industry Wastes to Produce Functional Ingredients*; Kanauchi, M., Ed.; IntechOpen: Rijeka, Croatia, 2017; Chapter 7; ISBN 978-953-51-3342-1.
8. Podpora, B.; Swiderski, F.; Sadowska, A.; Rakowska, R.; Wasiak-Zys, G. Spent Brewer's Yeast Extracts as a New Component of Functional Food. *Czech J. Food Sci.* **2016**, *34*, 554–563. [CrossRef]
9. Parapouli, M.; Vasileiadis, A.; Afendra, A.S.; Hatziloukas, E. Saccharomyces Cerevisiae and Its Industrial Applications. *AIMS Microbiol.* **2020**, *6*, 1–31. [CrossRef]
10. Liu, X.; Li, Y.; Zhang, R.; Huangfu, L.; Du, G.; Xiang, Q. Inactivation Effects and Mechanisms of Plasma-Activated Water Combined with Sodium Laureth Sulfate (SLES) against Saccharomyces Cerevisiae. *Appl. Microbiol. Biotechnol.* **2021**, *105*, 2855–2865. [CrossRef] [PubMed]
11. Chuang, W.Y.; Hsieh, Y.C.; Lee, T.T. The Effects of Fungal Feed Additives in Animals: A Review. *Animals* **2020**, *10*, 805. [CrossRef]
12. Rana, B.; Bhattacharyya, M.; Patni, B.; Arya, M.; Joshi, G.K. The Realm of Microbial Pigments in the Food Color Market. *Front. Sustain. Food Syst.* **2021**, *5*, 603892. [CrossRef]
13. Mussagy, C.U.; Khan, S.; Kot, A.M. Current Developments on the Application of Microbial Carotenoids as an Alternative to Synthetic Pigments. *Crit. Rev. Food Sci. Nutr.* **2022**, *62*, 6932–6946. [CrossRef] [PubMed]
14. Aman Mohammadi, M.; Ahangari, H.; Mousazadeh, S.; Hosseini, S.M.; Dufossé, L. Microbial Pigments as an Alternative to Synthetic Dyes and Food Additives: A Brief Review of Recent Studies. *Bioprocess Biosyst. Eng.* **2022**, *45*, 1–12. [CrossRef] [PubMed]
15. Kot, A.M.; Błażejak, S.; Kieliszek, M.; Gientka, I.; Bryś, J. Simultaneous Production of Lipids and Carotenoids by the Red Yeast Rhodotorula from Waste Glycerol Fraction and Potato Wastewater. *Appl. Biochem. Biotechnol.* **2019**, *189*, 589–607. [CrossRef]
16. Sibirny, A.A. *Biotechnology of Yeasts and Filamentous Fungi*; Springer: Cham, Switzerland, 2018; ISBN 9783319588292.
17. Paryad, A.; Mahmoudi, M. Effect of Different Levels of Supplemental Yeast (Saccharomyces Cerevisiae) on Performance, Blood Constituents and Carcass Characteristics of Broiler Chicks. *Afr. J. Agric. Res* **2008**, *3*, 835–842.
18. Grigore, D.M.; Ciurescu, G.; Radu, N.; Babeanu, N. Health Status, Performance and Carcass Caracteristics of Broiler Chicks Supplemented With Yeasts Bioproducts. *Anim. Sci.* **2022**, *65*, 140–149.
19. Sokoya, O.O.; Babajide, J.M.; Shittu, T.A.; Sanwo, K.A.; Adegbite, J.A. Chemical and Color Characterization of Breast Meat from FUNAAB Indigenous and Marshal Broiler Chickens. *Trop. Anim. Health Prod.* **2019**, *51*, 2575–2582. [CrossRef]
20. Karaoğlu, M.; Aksu, M.I.; Esenbuga, N.; Macit, M.; Durdağ, H. PH and Colour Characteristics of Carcasses of Broilers Fed with Dietary Probiotics and Slaughtered at Different Ages. *Asian-Australas. J. Anim. Sci.* **2006**, *19*, 605–610. [CrossRef]
21. Kanwugu, O.N.; Ranga Rao, A.; Ravishankar, G.A.; Glukhareva, T.V.; Kovaleva, E.G. Chapter 31—Astaxanthin from Bacteria as a Feed Supplement for Animals. In *Global Perspectives on Astaxanthin*; Ravishankar, G.A., Ranga Rao, A., Eds.; Academic Press: Cambridge, MA, USA, 2021; pp. 647–667. ISBN 978-0-12-823304-7.
22. Sun, J.; Li, M.; Tang, Z.; Zhang, X.; Chen, J.; Sun, Z. Effects of Rhodotorula Mucilaginosa Fermentation Product on the Laying Performance, Egg Quality, Jejunal Mucosal Morphology and Intestinal Microbiota of Hens. *J. Appl. Microbiol.* **2020**, *128*, 54–64. [CrossRef]
23. He, C.; Geng, H.; Qin, Y.; Yang, P.; Wang, W.; Mai, K.; Song, F. Dietary Xanthophyll Improved Growth, Antioxidant, Pigmentation and Meat Quality in the Southern Catfish (Silurus Soldatovi Meridionalis Chen). *Anim. Nutr.* **2022**, in press. [CrossRef]
24. Hu, P.; Mao, J.; Zeng, Y.; Sun, Z.; Deng, H.; Chen, C.; Sun, W.; Tang, Z. Isolation, Identification, and Function of Rhodotorula Mucilaginosa TZR2014 and Its Effects on the Growth and Health of Weaned Piglets. *Front. Microbiol.* **2022**, *13*, 922136. [CrossRef]

25. De Mello, J.L.M.; de Souza, R.A.; Ferrari, F.B.; Cavalcanti, E.N.F.; de Oliveira, R.F.; de Almeida Fidelis, H.; Pereira, M.R.; Villegas-Cayllahua, E.A.; Giampietro-Ganeco, A.; Dutra, D.R.; et al. Quality of Breast Meat from Broiler Chickens Raised in Brazil Affected by White Striping Myopathy. *Res. Soc. Dev.* **2021**, *10*, e42210212637. [CrossRef]
26. Paredes, J.; Cortizo-Lacalle, D.; Imaz, A.M.; Aldazabal, J.; Vila, M. Application of Texture Analysis Methods for the Characterization of Cultured Meat. *Sci. Rep.* **2022**, *12*, 3898. [CrossRef] [PubMed]
27. Abdel-Raheem, S.M.; Abd-Allah, S.M.S. The Effect of Single or Combined Dietary Supplementation of Mannan Oligosacharide and Probiotics on Performance and Slaughter Characteristics of Broilers. *Int. J. Poult. Sci.* **2011**, *10*, 854–862. [CrossRef]
28. Yasar, S.; Boselli, E.; Rossetti, F.; Gok, M.S. Effect of Fermented Cereals, Probiotics, and Phytase on the Sensory Quality of Poultry Meat. *Sci. Agric. Bohem.* **2018**, *49*, 225–235. [CrossRef]
29. Ciurescu, G.; Idriceanu, L.; Gheorghe, A.; Ropotă, M.; Drăghici, R. Meat Quality in Broiler Chickens Fed on Cowpea (Vigna Unguiculata [L.] Walp) Seeds. *Sci. Rep.* **2022**, *12*, 9685. [CrossRef] [PubMed]
30. Grigore, D.-M.; Ciurescu, G.; Mironeasa, S.; Iuga, M.; Băbeanu, N. Effect of Low-Fiber Sunflower Meal and Phytase Addition on Broiler Carcass Traits, and Meat Quality. *Arch. Zooteh.* **2019**, *22*, 32–47.
31. Fernandes, R.T.V.; De Arruda, A.M.V.; Melo, A.D.S.; Marinho, J.B.M.; Fernandes, R.T.V.; De Figueiredo, L.C. Chemical Composition and PH of the Meat of Broilers Submitted to Pre-Slaughter Heat Stress. *J. Anim. Behav. Biometeorol.* **2016**, *4*, 93–95. [CrossRef]
32. Fletcher, D.L.; Qiao, M.; Smith, D.P. The Relationship of Raw Broiler Breast Meat Color and PH to Cooked Meat Color and PH. *Poult. Sci.* **2000**, *79*, 784–788. [CrossRef]
33. De Araújo, P.D.; Araújo, W.M.C.; Patarata, L.; Fraqueza, M.J. Understanding the Main Factors That Influence Consumer Quality Perception and Attitude towards Meat and Processed Meat Products. *Meat Sci.* **2022**, *193*, 108952. [CrossRef]
34. Baéza, E.; Guillier, L.; Petracci, M. Review: Production Factors Affecting Poultry Carcass and Meat Quality Attributes. *Animal* **2022**, *16*, 100331. [CrossRef]
35. Hou, L.; Qiu, H.; Sun, P.; Zhu, L.; Chen, F.; Qin, S. Selenium-Enriched Saccharomyces Cerevisiae Improves the Meat Quality of Broiler Chickens via Activation of the Glutathione and Thioredoxin Systems. *Poult. Sci.* **2020**, *99*, 6045–6054. [CrossRef] [PubMed]
36. Koltovaya, N.A. DNA Damage-Independent Cell Cycle Stop in Saccharomyces Cerevisiae Yeast. *Radiat. Biol. Radioecol.* **2018**, *58*, 5–14. [CrossRef]
37. Ao, X.; Kim, I.H. Effects of Astaxanthin Produced by Phaffia Rhodozyma on Growth Performance, Antioxidant Activities, and Meat Quality in Pekin Ducks. *Poult. Sci.* **2019**, *98*, 4954–4960. [CrossRef] [PubMed]
38. Choe, J.; Kim, J.; Kim, B.; Park, H.; Kim, G.; Kim, H. Physicochemical Properties of Chicken Thigh Meat Sausage Manufactured with Red Yeast Rice Powder. *J. Korean Poult. Soc.* **2019**, *46*, 46–50.
39. Poorniammal, R.; Prabhu, S.; Dufossé, L.; Kannan, J. Safety Evaluation of Fungal Pigments for Food Applications. *J. Fungi* **2021**, *7*, 692. [CrossRef] [PubMed]
40. Marounek, M.; Pebriansyah, A. Use of Carotenoids in Feed Mixtures for Poultry: A Review. *Agric. Trop. Subtrop.* **2018**, *51*, 107–111. [CrossRef]
41. Meléndez-Martínez, A.J.; Mandić, A.I.; Bantis, F.; Böhm, V.; Borge, G.I.A.; Brnčić, M.; Bysted, A.; Cano, M.P.; Dias, M.G.; Elgersma, A.; et al. A Comprehensive Review on Carotenoids in Foods and Feeds: Status Quo, Applications, Patents, and Research Needs. *Crit. Rev. Food Sci. Nutr.* **2022**, *62*, 1999–2049. [CrossRef] [PubMed]
42. Satyanarayana, T.; Kunze, G. *Yeast Biotechnology: Diversity and Applications*; Springer: Dordrecht, The Netherlands, 2010; 744p. [CrossRef]
43. Puerta, M.L.; Shukla, V.; Dalle Carbonare, L.; Weits, D.A.; Perata, P.; Licausi, F.; Giuntoli, B. A Ratiometric Sensor Based on Plant N-Terminal Degrons Able to Report Oxygen Dynamics in Saccharomyces Cerevisiae. *J. Mol. Biol.* **2019**, *431*, 2810–2820. [CrossRef] [PubMed]
44. Nalunga, A.; Komakech, A.J.; Jjagwe, J.; Magala, H.; Lederer, J. Growth Characteristics and Meat Quality of Broiler Chickens Fed Earthworm Meal from Eudrilus Eugeniae as a Protein Source. *Livest. Sci.* **2021**, *245*, 104394. [CrossRef]
45. Ismail, I.; Joo, S.-T. Poultry Meat Quality in Relation to Muscle Growth and Muscle Fiber Characteristics. *Korean J. Food Sci. Anim. Resour.* **2017**, *37*, 873–883. [CrossRef]
46. Shin, D.-J.; Lee, H.J.; Lee, D.; Jo, C.; Choe, J. Fat Replacement in Chicken Sausages Manufactured with Broiler and Old Laying Hens by Different Vegetable Oils. *Poult. Sci.* **2020**, *99*, 2811–2818. [CrossRef] [PubMed]
47. Nusairat, B.; Tellez-Isaias, G.; Qudsieh, R. *An Overview of Poultry Meat Quality and Myopathies*; Tellez-Isaias, G., Latorre, J.D., Martínez-Aguilar, Y., Eds.; IntechOpen: Rijeka, Croatia, 2022; Chapter 5; ISBN 978-1-80355-328-3.
48. Godschalk-Broers, L.; Sala, G.; Scholten, E. Meat Analogues: Relating Structure to Texture and Sensory Perception. *Foods* **2022**, *11*, 2227. [CrossRef] [PubMed]
49. Von Lintig, J.; Moon, J.; Lee, J.; Ramkumar, S. Carotenoid Metabolism at the Intestinal Barrier. *Biochim. Biophys. Acta Mol. Cell Biol. Lipids* **2020**, *1865*, 158580. [CrossRef] [PubMed]
50. Surai, P.F.; Kochish, I.I. *Carotenoids in Aviculture BT—Pigments from Microalgae Handbook*; Jacob-Lopes, E., Queiroz, M.I., Zepka, L.Q., Eds.; Springer: Cham, Switzerland, 2020; pp. 515–540. ISBN 978-3-030-50971-2.

51. Ma, K.K.; Nolden, A.A.; Kinchla, A.J. *Sensory Qualities of Plant Protein Foods BT—Plant Protein Foods*; Manickavasagan, A., Lim, L.-T., Ali, A., Eds.; Springer: Cham, Switzerland, 2022; pp. 313–342. ISBN 978-3-030-91206-2.
52. Ienczak, J.L.; de Oliveira Pereira, I.; da Silveira, J.M. *Utilization of Saccharomyces Cerevisiae as a Source of Natural Food Additives BT—Natural Additives in Foods*; Valencia, G.A., Ed.; Springer: Cham, Switzerland, 2023; pp. 185–214. ISBN 978-3-031-17346-2.

applied sciences

MDPI

Article

Nutritive Profile of Canned Goat Meat Food with Added Carrot

Urishbay Chomanov, Gulmira Kenenbay, Alibek Tursynov, Torgyn Zhumalieva, Nurzhan Tultabayev and Anuarbek Suychinov *

Kazakh Research Institute of Processing and Food Industry, Almaty 050060, Kazakhstan
* Correspondence: asuychinov@gmail.com or a.suichinov@rpf.kz

Abstract: This study describes the effect of adding carrot on the nutritional and biological value of canned goat meat. Four batches of canned goat meat were produced: control (without carrot), and three experimental treatments T10, T20, and T30 that was canned goat meat with the addition of 10, 20, and 30% carrot, respectively. Canned goat meat with 30% added carrot had the lowest fat (5.76%) and protein (17.08%) content. The mass fraction of fiber was significantly increased, up to 1.96% in T10, 2.33% in T20, and 2.71% in T30. The same trend was observed for β-carotene content (from 0.78 mg/100 g in the control sample to 1.91 mg/100 g in T30). Among the amino acids, histidine was significantly increased in T30 ($p < 0.05$), and lysine in T20 and T30 ($p < 0.05$). There was also evidence of increased vitamin A, B_6 (pyridoxine), B_9 (folic acid), and B_5 (pantothenic acid) content in experimental samples than in the control sample. Among the minerals, the most significant increase was in potassium, magnesium, and phosphorus levels in samples with carrot. Organoleptic evaluation showed differences in color, taste, consistency, and odor. Thus, the addition of carrots increases the nutritive value and improves the palatability of canned goat meat.

Keywords: goat meat; carrots; nutritional value; carotene; canned food

1. Introduction

Meat products are considered one of the most important components of the human diet. The global production of meat products is increasing, which is not only associated with an increasing world population, but also with an increasing level of consumption of meat products in developed and developing countries. In Asia, for example, meat consumption has increased 15 times over the last 50 years, reaching almost half of the global meat consumption [1]. Experts predict that if dietary habits in developing countries continue to evolve as in richer countries, global meat consumption will increase dramatically in this century [1,2].

Meat products are rich in nutrients and vitamins, which are essential for the human body. Despite the popular concepts of healthy eating and veganism, meat is still the basis of nutrition. It contains a high concentration of essential minerals, trace elements, and amino acids [3,4]. Among the variety of types of meat products, canned meat is in high demand. Canned meat is a meat product that has been hermetically sealed in tin or glass jars. It is exposed to high temperatures for destruction of pathogenic microflora to be made storage-friendly. Canned meat is characterized by its high nutritional value, long shelf life, and easy transportation [5,6]. In addition to traditional types of meat (beef, pork) used in the production of canned meat products, horse meat, mutton, and goat meat are widely used.

Goat breeding is widespread in Central Asian countries. Goats were domesticated by humans a long time ago; their milk, hide, and meat have been used since ancient times. Goats are unevenly distributed in Kazakhstan. Their highest numbers are in the southern regions of Kazakhstan, where a large part of the year has warm temperatures, allowing year-round maintenance of animals on pastures [7,8].

Goat meat has a positive perception compared with other meats in terms of health. The low cholesterol content of goat meat is one of its many health benefits. This makes it an ideal food for dietary or therapeutic purposes. In addition, goat meat contains iron and various antioxidants [9]. Consumption of this product helps to reduce cholesterol and supplies the body with important nutrients [10]. Goat meat is easily digestible and absorbed by the body. One of the main advantages of goat meat is hypoallergenicity; it can even be used in the production of baby food [11].

Goat meat is considered a delicacy in most countries. The chemical composition of goat meat is as follows: 74–76% moisture, 20–22% protein, 0.6–2.6% fat, and 1% ash [12]. Goat meat is traditionally considered a lean red meat with dietary properties, characterized by a slightly darker red color, a coarser texture, and a different taste and aroma from mutton [13,14]. The results of sensory studies show that goat meat is different, but certainly not inferior to mutton [15]. Goat meat is a valuable source of essential amino acids and minerals (potassium, zinc, iron, etc.). Goat meat and its meat products also tend to be less succulent than lamb, mainly due to their lower fat content [16,17].

These days, an issue of technology improvement and expansion of the assortment of meat products receives special urgency. Developing meat product recipes that involves replacing a part of the raw animal material with vegetable additives is a promising way to solve the problem of increasing the availability of meat products, expanding their assortment, and increasing their nutritional value.

It is known that one of the most effective methods of normalizing the nutritional status of human beings is the use of protein from raw vegetable material in the diet, which makes it possible to enrich food with necessary micronutrients and compensate for a lack of animal protein. Enrichment of meat products with useful nutrients from raw plant material helps to compensate for the deficit of vitamins, minerals, dietary fibers, and other useful substances in the human body. This could reduce the presence of nutrient deficiencies in the diet of the population [18,19].

Carrot (*Daucus carota subsp. Sativus*) is a vegetable whose root is widely used in the human diet due to its global distribution, nutritional value, health benefits, and technological properties. It is a rich source of vitamin A (beta-carotene) and dietary fiber [20]. Beta-carotene is correlated with improved cognitive health because of its antioxidant effects. Beta-carotene is also potentially useful for treating neurodegenerative diseases such as Alzheimer's disease [21].

Carrots help reduce cholesterol level and risk of cancer and increase immune function. Carrots also support the immune system with antiseptic and antibacterial properties that help the body's immune reactions to infections and disease [22]. In addition, vitamin A works together with vitamin C, another powerful antioxidant, to boost immunity and prevent cell damage caused by free radicals. Carrots are rich in potassium. Potassium helps lower blood pressure through its interaction with kidney and sodium [23]. Regular consumption of carrots may significantly reduce the risk of cardiovascular disease. It contains the antioxidant luteolin, which prevents arterial inflammation caused by cholesterol clots [24].

Therefore, the combination of raw animal and plant material makes it possible to supplement them with biologically active substances and give the final food products functional, therapeutic, and prophylactic properties [25,26]. The aim of this study is to evaluate the effect that adding carrot has on the nutritional and biological value of canned goat meat.

2. Materials and Methods

2.1. Materials

Raw meat: the butchering of 35 goats of the Zaanen breed (age: 8 months, weight of carcass: 11.58 ± 1.24 kg) was carried out in a slaughter shop of FE called "Aigerim" (South Kazakhstan region, the Republic of Kazakhstan). Carcasses were kept for 24 h at −18 °C, subjected to deboning, roasting, and subsequent technological processing. Cuts of meat samples were from the loin side. The back part of the goat carcass was cut lengthwise along the vertebrae into two loin parts (ham). Three basic techniques were used for deboning the hind leg: the pelvic bone was separated, then the tibia and femur bones. The flesh of the loin mass was cut into small pieces for grinding in the meat grinder.

Raw vegetable material: carrots, onions, beans, salt, and spices (black pepper) were purchased from the supermarket network called "Magnum" (Republic of Kazakhstan, Almaty).

2.2. Preparation of Control and Experimental Canned Meat

Four batches of canned meat were prepared: one control and three experimental batches, with the addition of carrots from 10 to 30% instead of goat meat (Table 1). Each batch consisted of 12 cans (Figure 1). The weight of one can with product before autoclaving was 350 g.

(a) Hand canning

(b) Ready canned goat meat

(c) packaged canned food

Figure 1. Canned goat meat.

Table 1. Composition of canned meat with different carrot content (g/kg).

Ingredients	Control	Treatment		
		T10	T20	T30
Goat meat	846	761.4	676.8	592.2
Carrots	0	84.6	169.2	253.8
White beans	132.58	132.58	132.58	132.58
Onion	9.22	9.22	9.22	9.22
Black pepper	0.2	0.2	0.2	0.2
Table salt	12	12	12	12
Total	1000	1000	1000	1000

The process of canned meat production was carried out according to the patent [27]. To produce canned meat, the goat meat was cut using Cutter ZB-40 (Hualian Machinery, China); the size of the pieces after cutting was 10–15 mm. All raw vegetable material were cleaned and washed. Defective or rotten roots were removed. Onions were cut by a cutter into pieces 5 mm in size, carrots were sliced by a Torgmash OM-350-01P vegetable slicer (Torgmash Plant, Belarus) into pieces 10–12 mm in size. Sliced vegetables were sauteed at 105 °C in vegetable oil. The beans were half-cooked.

For canning, a manual canning machine MZ.04 (Forcom LLC, Moscow, Russia) was used. The canned food was sterilized using a Malysh Nerzh autoclave (Forcom LLC, Moscow, Russia) under the following regime: heating for 20 min, sterilization for 60 min at a pressure of 0.25 MPa and sterilization temperature of 115 °C, and gradually decreasing the pressure and temperature down to 35 °C for 30 min. All ingredients were weighed according to the recipe presented in Table 1.

2.3. Determination of the Chemical Composition

The moisture content was determined on an MX-50 moisture meter (A&D Co., LTD, Tokyo, Japan). All samples for moisture content determination were weighed at 5 g each and evenly distributed inside the device cup.

Moisture measurement was based on the method described in [28]. After determining moisture content, each dried sample was used for fat determination according to [29]. The samples were calcined in a muffle furnace (500–600 °C) to measure the ash content [28]. The protein content was analyzed according to [30].

2.4. Calculation of Energy Value of Canned Food

The energy value of canned food was determined by a formula based on the values of protein, fat, and carbohydrates and their caloric index in the finished product according to the method described in [4]:

$$CV = 4(P + C) + 9F \quad (1)$$

where CV—caloric value, kCal; P—protein content, g; F—fat content, g; C—carbohydrate content, g; 4—caloric index for protein and carbohydrate; and 9—caloric index for fat.

2.5. Determination of β-Carotene Content

Our method was based on the extraction of carotenoids from precipitate, previously obtained by treating the sample with Carrez I and Carrez II solutions, followed by purification of the isolated substance with petroleum ether and spectrophotometric determination of the mass concentration of β-carotene [31].

The obtained extract was used for spectrometric determination of total carotenoids (spectrophotometer SF-56 (OKB Spectr, Sankt-Peterburg, Russia), 450 nm, optical glass cuvettes, optical path length 1 cm). Petroleum ether was used as a comparative solution.

The mass concentration of carotenoids C, mg/dm3, is calculated by Formula (2):

$$C = 4.0 \cdot A \cdot F \quad (2)$$

where 4.0—the optical density conversion factor;

A—measured optical density value;

F—dilution factor (ratio of extract volume in petroleum ether, or volumes of different fractions to the volume of sample taken for analysis).

2.6. Determination of Fiber Content

Mass fraction of dietary fiber in the product was determined by the gravimetric method, according to GOST R 54014-2010. The method was based on the enzymatic hydrolysis of starch and non-starch compounds using α-amylase, protease, and amyloglucosidase to mono-, di-, and oligosaccharides and peptides. Dietary fibers were precipitated with ethyl alcohol, dried, and gravimetrically determined. The tests were performed on the Velp FIWE-3 fiber analyzer ("VELP", Usmate, Italy). The total mass fraction of dietary fiber is expressed as a percentage (g/100 g) [32].

2.7. Determination of Salt Content

The amount of salt in the canned food was determined by GOST 9957-2015 [33]. Samples from each batch were ground, then mixed for sample homogeneity. An amount of 5 g of the sample was placed in a beaker and 100 cm^3 of distilled water was added. After 40 min of infusion (with periodic stirring with a glass rod), the aqueous extract was filtered through a filter paper.

After cooling to room temperature, 5 cm^3 of the filtrate was titrated with 0.05 mol/dm^3 silver nitrate solution in the presence of 0.5 cm^3 potassium chromate solution until the color changed to orange.

The mass fraction of sodium chloride X, %, was calculated by Formula (3), where:

$$X = \frac{0.00292 \times K \times v \times 100 \times 100}{v_1 \times m} \quad (3)$$

0.00292—amount of sodium chloride equivalent to 1 cm of 0.05 mol/dm^3 silver nitrate solution, g;

K—correction for the titer of 0.05 mol/dm^3 silver nitrate solution;

v—amount of 0.05 mol/dm^3 silver nitrate solution used for the titration of the test solution, cm^3

v_1—amount of aqueous extract taken for titration, cm^3;

m—weight of the sample, g;

100—volume to which the analyzed sample is diluted, cm^3;

100—conversion coefficient to percent.

2.8. Determination of Amino Acid Composition

The determination of amino acid composition was performed on a SHIMADZU LC-20 Prominence HPLC instrument (Kyoto, Japan) with fluorimetric and spectrophotometric detectors. We used a 25 cm by 4.6 mm SUPELCO C18 chromatographic column, 5 in front of the column to protect the main column from impurities. Chromatographic analysis was performed in eluent gradient mode at a flow rate of 1.2 mL min^{-1} and a column thermostat temperature of 40 °C in reversed phase with fluorimetric and spectrophotometric detectors at 246 and 260 nm, using acid hydrolysis and amino acid modification with phenylisothiocyanate solution in isopropanol to obtain phenylthiohydantoins. A mixture of 6.0 mm of CH3 SONA solution at pH 5.5 (component A), 1% isopropanol in acetonitrile solution (component B), and 6.0 mm of CH3 SONA solution at pH 4.05 (component C) was used as the mobile phase. We used standard amino acid samples produced by Sigma Aldrich (Burghausen, Germany) [4].

2.9. Determination of Mineral Content

The mineral content of the samples was determined according to the method of AOAC (2000) [34]. All samples for determining the mineral content weighed 5 g each, being placed on a container and incinerated in a microwave muffle for 12 h to a final temperature of 600 °C. After microwave splitting, the samples were diluted with a 10 ml solution of hydrochloric acid (HCl) in distilled water (1:1), mixed with a glass rod, and passed through a paper filter. Mineral elements were determined on an atomic emission spectrometer ICP-OES (Spectro, Boschstr, Burghausen, Germany).

2.10. Determination of Vitamin Composition

Vitamin content was determined according to GOST 55482-2013 [35] for water-soluble vitamins using a single-substituted potassium phosphate buffer solution and GOST 32307-2013 for fat-soluble vitamins [36]. The analysis was performed on a Dionex Ultimate 3000 chromotograph ("Dionex", Sunnyvale, CA, USA) with a Supelco SUPELCOSIL LC-PAH 5 µm 4.6 × 150 mm column. The eluent composition was a mixture of methanol and distilled water (92:8).

2.11. Organoleptic Evaluation

Organoleptic evaluation of canned meat was carried out according to the requirements of state standard GOST 33741-2015 [37]. Organoleptic characteristics of canned meat were evaluated by pre-trained and instructed tasters consisting of 7 people from the Kazakh Research Institute of Processing and Food Industry. The appearance of cans was evaluated by external signs of the container, the presence or absence of leakage, deformation and rust, etc. Canned food was heated before tasting; the can was placed in a water bath for 20 min. Tasters compared opinions on the appearance, color, odor, consistency, and taste of each product. Organoleptic indicators of canned products were determined in the following sequence: appearance, color, odor, consistency, and taste. When assessing the appearance of canned products, tasters assessed the degree of chopping, whether shape was maintained by chopped or molded ingredients, the state of the broth, and the presence of foreign impurities. When assessing color, various deviations from the color typical for a given type of canned food were determined. When assessing odor, tasters determined whether the aroma was typical, harmonious, and whether there was a presence of an extraneous odor. When assessing the consistency, tasters looked at tenderness, fibrousness, coarseness, crumbliness, uniformity, chewability, presence of rigid structural components, etc.. When evaluating taste, tasters determined the characteristic taste for this type of canned food and established the presence of specific unfavorable flavors and other extraneous flavors.

2.12. Statistical Analysis

The experiments were carried out in triplicate. Standard deviation values are given for all measurements. Differences between the experimental and control groups were calculated using a one-way ANOVA with Tukey test. $p < 0.05$ was considered as significant.

3. Results and Discussion

3.1. Influence of Carrots on the Chemical Composition of Canned Food

The results of the study of the chemical composition of canned meat under the standard recipe (control) and with the replacement of goat meat with carrots at 10% (T10), 20% (T20), and 30% (T30) are presented in Table 2. The experimental samples significantly differed from the control ($p < 0.05$) in moisture, carbohydrates, ash, carotenoids, and fiber content.

The highest mass fraction of fat was found in the control samples and gradually decreased as the proportion of carrots in the product increased. This confirms the findings of Roccetti et al. who showed a decrease in fat when substituting vegetable ingredients—carrots (*Daucus Carota* L.)—for turkey meat in sausages in amounts of 10 to 30% [38], as well as Corvalho et al., who added spinach to chicken burgers [39]. Roccetti et al. attributed this effect to the low fat content of vegetable ingredients [38].

Table 2. Chemical composition of canned goat meat containing different amounts of carrot.

Indicators	Canned Goat Meat Samples			
	Control	T10	T20	T30
Protein, %	17.82 ± 0.26	17.82 ± 0.26	17.11 ± 0.29	17.08 ± 0.19
Fat, %	6.70 ± 0.09 ^{b}	6.15 ± 0.17 ^{b}	6.02 ± 0.14 ^{b}	5.76 ± 0.13 ^{a}
Carbohydrates, %	11.78 ± 0.09 ^{a}	13.03 ± 0.02 ^{b}	13.77 ± 0.02 ^{b}	14.36 ± 0.13 ^{b}
Moisture, %	64.7 ± 1.61 ^{b}	63.0 ± 1.57 ^{a}	63.1 ± 1.59 ^{a}	62.8 ± 1.60 ^{a}
Ash, %	1.81 ± 0.03 ^{a}	2.38 ± 0.02 ^{b}	2.06 ± 0.03 ^{b}	2.06 ± 0.01 ^{b}
Energy value, kCal/100 g	178.7	178.7	177.7	177.6

^{a,b} means within the same row, with different letters meaning there is a significant difference among different samples of canned meat ($p < 0.05$).

The protein content of the experimental samples slightly decreased ($p > 0.05$), which in turn agrees with the experimental data of other studies [40,41]. The carbohydrate content significantly increased in the experimental samples with the addition of carrots ($p < 0.05$). Thus, in sample T30, the carbohydrate content was the highest and amounted to 14.36%. The proportion of carbohydrates in samples T20 and T10 increased to 13.77 and 13.03% compared with the control (11.78%). The increase in the proportion of carbohydrates was expected, as carrot composition contain up to 9% of carbohydrates [42]. The salt content in all samples was 1.63%. By lowering the fat content in the experimental samples, the energy value of canned meat slightly decreased. The salt content was the same in all samples (1.63%).

In terms of the mass fraction of fiber and content of β-carotene, the experimental samples significantly exceeded the control samples ($p < 0.05$), with maximum values obtained when 30% of goat meat was replaced with carrots(Figures 2 and 3). It is known that β-carotene has diverse functions, such as protecting against cancer, increasing immune response, possessing antioxidant properties, and other health benefits including anti-ulcer and anti-aging activities [43,44]. Carrots contain dietary fiber, which activates the intestines. Fibers activate the cleansing process of the body and help recover intestinal functions. Due to this, the metabolism is normalized, and the skin, hair, and nails become healthier [45,46].

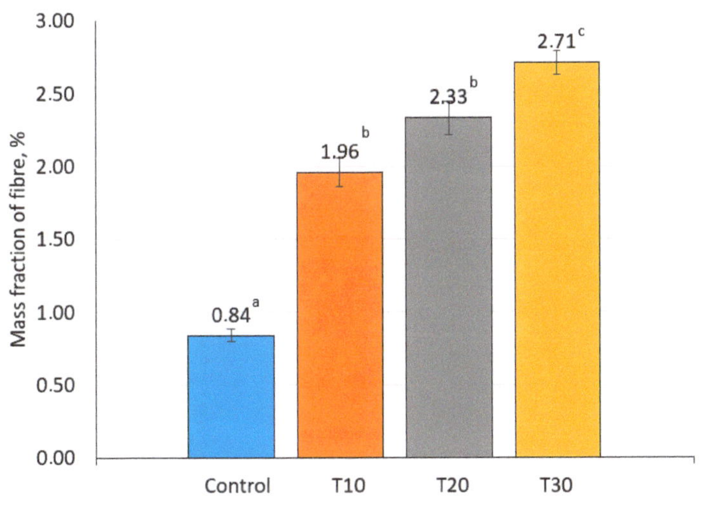

Figure 2. Fiber content in canned goat meat samples (the values for bars sharing different letters (a–c) are significantly different ($p < 0.05$)).

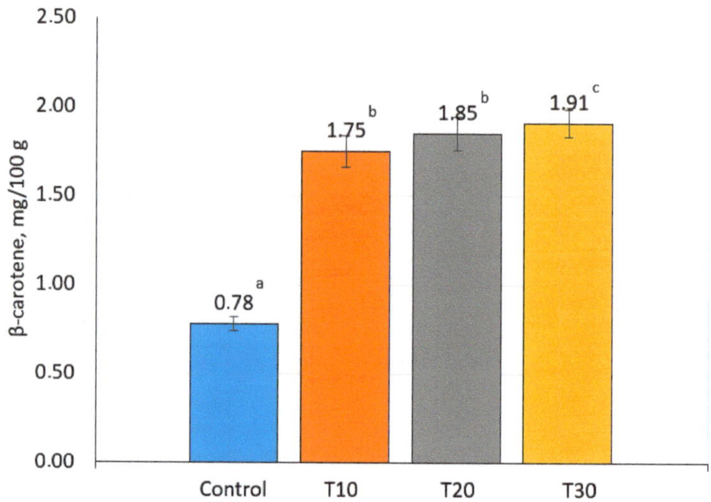

Figure 3. β-carotene content in canned goat meat samples (the values for bars sharing different letters (a–c) are significantly different ($p < 0.05$)).

3.2. Amino Acid Composition of Canned Goat Meat

Analysis of amino acid composition of the control and experimental canned samples showed that replacement of 20 and 30% of goat meat with carrots significantly increased the histidine and lysine content in the T30 sample (Table 3). Also, in T30 samples, there was a significant increase in levels of leucine + isoleucine from 1737 (control) to 1822 mg/100 g (T30). However, levels of other amino acids were decreased, which is expected due to a decrease in protein content. These data are consistent with past experimental findings [38,40,41]. Among the samples with added carrot, the highest total amino acid content (8508 mg/100 g) was identified in the T20 sample, but T30 showed the highest total content for essential amino acids (5956 mg/100 g), which is important for complete nutrition.

Table 3. Amino acid composition of canned goat meat, (mg/100 g).

Name of Amino Acids	Canned Goat Meat Samples				p Value
	Control	T10	T20	T30	
Non-essential	3061 ± 109 [b]	2506 ± 89.6 [a]	2561 ± 91.5 [a]	2311 ± 82.5 [a]	<0.05
Alanine	718 ± 7.71 [c]	583 ± 8.72 [a]	602 ±10.1 [b]	544 ± 6.41 [a]	<0.05
Glycine	702 ± 9.12 [b]	583 ± 7.38 [a]	586 ± 7.65 [a]	528 ± 6.73 [a]	<0.05
Tyrosine	483 ± 9.20 [c]	405 ± 5.88 [b]	402 ± 5.63 [b]	362 ± 3.92 [a]	<0.05
Proline	603 ± 11.2 [c]	481 ± 7.30 [b]	506 ± 7.15 [b]	458 ± 6.51 [a]	<0.05
Serin	555 ± 10.5 [b]	454 ± 6.55 [b]	465 ± 9.40 [b]	419 ± 8.36 [a]	<0.05
Essential	6199 ± 112 [c]	2480 ± 91.2 [a]	5947 ± 98.7 [b]	5956 ± 90.5 [b]	<0.05
Arginine	741 ± 14.9 [c]	622 ± 6.43 [b]	617 ±8.08 [b]	555 ± 8.42 [a]	<0.05
Methionine	322 ± 5.05 [b]	297 ± 2.65 [a]	318 ± 4.71 [b]	316 ± 4.58 [b]	<0.05
Valine	754 ± 13.1 [c]	686 ± 7.15 [a]	738 ± 11.2 [b]	730 ± 9.76 [b]	<0.05
Histidine	409 ± 5.03 [b]	356 ± 4.26 [a]	427 ± 6.67 [c]	436 ± 6.87 [c]	<0.05
Threonine	533 ± 6.07 [c]	460 ± 5.44 [a]	496 ± 8.46 [b]	478 ± 6.20 [a]	<0.05
Leucine + isoleucine	1737 ± 33.5 [b]	1578 ± 26.08 [a]	1730 ± 29.8 [b]	1822 ± 22.45 [c]	<0.05

Table 3. *Cont.*

Name of Amino Acids	Canned Goat Meat Samples				p Value
	Control	T10	T20	T30	
Lysine	969 ± 9.04 [a]	898 ± 16.4 [a]	994 ± 12.8 [a]	1006 ± 13.2 [b]	<0.05
Phenylalanine	654 ± 10.7 [c]	583 ± 9.12 [a]	627 ± 7.04 [b]	613 ± 13.0 [b]	<0.05
Total	9250 ± 233 [c]	4986 ± 142 [a]	8508 ± 107 [b]	8267 ± 180 [b]	<0.05

[a–c] means within the same row with different letters mean there are significant differences among different samples of canned meat ($p < 0.05$).

3.3. Vitamin and Mineral Content of Canned Goat Meat

A study of the vitamin composition of canned meat products showed a trend towards an increase in the content of both fat-soluble and water-soluble groups of vitamins. As carrots are rich in vitamin A, the levels of vitamin A increased to 212.08 and 193.77 µg/100 g in samples T20 and T30, respectively, compared with 73.46 µg/100 g in the control.

It should be noted that 100 g of the product with 20 and 30% carrot substitution for meat ingredients satisfied 23.6 and 21.53% of the daily recommended vitamin A intake (by the WHO) [47], whereas control samples showed only 8.16% of the daily recommended intake. There was also an increase in vitamins B_6 (pyridoxine), B_9 (folic acid), and B_5 (pantothenic acid) (Table 4).

Table 4. Vitamin content of canned goat meat, mg/100 g.

Name of Vitamins	Canned Goat Meat Samples				Recommended Rates, mg/day [47]
	Control	T10	T20	T30	
Vitamin A, µg	73.46 ± 1.9 [b]	60.05 ± 1.18 [a]	212.08 ± 4.24 [c]	193.77 ± 7.75 [c]	900
Vitamin E	0.36 ± 0.002 [a]	0.56 ± 6.41 [b]	0.51 ± 0.01 [b]	0.48 ± 0.01 [b]	15
VitaminB_1	0.075 ± 0.001 [a]	0.086 ± 0.017 [b]	0.18 ± 0.036 [c]	0.172 ± 0.034 [c]	1.5
VitaminB_2	0.650 ± 0.27 [b]	0.587 ± 0.246 [a]	0.687 ± 0.289 [b]	0.655 ± 0.27 [b]	1.8
Vitamin B_6	0.009 ± 0.001 [a]	0.014 ± 0.002 [b]	0.026 ± 0.005 [c]	0.025 ± 0.005 [c]	2
Vitamin B_3	4.22 ± 0.84 [b]	3.35 ± 0.67 [a]	4.353 ± 0.87 [b]	4.146 ± 0.83 [b]	15–20
Vitamin B_5	0.021 ± 0.007 [a]	0.040 ± 0.007 [b]	0.036 ± 0.007 [b]	0.035 ± 0.006 [b]	5
Vitamin B_9	0.006 ± 0.001 [a]	0.009 ± 0.001 [b]	0.015 ± 0.003 [b]	0.014 ± 0.003 [b]	0.4
Vitamin C	0.81 ± 0.27 [c]	0.371 ± 0.13 [b]	0.189 ± 0.064 [a]	0.18 ± 0.06 [a]	90

[a–c] means within the same row with different letters mean there is a significant difference among different samples of canned meat ($p < 0.05$).

The analysis of mineral composition in canned products showed that the increase in ash content in the experimental samples corresponded to an increase in the content of almost all the studied mineral elements. The most significant increases were in potassium, magnesium, and phosphorus content (Table 5). It should be noted that in samples T20 and T30, the iron content satisfies up to 32 and 42% of the daily requirement in men and up to 17 and 22% of the daily requirement in women. The control sample satisfies 28 (for men) and 14.9% (for women) [47]. It should also be noted that excessive intake of a number of elements can lead to serious diseases in humans [48,49].

Table 5. Mineral content of canned goat meat (mg/100 g).

Name of Elements	Canned Goat Meat Samples				Recommended, mg/day [47]
	Control	T10	T20	T30	
Potassium	362.58 ± 5.43 [a]	438.54 ± 13.15 [b]	397.69 ± 11.93 [b]	526.29 ± 7.89 [c]	1500–2500
Calcium	20.06 ± 0.59 [a]	36.48 ± 0.19 [b]	36.7 ± 1.09 [c]	29.78 ± 0.89 [b]	800–1200
Sodium	67.17 ± 2.01 [a]	67.31 ± 2.03 [a]	79.03 ± 2.37 [b]	74.24 ± 2.22 [b]	3000–5000

Table 5. *Cont.*

Name of Elements	Canned Goat Meat Samples				Recommended, mg/day [47]
	Control	T10	T20	T30	
Magnesium	15.33 ± 0.30 [a]	25.04 ± 0.75 [b]	52.12 ± 2.08 [d]	38.21 ± 1.14 [c]	300–400
Iron	2.24 ± 0.06 [b]	1.05 ± 0.03 [a]	2.57 ± 0.07 [b]	3.41 ± 0.10 [c]	8–10 (man) 15–20 (woman)
Copper	1.56 ± 0.04 [b]	0.72 ± 0.01 [a]	2.21 ± 0.04 [b]	2.28 ± 0.08 [b]	0.9–2.3
Zinc	4.19 ± 0.10 [b]	1.63 ± 0.04 [a]	3.85 ± 0.11 [b]	3.18 ± 0.09 [b]	10–15 (man) 10–12 (woman)
Phosphorus	188.23 ± 3.76 [a]	179.54 ± 7.18 [a]	247.08 ±7.42 [b]	234.05 ± 7.02 [b]	550–1400
Iodine	0.78 ± 0.01 [a]	0.85 ± 0.01 [a]	1.71 ± 0.05 [b]	1.48 ± 0.04 [b]	0.15
Selenium	17.11 ± 0.51 [b]	12.55 ± 0.29 [a]	17.92 ± 0.34 [b]	20.62 ± 0.82 [c]	55
Chlorine	1.58 ± 0.03 [a]	4.25 ± 0.07 [b]	5.09 ± 0.13 [b]	5.62 ± 0.17 [b]	-

[a–c] means within the same row with different letters mean there is a significant difference among different samples of canned meat ($p < 0.05$).

3.4. Organoleptic Characteristics

Organoleptic characteristics of food products are one of the main criteria of choice for consumers. We evaluated the appearance, appearance of the product packaging, appearance on the cut, color, recipe composition, etc. (Figure 4).

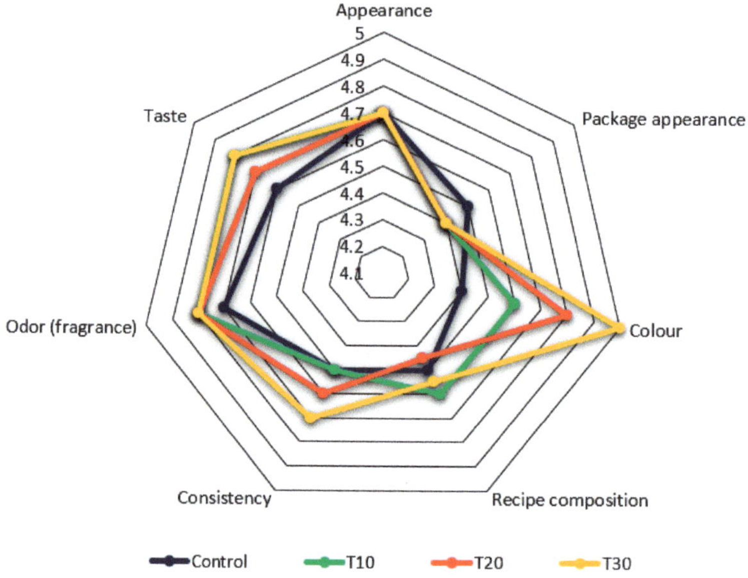

Figure 4. Organoleptic characteristics of the control and experimental samples of canned meat.

According to the results of organoleptic evaluation (Figure 4) all experimental and control samples had no critical differences in the type of packaging and labels, odor, formulation composition, and appearance of canned food. Differences were noticeable in color assessment: the control sample (4.4 points) and sample T10 (4.6 points) had a pale color, samples T30 (4.8 points) and T20 (5.0 points) had a richer color due to carrot inclusions. The color scores were higher in the experimental samples compared with the control. The highest score was in the T-30 sample with 30% carrot inclusion. This effect may be associated with the presence of orange pigment and carotenoids affecting color in carrot. Similar findings were also reported by [50] in chicken nuggets and by [51] in turkey meat sausages.

Canned samples T30 (4.6 points) and T20 (4.7 points) had a more tender and juicy consistency. In evaluating the taste of the experimental samples, T30 and T20 also received the highest score (4.7 and 4.8 points, respectively) and were described as typical of canned meat, without specific flavors. Organoleptic evaluation showed improved consistency in samples T-20 (4.6 points) and T-30 (4.7 points). Similar results were obtained by [51] in a study of turkey meat sausages with the inclusion of carrot paste. Improving juiciness and tenderness of the meat product by increasing the proportion of carrots is due to the presence of natural polysaccharides and dietary fiber, which join with the water–protein–fat matrix to form a more gel-like consistency [52,53].

Any change in the chemical composition of the finished product due to the partial replacement of one ingredient by another, consequently, can affect the sensory characteristics and consumer qualities of the product. As supposed by [54], the amount and type of amino acids in the composition of finished meat and meat products can affect its flavor characteristics and sensorimetric score. For example, alanine, glycine, proline, serine, and threonine can contribute to a sweet taste; histidine, allo-isoleucine, leucine, methionine, pheninalanine, tryptophan, and valine can add a bitter taste to products [55]. The maximum content of amino acids responsible for sweet taste was found in the control and T-20 sample. The highest content of the amino acids contributing to bitter taste corresponded to the control and T-30 sample. The highest grades of taste in organoleptic evaluation were given to the T-10 and T-30 samples. These findings differ from those of [56], on the reduction of taste characteristics in meat bread made of buffalo liver and vegetables with the inclusion of carrot paste, and with the results of [52], in a study on beef frankfurters with carrot paste (3,5, and 10%). This difference in results may be explained by differences in the type of finished products, the raw meat, and the formulation.

4. Conclusions

To summarize the results of the study, we can conclude that replacing part of the goat meat with carrot had a significant effect on the chemical, amino acid, vitamin, and mineral composition of canned food. In particular, some parameters significantly increased ($p < 0.05$), depending on the amount of carrot addition: the mass content of moisture, ash, carotenoids, and fiber, with a simultaneous decrease in fat content. The vitamin and mineral content were significantly increased ($p < 0.05$), according to our tests. Organoleptic evaluation showed differences in color, taste, consistency, and odor. Thus, the addition of carrots could increase the biological value and improve the palatability of canned goat meat.

Author Contributions: Conceptualization, U.C. and G.K.; data curation, A.T. and N.T.; investigation, A.T. and N.T.; methodology, T.Z.; project administration, U.C.; resources, T.Z.; software, T.Z.; visualization, A.S.; writing—review and editing, G.K. and A.S. All authors have read and agreed to the published version of the manuscript.

Funding: This research was funded by the Ministry of Agriculture of Kazakhstan (grant number IRN BR10764970) and the Ministry of Education and Science of the Republic of Kazakhstan (grant number IRN AP09058083).

Institutional Review Board Statement: Not applicable.

Informed Consent Statement: Not applicable.

Data Availability Statement: The data presented in this study are available on request from the corresponding author.

Acknowledgments: The authors wish to thank laboratory staff of Kazakh Research Institute of Processing and Food Industry for meat sample analysis.

Conflicts of Interest: The authors declare no conflict of interest.

References

1. Bonnet, C.; Bouamra-Mechemache, Z.; Réquillart, V.; Treich, N. Regulating meat consumption to improve health, the environment and animal welfare. *Food Policy* **2020**, *10*, 101847. [CrossRef]
2. FAO. *Food Outlook—Biannual Report on Global Food Markets*; Food Outlook: Rome, Italy, 2021. [CrossRef]
3. Toldrá, F.; Reig, M. Innovations for healthier processed meats. *Trends Food Sci. Technol.* **2011**, *22*, 517–522. [CrossRef]
4. Yessimbekov, Z.; Kakimov, A.; Caporaso, N.; Suychinov, A.; Kabdylzhar, B.; Shariati, M.A.; Baikadamova, A.; Domínguez, R.; Lorenzo, J.M. Use of meat-bone paste to develop calcium-enriched liver pâté. *Foods* **2021**, *10*, 2042. [CrossRef] [PubMed]
5. Słowiński, M.; Miazek, J.; Chmiel, M. Influence of the dose and length of wheat fiber on the quality of model sterilized canned meat products. *Foods* **2020**, *9*, 1001. [CrossRef]
6. Ebeed, A.; Elsayed, E.; Nabil, M.; Yasser, G.; Safaa, H. Quality assurance of imported canned meat. *Glob. Vet.* **2015**, *14*, 511–516.
7. Kakimov, M.M.; Tokysheva, G.M.; Makangali, K.K. Perspectives for goat meat processing development in the Republic of Kazakhstan. *Sci. Technol. Kazakhstan* **2021**, *2*, 95–100.
8. Kazhybayeva, G.; Agibayeva, A.; Kuderinova, N.; Harlap, S.; Fedoseeva, N.; Usov, V.; Bakirova, L. Development of technology and assessment of nutritional value of a delicacy goat meat product. *Int. J. Innov. Technol. Explor. Eng.* **2019**, *8*, 239–242.
9. Nassu, R.T.; Gonçalves, L.A.G.; da Silva, M.A.A.P.; Beserra, F.J. Oxidative stability of fermented goat meat sausage with different levels of natural antioxidant. *Meat Sci.* **2003**, *63*, 43–49. [CrossRef]
10. Ali, M.; Park, J.Y.; Lee, S.Y.; Choi, Y.S.; Nam, K.C. Physicochemical and microbial characteristics of longissimus lumborum and biceps femoris muscles in Korean native black goat with wet-aging time. *J. Anim. Sci. Technol.* **2021**, *63*, 149. [CrossRef]
11. Tokysheva, G.; Makangali, K.; Uzakov, Y.; Kakimov, M.; Vostrikova, N.; Baiysbayeva, M.; Mashanova, N. The potential of goat meat as a nutrition source for schoolchildren. *Potravin. Slovak J. Food Sci.* **2022**, *16*, 398–410. [CrossRef]
12. Malekian, F.; Khachaturyan, M.; Gebrelul, S.; Henson, J. Nutritional characteristics and consumer acceptability of sausages with different combinations of goat and beef meats. *Funct. Foods Health Dis.* **2016**, *6*, 42–58. [CrossRef]
13. Teixeira, A.; Almeida, S.; Pereira, E.; Mangachaia, F.; Rodrigues, S. Physicochemical characteristics of sheep and goat pâtés. differences between fat sources and proportions. *Heliyon* **2019**, *5*, e02119. [CrossRef]
14. Babiker, S.A.; El Khider, I.A.; Shafie, S.A. Chemical composition and quality attributes of goat meat and lamb. *Meat Sci.* **1990**, *28*, 273–277. [CrossRef]
15. Rababah, T.M.; Ereifej, K.I.; Alhamad, M.N.; Al-Qudah, K.M.; Rousan, L.M.; Al-Mahasneh, M.A.; Al-u'datt, M.H.; Yang, W. Effects of green tea and grape seed and TBHQ on physicochemical properties of Baladi goat meats. *Int. J. Food Prop.* **2011**, *14*, 1208–1216. [CrossRef]
16. Madruga, M.S.; Bressan, M.C. Goat meats: Description, rational use, certification, processing and technological developments. *Small Rumin. Res.* **2011**, *98*, 39–45. [CrossRef]
17. Oliveira, A.N.D.; Selaive-Villarroel, A.B.; Monte, A.L.S.; Costa, R.G.; Costa, L.B.A. Evaluation of carcass characteristics of crossbred Anglo-Nubian, Boer and undefined breed goats. *Ciência Rural.* **2008**, *38*, 1073–1077. [CrossRef]
18. Assenova, B.; Okuskhanova, E.; Rebezov, M.; Zinina, O.; Baryshnikova, N.; Vaiscrobova, E.; Kasatkina, E.; Shariati, M.A.; Khan, M.U.; Ntsefong, G.N. Effect of germinated wheat (Triticum Aestivum) on chemical, amino acid and organoleptic properties of meat pate. *Potravin. Slovak J. Food Sci.* **2020**, *14*, 580–586. [CrossRef]
19. Zinina, O.; Merenkova, S.; Tazeddinova, D.; Rebezov, M.; Stuart, M.; Okuskhanova, E.; Yessimbekov, Z.; Baryshnikova, N. Enrichment of meat products with dietary fibers: A review. *Agron. Res.* **2019**, *17*, 1808–1822.
20. Sharma, K.D.; Karki, S.; Thakur, N.S.; Attri, S. Chemical composition, functional properties and processing of carrot—A review. *J. Food Sci. Technol.* **2012**, *49*, 22–32. [CrossRef]
21. Hira, S.; Saleem, U.; Anwar, F.; Sohail, M.F.; Raza, Z.; Ahmad, B. β-carotene: A natural compound improves cognitive impairment and oxidative stress in a mouse model of streptozotocin-induced Alzheimer's disease. *Biomolecules* **2019**, *9*, 441. [CrossRef]
22. Alvarado-Ramírez, M.; Santana-Gálvez, J.; Santacruz, A.; Carranza-Montealvo, L.D.; Ortega-Hernández, E.; Tirado-Escobosa, J.; Cisneros-Zevallos, L.; Jacobo-Velázquez, D.A. Using a functional carrot powder ingredient to produce sausages with high levels of nutraceuticals. *J. Food Sci.* **2018**, *83*, 2351–2361. [CrossRef] [PubMed]
23. Średnicka-Tober, D.; Kopczyńska, K.; Góralska-Walczak, R.; Hallmann, E.; Barański, M.; Marszałek, K.; Kazimierczak, R. Are organic certified carrots richer in health-promoting phenolics and carotenoids than the conventionally grown ones? *Molecules* **2022**, *27*, 4184. [CrossRef] [PubMed]
24. Pereira, R.; Costa, M.; Velasco, C.; Cunha, L.M.; Lima, R.C.; Baião, L.F.; Batista, S.; Marques, A.; Sá, T.; Campos, D.A.; et al. Comparative analysis between synthetic vitamin E and natural antioxidant sources from tomato, carrot and coriander in diets for market-sized Dicentrarchus labrax. *Antioxidants* **2022**, *11*, 636. [CrossRef] [PubMed]
25. Kamani, M.H.; Meera, M.S.; Bhaskar, N.; Modi, V.K. Partial and total replacement of meat by plant-based proteins in chicken sausage: Evaluation of mechanical, physico-chemical and sensory characteristics. *J. Food Sci. Technol.* **2019**, *56*, 2660–2669. [CrossRef] [PubMed]
26. Zheleuova, Z.S.; Uzakov, Y.M.; Shingisov, A.U.; Alibekov, R.S.; Khamitova, B.M. Development of halal cooked smoked beef and turkey sausage using a combined plant extracts. *J. Food Process. Preserv.* **2021**, *45*, 1. [CrossRef]
27. Kvasenkov, O.I. Method of Making Canned Goat Roast Spiked. Patent No. 2358573, 28 January 2008.
28. Antipova, L.V.; Glotova, I.A.; Rogov, I.A. *Meat and Meat Products Research Methods*; "KoloS" Publishing House: Moscow, Russia, 2001.

29. *GOST 23042-86*; Interstate Standard. Meat and Meat Products. Methods of Determination of Fat. Standartinfrom: Moscow, Russia, 2010.
30. *GOST 25011-81*; Interstate Standard. Meat and Meat Products. Methods of Determination of Protein. Standartinfrom: Moscow, Russia, 2010.
31. *GOST R 54058-2010*; Functional Food Products. Method for Determination of Carotenoids. Standardinform: Moscow, Russia, 2011.
32. *GOST R 54014-2010*; Functional Food. Determination of Soluble and Unsoluble Dietary Fiber by Enzymatic-Gravimetric Method. Standardinform: Moscow, Russia, 2011.
33. *GOST 9957-2015*; Meat and Meat Products. Methods of Determination of Sodium Chloride Content. Standardinform: Moscow, Russia, 2016.
34. AOAC. *Official Methods of Analysis*, 17th ed.; Official Methods of Analysis of AOAC International; CAB Publisher: Gaithersburg, MD, USA, 2000.
35. *GOST R 55482-2013*; Meat and Meat Products. Method for Determination of Water-Soluble Vitamins Content. Standardinform: Moscow, Russia, 2014.
36. *GOST 32307-2013*; Meat and Meat Products. Determination of Fat-Soluble Vitamins by High Performance Liquid Chromatography. Standardinform: Moscow, Russia, 2014.
37. *GOST-33741-2015*; Canned Meat and Meat-Containing Products. Methods for Determining Organoleptic Characteristics, Net Weight and Mass Portion of Constituent Parts. Standartinfrom: Moscow, Russia, 2016.
38. Rocchetti, G.; Pateiro, M.; Campagnol, P.C.; Barba, F.J.; Tomasevic, I.; Montesano, D.; Lucini, L.; Lorenzo, J.M. Effect of partial replacement of meat by carrot on physicochemical properties and fatty acid profile of fresh turkey sausages: A chemometric approach. *J. Sci. Food Agric.* **2020**, *100*, 4968–4977. [CrossRef]
39. Carvalho, F.A.L.; Pateiro, M.; Domínguez, R.; Barba-Orellana, S.; Mattar, J.; Rimac Brnčić, S.; Barba, F.J.; Lorenzo, J.M. Replacement of meat by spinach on physicochemical and nutritional properties of chicken burgers. *J. Food Process. Preserv.* **2019**, *43*, e13935. [CrossRef]
40. Zargar, F.A.; Kumar, S.; Bhat, Z.F.; Kumar, P. Effect of incorporation of carrot on the quality characteristics of chicken sausages. *Indian J. Poult. Sci.* **2017**, *52*, 91–95. [CrossRef]
41. Yadav, S.; Pathera, A.K.; Islam, R.U.; Malik, A.K.; Sharma, D.P. Effect of wheat bran and dried carrot pomace addition on quality characteristics of chicken sausage. *Asian-Australas. J. Anim. Sci.* **2018**, *31*, 729–737. [CrossRef]
42. Aleksashina, S.A.; Makarova, N.V. Study of the chemical composition and antioxidant activity of carrots, beets and pumpkins. *Storage Process. Farm Prod.* **2016**, *6*, 29–32.
43. Rao, A.V.; Rao, L.G. Carotenoids and human health. *Pharmacol. Res.* **2007**, *55*, 207–216. [CrossRef] [PubMed]
44. Zaini, R.; Clench, M.R.; Le Maitre, C.L. Bioactive chemicals from carrot (*Daucus carota*) juice extracts for the treatment of leukemia. *J. Med. Food* **2011**, *14*, 1303–1312. [CrossRef] [PubMed]
45. Guan, Z.-W.; Yu, E.-Z.; Feng, Q. Soluble dietary fiber, one of the most important nutrients for the gut microbiota. *Molecules* **2021**, *26*, 6802. [CrossRef]
46. Yang, X.; Dai, J.; Zhong, Y.; Wei, X.; Wu, M.; Zhang, Y.; Huang, A.; Wang, L.; Huang, Y.; Zhang, C.; et al. Characterization of insoluble dietary fiber from three food sources and their potential hypoglycemic and hypolipidemic effects. *Food Funct.* **2021**, *12*, 6576–6587. [CrossRef] [PubMed]
47. Human Vitamin and Mineral Requirements. *Report of a Joint FAO/WHO Expert Consultation Bangkok, Thailand*; World Health Organization: Geneva, Switzerland; Food and Agriculture Organization of the United Nations: Rome, Italy, 2002.
48. Sitalakshmi, R.; Kumar, S.P. Trace Elements in Health and Disease: A Review. *Res. J. Pharm. Biol. Chem. Sci.* **2014**, *5*, 450–455.
49. Khanvilkar, D.G.; Nagarjee, S.; Noah, A.A.; Adedeji, M.A.; Jabeen, N.M.; Kumar, G.P.; Devanna, N.; Manjunath, S.S.; Jithendran, L.; Kalpana, C.A. Trace element status and human endocrine health: A perspective. *Int. J. Food Nutr. Sci.* **2021**, *10*, 1–5. [CrossRef]
50. Bhosale, S.S.; Biswas, A.K.; Sahoo, J.; Chatli, M.K.; Sharma, D.K.; Sikka, S.S. Quality evaluation of functional chicken nuggets incorporated with ground carrot and mashed sweet potato. *FSTI* **2011**, *17*, 233–239. [CrossRef]
51. Reddy, M.N.K.; Kumar, M.S.; Reddy, G.V.B.; Reddy, N.A.; Rao, V.K. Quality evaluation of turkey meat sausages incorporated with ground carrot. *Pharm. Innov. J.* **2018**, *7*, 773–777.
52. Sam, F.E.; Ma, T.Z.; Atuna, R.A.; Salifu, R.; Nubalanaan, B.A.; Amagloh, F.K.; Han, S.Y. Physicochemical, Oxidative Stability and Sensory Properties of Frankfurter-Type Sausage as Influenced by the Addition of Carrot (*Daucus carota*) Paste. *Foods* **2021**, *10*, 3032. [CrossRef]
53. Slima, S.B.; Ktari, N.; Triki, M.; Trabelsi, I.; Moussa, H.; Makni, S.; Abdeslam, A.; Herrero, A.M.; Jiménez-Colmenero, F.; Ruiz-Capillas, C.; et al. Effects of two fibers used separately and in combination on physico-chemical, textural, nutritional and sensory properties of beef fresh sausage. *Br. Food J.* **2019**, *121*, 1428–1440. [CrossRef]
54. Zamuz, S.; Purriños, L.; Galvez, F.; Zdolec, N.; Muchenje, V.; Barba, F.J.; Lorenzo, J.M. Influence of the addition of different origin sources of protein on meat products sensory acceptance. *J. Food Process. Preserv.* **2019**, *43*, e13940. [CrossRef]
55. Food Ingredients Brazil. Os Aminoácidos e o Sabor. Food Ingredients Brazil, 2017, 31. Available online: www.revista-fi.com (accessed on 24 August 2022).
56. Devatkal, S.; Mendiratta, S.K.; Kondaiah, N. Quality Characteristics of Loaves from Buffalo Meat, Liver and Vegetables. *Meat Sci.* **2004**, *67*, 377–383. [CrossRef] [PubMed]

applied sciences

MDPI

Article

Development and Characterization of a Low-Fat Mayonnaise Salad Dressing Based on *Arthrospira platensis* Protein Concentrate and Sodium Alginate

Jorge Metri-Ojeda [1], Milena Ramírez-Rodrigues [2], Lizbeth Rosas-Ordoñez [1] and Diana Baigts-Allende [3,*]

[1] Departamento de Ingeniería Química, Ambiental y Alimentos, Universidad de las Américas Puebla, San Andrés Cholula, Puebla 72810, Mexico; jorge.metrioa@udlap.mx (J.M.-O.); lizbeth.rosasoz@udlap.mx (L.R.-O.)
[2] Food Analysis Laboratory, Intema S.A. de C.V., 31 Sur 2901, Col. Santa Cruz Los Ángeles, Puebla 72400, Mexico; mm.ramirez@intema.com.mx
[3] Faculty of Agrobiology, Food and Natural Resources, Czech University of Life Sciences Prague, Kamýcká 129, 165 00 Prague, Czech Republic
* Correspondence: baigts_allende@af.czu.cz

Featured Application: This work presents an alternative approach to produce low-fat mayonnaises with good acceptability and using clean label ingredients, useful for developing new formulations in this type of products.

Abstract: The food industry is constantly reformulating different foods to fulfill the demands of the consumers (natural ingredients and good sensory quality). The present work aimed to produce low-fat mayonnaises using 30.0, 22.5, and 15.0% oil, 1% soy protein isolate (SPI) or spirulina (*Arthrospira platensis*) protein concentrate (SPC), and 2% sodium alginate. The physical properties (thermal stability, rheological behavior, and particle size), the sensory attributes (appearance, texture, taste, and acceptability), the purchase probability, and amino acid availability (after a simulated digestion) were evaluated. The mayonnaises demonstrated good thermal stability (>90%) using 22.5 and 15% oil, all products showed shear-thinning behavior and a consistency index of 20–66 Pa·s. The reduction of oil from 30 to 15% increased the particle size from 6–9 μm to 10–38 μm. The most acceptable product was the formulated with SPI and 22.5% oil (8.3 of acceptability and 79% of purchase probability). Finally, the addition of proteins improved the total essential amino acids compared to a commercial product (28 and 5 mg/25 g, respectively). In summary, it was possible to obtain well accepted products with high purchase probability using low concentrations of oil and vegetable proteins.

Keywords: salad dressing; sensory evaluation; acceptability; spirulina protein concentrate; soy protein isolate

1. Introduction

In recent decades, the demand for healthier foods has risen, boosting the interest of the food industry to reformulate a wide variety of products. This reformulation includes reducing a specific component (such as sugar or fat content) or the use of naturally derived ingredients [1], but without compromising the food safety and the sensory quality. Fulfilling the latter requirements is especially challenging in food products where fat is the primary component, for example, in ice creams, vegetable spreads (i.e., peanut butter), and salad dressings (i.e., mayonnaise), because the fat plays a crucial role as structuring agent, contributing to the development of texture [2].

Mayonnaise is a worldwide-consumed seasoning used in several types of foods, for instance, sandwiches, salads, hamburgers, hot dogs, seafood, and others. The traditional formulation contains approximately 70–80% of oil, vinegar, and spices; however, the consumers usually perceive its high oil content as potentially unhealthy [3]. This conception led to the elaboration of low-fat mayonnaise, which contains 20–40% oil, this has been possible by using thickening agents such as modified starches or gums to develop texture and physical stability in the product [4–6]. However, the use of the latter ingredients might have some sensory disadvantages. For instance, the texture or taste of the product could be negatively affected [7,8], and some of these ingredients (modified starches) could be perceived as potentially harmful [9].

One alternative to overcome the negative texture impact of reducing the fat content in mayonnaises is the use of Emulsion-Filled Gels (EFG). The EFGs are gel matrices in which emulsified oil droplets replace the water phase of the gel. These systems are stable against creaming, flocculation, and coalescence, have a gel-like behavior, and they can be easily produced with a wide range of biopolymers such as gums (pectin, tragacanth gum, gellan gum, alginates, carrageenan) and proteins [10,11]. To date, the application of EFGs in mayonnaise has been demonstrated to mimic the sensory properties (texture and mouthfeel) of the regular product using egg yolk (as emulsifier), tragacanth gum, and sodium alginate (as a thickening and gelling agents) [12,13], fulfilling the actual demands of the consumers (low-fat and naturally derived ingredients).

There are many sources for obtaining natural derived biopolymers, for instance, cereals, tubers, oilseeds, and algae. These latter are comprised by macroalgae and microalgae, where the first one is a common source of sodium alginate; this biopolymer is used as thickening and gelling agent, and one of the most important characteristics of this compound is the irreversible gelation when is combined with ionic salts, for example, calcium chloride [14]. In the case of microalgae, the *Arthrospira platensis*, commonly known as spirulina, is nowadays an important raw material to extract high valuable ingredients, because it contains unsaturated fatty acids (omega 3), pigments (phycocyanin), and a high concentration of proteins (40–50%) [15,16] with promising applications due to its emulsion capacity (approximately 60%) and stability (ranging from 80–100%) [17–20].

In our previous work [21], it was observed that the consumers of mayonnaise (in Mexico) perceived the proteins as natural ingredients and have good acceptability, hence, it was decided to produce mayonnaises using egg yolk, vegetable proteins, and sodium alginate. Therefore, the present study aimed to exploit the techno-functionality (i.e., emulsifying properties and water and oil holding capacities) of the *A. platensis* protein in combination with sodium alginate to produce low-fat mayonnaises (using an EFG approach) with comparable physical and sensory characteristics than a commercial low-fat mayonnaise.

2. Materials and Methods

Microalgae (*A. platensis*) was purchased from Galtec Algae Technologies (Guadalupe, Mexico). Soy protein isolate containing 91.9% *w*/*w* of protein was purchased from Food Technologist Trading S.A de C.V. (Atizapan de Zaragoza, Mexico). Soybean oil, salt, sugar, vinegar, and low-fat commercial mayonnaise (41% of total fat) used as reference were purchased from local groceries in Puebla, Mexico. Sodium alginate and calcium chloride were obtained from Sigma-Aldrich Co. (St. Louis, MO, USA) and J.T Baker (Phillipsburg, NJ, USA), respectively. All reagents used for the analyses (amino acid mix, *o*-phthaldehyde (OPA) and 9-fluorenylmethylchloroformate (FMOC)) were analytical grade obtained from Sigma-Aldrich Co. (St. Louis, MO, USA).

2.1. Spirulina Conditioning and Protein Extraction

The biomass (*A. platensis*) was freeze-dried, grind, and sieved using a 300 μm mesh. The freeze-dried spirulina was defatted twice with n-hexane at room temperature (for 30 min), at a ratio of 1:5 (powder: solvent *w*/*v*), then the n-hexane was decanted, and the remaining solvent was allowed to evaporate overnight at room temperature.

To obtain the Spirulina Protein Concentrate (SPC), the defatted biomass was subjected to alkaline solubilization for the protein extraction at pH 12 using a NaOH solution at 0.05 N in a ratio of 1:10 (*w*/*v*) for 1 h (at 25 °C). Afterwards, the sample was centrifuged (4500 rpm, 4 °C, 10 min), and the pH of the supernatant was adjusted to 3.5 to precipitate the proteins. The precipitate was collected by centrifugation (at the same conditions), rinsed with distilled water, and centrifuged again [17]. The protein concentrates were freeze-dried at −50 °C for approximately 48 h and stored in sealed flasks.

2.2. Protein Concentrates Characterization

2.2.1. Protein Content

The protein content was determined spectrophotometrically using the Total Protein Kit, Micro Lowry, Peterson's Modification (Sigma-Aldrich Co., St. Louis, MO, USA). This method was selected to avoid the interaction of the polyphenolic compounds of the spirulina. First, 500 µL of solubilized protein (at pH 13 to promote full solubilization) was mixed with 50 µL a deoxycholate solution (0.15%) and followed by 50 µL of trichloroacetic acid (72%), the sample was allowed to stand for 10 min after the addition of each reagent. These solutions promote the protein precipitation, which was further collected by centrifugation (5000 rpm at 25 °C for 5 min). The protein pellet was dissolved with 500 µL of Lowry's reagent and 500 µL of water. After 20 min, 250 µL of Folin Ciocalteau reagent was added and allowed to react for 30 min. Finally, 200 µL of the samples were transferred to a 96 wells microplate and the absorbance was measured using a UV-Vis Multiskan Sky Microplate (Thermo Fischer Scientific, Walthman, MA, USA) spectrophotometer at 720 nm. A solution of bovine serum albumin (0–1000 µg/mL) was used as standard following the same procedure. The protein content was calculated using Equation (1) and expressed as a percentage (%).

$$\text{Protein content } (\%) = \frac{P_S}{S_w} \times 100, \tag{1}$$

where P_S is the total protein quantified in the sample and S_W is the initial sample weight.

2.2.2. Water Holding Capacity (WHC) and Oil Holding Capacity (OHC)

Water and oil holding capacities (WHC, OHC, respectively) were measured by weighing 0.25 g of protein in a 50 mL Falcon tube. Then, 10 mL of distilled water or oil was added, and the sample was stirred using a vortex for 10 s every 5 min (for 30 min). Afterwards, the samples were centrifuged (at 4500 rpm, 4 °C, for 10 min), the supernatants were discarded, and the WHC and OHC were calculated by weight difference as is shown in Equation (2) [22].

$$\text{WHC or OHC} = \frac{\text{Final weight} - \text{Initial weight}}{\text{Initial weight}}, \tag{2}$$

2.3. Mayonnaise Preparation

The continuous phase, constituted by an aqueous solution added with water, vinegar, sugar, salt, and protein (soy or spirulina), was homogenized (Ultraturrax (IKA, Staufen, Germany) with the dispersed phase (egg yolk and oil) at 7500 rpm (60 s). A double emulsion was prepared by adding a solution of sodium alginate (2%) to the primary emulsion at a 1:1 ratio. Finally, for gelation of sodium alginate in the secondary emulsion, 1 mL of $CaCl_2$ (0.5 % w/v) was dropped in constant stirring at 7500 rpm [13]. Finally, we obtained nine different formulations of low-fat mayonnaise (Table 1).

2.4. Mayonnaise Characterization

2.4.1. Stability Analysis

The thermal stability of the samples was studied by pouring approximately 20 g of mayonnaise into a sealed Falcon flask (50 mL of capacity) and heating them in a water bath (85 °C) for 30 min. Afterwards, the samples were centrifuged at 4000 rpm for 10 min to induce phase separation and the stability was calculated with Equation (3) [23].

Table 1. The concentration of the ingredients in the low-fat mayonnaise.

Sample	Oil (%)	Aqueous Phase						
		Egg Yolk (%)	Water (%)	Vinegar (%)	Sugar (%)	Salt (%)	Protein (Type; %)	Sodium Alginate (%)
A	30.0	10	22	4.5	2	1.5	NA	30
B	22.5	10	26	4.5	2	1.5	NA	34
C	15.0	10	30	4.5	2	1.5	NA	38
D	30.0	10	21	4.5	2	1.5	SPI; 1	30
E	22.5	10	25	4.5	2	1.5	SPI; 1	34
F	15.0	10	29	4.5	2	1.5	SPI; 1	38
D′	30.0	10	21	4.5	2	1.5	SPC; 1	30
E′	22.5	10	25	4.5	2	1.5	SPC; 1	34
F′	15.0	10	29	4.5	2	1.5	SPC; 1	38

SPI = soy protein isolate. SPC = spirulina protein concentrate.

$$\text{Mayonnaise stability (ES)} = \frac{\text{Remaining emulsified layer (cm)}}{\text{Initial emulsified layer (cm)}} \times 100 \quad (3)$$

2.4.2. Particle Size Analysis

The particle size was measured using dynamic light scattering with a Bluewave Nanotrac Wave II (Microtrac Inc., Montgomeryville, PA, USA) analyzer. All samples were diluted 1:100 (v/v) with distilled water and homogenized with a Vortex at full speed for 15 s. The refractive index was set at 1.47 for oil droplets and 1.33 for water [24], the results were expressed as mean particle size (μm) and the span was calculated using Equation (4).

$$\text{Span} = \frac{d_{90} - d_{10}}{d_{50}}, \quad (4)$$

where d_{90}, d_{50}, and d_{10} represents the diameter of the 90%, 50%, and 10% of all particles measured.

2.4.3. Rheological Characterization

The viscosity was determined at 25 °C using a Brookfield Viscometer DV II (AMETEK Inc., Devon-Berwyn, PA, USA) equipped with an LV-3 spindle at 0.25, 0.5, 1, 2, 5, 10 and 16.67 s^{-1} for 45 s. The data were fitted to the power law model [25], this mathematical model allows a simple and easy interpretation of the flow behavior of a non-Newtonian fluid. This process was done using Microsoft Office Excel 365 solver to obtain the consistency index (K) and power law index (n) using the following expression.

$$\sigma = K\gamma^n, \quad (5)$$

afterwards, the loss of consistency index (ΔK) was calculated using Equation (6) to assess the effect of oil concentration and the vegetable proteins in the loss of viscosity of the systems.

$$\Delta K = |K_{30} - K_i|, \quad (6)$$

where K_{30} is the consistency index of the sample with 30% oil and K_i is the consistency index of the sample.

2.4.4. Sensory Analysis

A total of 50 non-trained panelists were recruited in "Universidad De Las Américas Puebla". The samples were spread on a wheat cookie and labeled using three random digits. The sensory evaluation of the mayonnaises was studied using a structured scale from 1 to 10 (1 = lowest score and 10 = highest score) for assessing the general appearance, taste, texture, and acceptability. Furthermore, the panelists were asked for their purchase intention (possible answers: Yes/No) for each sample after its evaluation; the results were expressed as percentage (%) [26].

2.4.5. Amino Acid Bioavailability

To investigate the nutritional enhancement of proteins (bioavailability of amino acids) in the low-fat mayonnaise, the products with the best purchase intention with one of each vegetable protein (SPI and SPC) were selected for an in vitro gastric digestion simulation (GDS). First, artificial saliva (AS) and gastric juices (GJ) were prepared to mimic the buccal and gastric environment. The AS was prepared by solubilization of α-amylase (0.2 mg/mL) in phosphate buffer at 20 mM (pH 7) and GJ consisted of pepsin enzyme in HCL solution (0.1 M) at 3.2 mg/mL. The digestion in the mouth was simulated by homogenizing mayonnaise (25 g) and 10 mL of AS using a magnetic stirrer at 150 rpm for 3 min (at 37 °C). Afterward, the gastric digestion was simulated by adjusting the pH to 2.0 using concentrated HCl (4 M) and adding 15 mL of GJ using the same stirring conditions as above; this digestion was done for 60 min. The conditions of the experiment were selected according to the study of Rui et al. [27], which selected the appropriate food to digestive juices ratios based on the human physiology (food to AS = 2.5:1; food to GJ = 1.5:1) [28].

The sample collection was done after the buccal digestion (time 1) and gastric digestion (time 2), whereas the fresh sample was considered as time 0. For each sample, the amino acid profile was determined (time 0, 1, and 2) with the Agilent amino acid analysis protocol [29] in an Agilent 1290 Infinity II (Agilent Technologies Inc., Santa Clara, CA, USA) HPLC coupled with a diode array detector (DAD). The column used was a ZORBAX Eclipse Plus C18 (Agilent Technologies Inc., Santa Clara, CA, USA), the mobile phase A was a 10 mM of Na_2HPO_4:10 mM $Na_2B_4O_7$ (pH 8.4) buffer, and the mobile phase B was a solution of acetonitrile: methanol: water (45:45:10). The column temperature was at 40 °C, and the flow rate was 1.5 mL/min. The total time of the assay was 16 min, with 98% of mobile phase A (at 0.35 min), 100% of mobile phase B at 13.5 min, and 98% of mobile phase A used at 16 min. Amino acids (primary and secondary) were derivatized using *o*-phthalaldehyde (OPA) and 9-fluorenylmethyl chloroformate (FMOC). The amino acids detection was done at 263 and 338 nm and quantified using norvaline (internal standard) and a calibration curve (amino acid mix).

2.5. Statistical Analysis

All experiments were done in triplicate (n = 3). The data was expressed as mean with standard deviation, median with interquartile ranges, and percentage where appropriate. The statistical differences were analyzed using ANOVA ($p < 0.05$). Afterward, a pairwise *t*-test was used to determine the sample's mean differences ($p < 0.05$). The relationship between mayonnaise properties and the sensory scores was determined by a correlation coefficient analysis and expressed as Correlation Index (CI). All statistical analyses were done in Microsoft Office Excel 365 using the open-source Real Statistics Resource Pack [30]. Artwork (Figures) was processed in Python 3.9 using Matplotlib (v. 3.4.2) library in Spyder (v. 5.1.1) IDE.

3. Results and Discussion

3.1. Protein Concentrates Characterization and Techno-Functional Properties

The content of protein (dry basis) found in SPC was 66.1 $\pm$ 1.5%, this concentration was slightly lower to others reports on *A. platensis* protein concentrates (69–75%) [19,31]. The difference in protein concentration may be due to the biomass growth conditions or the protein extraction process, in this regard, the use of pretreatments (sonication or high pressures) for cell disruption as used in [19,32] and [33] could increase the protein concentration up to 73–85% in the concentrates.

The WHC and OHC properties of SPC were 0.49 $\pm$ 0.029 g/g and 3.25 $\pm$ 0.218 g/g, respectively. The WHC observed for the SPC was very lower than previous reports (3–5 g/g) [17,20] but similar than the values found by Bleakley and Hayer [32]. In the case of the OHC, it was slightly higher than the reported by Benelhadj et al. (2.5 g/g) [17] but lower than the OHC of the SPC in [20,32], these studies demonstrated around 6–8 g/g, this enhanced OHC could be attributed to a partial unfolding of the protein structure due to the sonication pretreatment for the protein extraction.

For soy protein (control) the WHC (18.77 ± 1.091 g/g) and OHC (1.48 ± 0.126 g/g) were significantly higher and lower ($p < 0.05$), respectively, compared to the results of SPC. In related works [34–36], the OHC of soy protein (1.0–5.5 g/g) agrees with the results of this study. In contrast, the WHC determined herein was higher than previous studies (3.5–6.9 g/g) [35–37]. The difference in WHC may be due to the processing of the protein or drying methods.

3.2. *Mayonnaise Characterization*

3.2.1. Stability Analysis

The physical stability of the mayonnaise is an important quality attribute, but also thermal stability for applications in hot dishes such as hamburgers, hot dogs, and grilled sandwiches, among others. The low-fat mayonnaises showed excellent physical stability (100% of stability and any syneresis was observed) during storage (30 days) at 4 °C. This result was consistent with previous reports of similar systems; for example, low-fat mayonnaises developed with *A. platensis* and *Dunaliella* proteins and starch showed a low reduction of G′ and G″ during 60 days of storage, which was indicative of a physically stable network [38].

In the case of the thermal stability (Figure 1), this property was negatively affected by the increase in the oil concentration ($p < 0.05$), for instance, the stability was in the range of 81–90% at 30% of oil, where the highest value was found in D′, E′, and F′ systems. The mayonnaise formulated with 15% oil showed similar stability and presented the best stability (>94%) compared to all systems elaborated. Finally, as was expected, the commercial product did not show instability against high temperature (due to the high content of stabilizers), and it was significantly higher ($p < 0.05$) than all our samples.

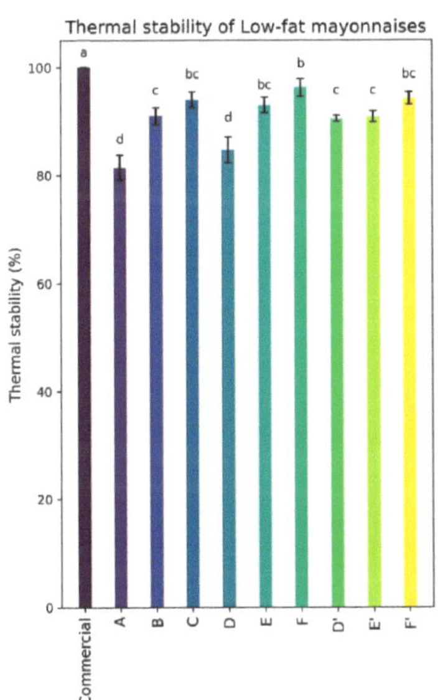

Figure 1. Thermal stability of low-fat mayonnaises. A, B, C = sodium alginate low fat mayonnaises with 30, 22.5 and 15% oil. D, E, F = low fat mayonnaises with 30, 22.5 and 15% oil and soy protein isolate. D′, E′, F′ = low fat mayonnaises with 30, 22.5% oil and spirulina protein concentrate. Different superscript letters indicate significant difference ($\alpha = 0.05$).

Mayonnaises are physically stable emulsions due to the high viscosity of the system; however, the stability could be affected by ingredient interactions (proteins and polysaccharides). For instance, pea protein used in the low-fat mayonnaises decreased the product's thermal stability (62%) [39]. In contrast, using starches or cereals flour has enhanced thermal stability (~99%) [7,40]. The products' stability in this study agrees with gelled emulsions with sodium alginate (previously reported), which demonstrated ~95% stability at pH 2 and 4 [41]. Moreover, Yang et al. [42] and Li et al. [12] developed very similar mayonnaises to the present study (containing sodium alginate) with good thermal stability due to minor microstructural changes (observed by Confocal Laser Scanning Microscopy). However, these results cannot be compared with those obtained in the present work. A possible synergistic mechanism among the polysaccharides and proteins in our products probably helped produce systems with good thermal stability. In this regard, sodium alginate could increase the stability due to the irreversible gelation of the continuous phase, preventing the mobility of oil droplets [13,43]. Meanwhile, the WHC and the OHC could favor the interactions between the protein hydrophilic residues with the hydrogel and the hydrophobic protein surface with the oil.

3.2.2. Particle Size Analysis

The particle size in the samples increased when the oil concentration decreased and was significantly higher than the commercial product ($p < 0.05$) (Figure 2). All samples with 30% oil (A, D, and D') demonstrated a mean particle size in a range of 6–9 μm; however, the presence of protein showed a slight reduction in particle size (6–7 μm). In samples with lower oil content (22.5%), there was an increase in the particle size for A, B, C, D, E, and F (~10 μm), while the systems with SPC (D', E,' and F') showed the twice size compared to the samples with 30% oil. In the case of the mayonnaises with the lowest content of oil (15%), D', E,' and F' showed smaller particle sizes (12 μm) in comparison to the rest of the samples where the particles increased up to 30 μm. Two possible mechanisms for this increase in particle size could occur. First, A, B, and C behavior could be due to the absence of a second emulsifier to help reduce the interfacial tension and reach smaller droplet sizes in emulsions [44]. In the case of D, E, and F, the SPI could contribute to a larger droplet size because the mayonnaise pH is close to this protein's isoelectric point (4–4.5). Hence, when the net charge is near or equal to zero (at the isoelectric point of the protein) leads to poor emulsifying activity and the generation of larger droplet size [45].

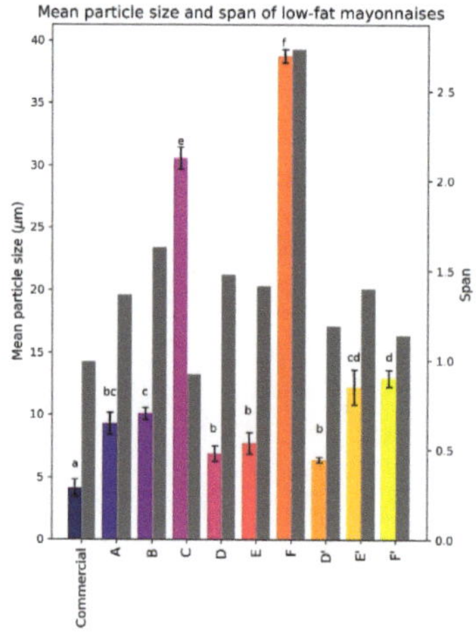

Figure 2. Mean particle size (color bars) and span (gray bars) of the low-fat mayonnaises. A, B, C = sodium alginate low fat mayonnaise with 30, 22.5 and 15% oil. D, E, F = low fat mayonnaises with 30, 22.5 and 15% oil and soy protein isolate. D', E', F' = low fat mayonnaises with 30, 22.5 and 15% oil and spirulina protein concentrate. The different superscript letter indicates a significant difference ($p < 0.05$).

Our products showed larger particle sizes than related studies of low-fat mayonnaises. For example, Sun et al. [46] developed mayonnaises with microparticulate Whey Protein Isolate (M-WPI), and the systems had approximately 2 μm in diameter. The difference in the particle size could be because the M-WPI have a tiny particle size (~8 μm), contributing to reducing the lipid droplets size in their products [46]. Nevertheless, the particle size determined in the samples with more than 22.5% oil showed comparable results to a previous report of low-fat mayonnaise (2–12 μm) elaborated with modified starches [7].

The span measurement indicates the width of the particle size distribution. Higher span values indicate more heterogeneous particle sizes within the system. The lowest span values (Figure 2) were found for the commercial low-fat mayonnaise and C (~0.9). In contrast, the rest of the samples showed values of 1.1–1.6, and the highest span (span = 2.7) was recorded in F′. The particle size distribution (Figure 3) in commercial mayonnaise and the samples with 22.5 and 30% oil (A, B, D, E, D′, and E′) was monomodal; on the contrary, the mayonnaises with 15% oil content (C, F, and F′) showed a bimodal or multimodal distribution. Usually, systems with high viscosity and homogeneous particle size distribution (monomodal) have been related to better stability. In contrast, multimodal distributions are usually observed in low viscosity and unstable systems [47]. In our low-fat mayonnaises, only the latter behavior was detected (higher viscosity for monomodal systems), and even though the particle size distribution (multimodal) could induce instability, the presence of the sodium alginate probably helped to prevent the oil droplet movement and their further coalescence in the products [10].

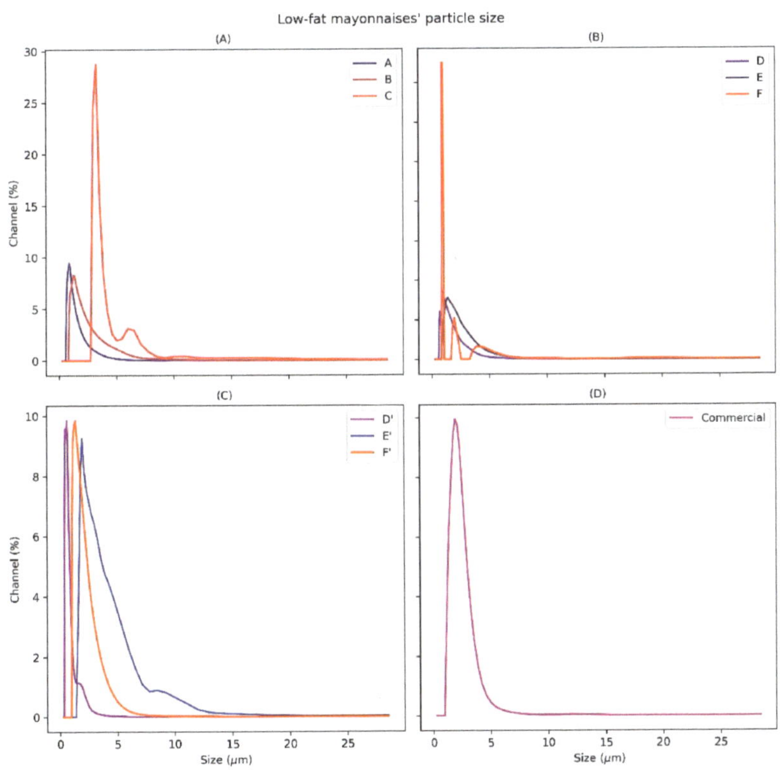

Figure 3. Particle size distribution of different mayonnaises. (**A**) Sodium alginate low-fat mayonnaises (A, B, C = 30, 22.5, 15% oil, respectively). (**B**) Low-fat mayonnaises with soy protein isolate (D, E, F = 30, 22.5, 15% oil, respectively). (**C**) Low-fat mayonnaises with spirulina protein concentrate (D′, E′, F′ = 30, 22.5, 15 oil%, respectively). (**D**) Commercial low-fat mayonnaise.

3.2.3. Rheological Characterization

Regarding the rheological behavior, the low-fat mayonnaises fit well with the power law model (r^2 = 0.998–0.999 and RMSE = 0.284–1.442). All samples showed a power law index in a range of 0.28–0.33 (Table 2), indicating a shear-thinning fluid (pseudoplastic fluid) behavior (power law index < 1). The results herein agree with previous reports of low-fat mayonnaises where this parameter was 0.28–0.52 [25,48].

Table 2. Rheological behavior of low-fat mayonnaise samples using different oil concentrations (%).

Mayonnaise Sample	K (Pa·s^n)	ΔK	n	RMSE	r^2
Commercial	43.284	ND	0.297	0.583	0.999
A	66.262	0.000	0.281	1.442	0.998
B	35.008	31.223	0.260	1.397	0.999
C	20.922	45.291	0.335	0.915	0.999
D	35.284	0.000	0.281	0.883	0.999
E	31.524	3.524	0.299	1.062	0.998
F	25.876	9.408	0.329	0.380	0.999
D′	49.552	0.000	0.308	1.324	0.998
E′	43.879	5.673	0.305	0.839	0.999
F′	35.013	14.539	0.308	0.284	0.999

K (consistency index), n (power law index), RMSE (Root Mean Square Error), r^2 (Coefficient of Determination).

On the other hand, the consistency index (K) represents the viscosity of the sample; the highest K values were found in samples with 30% oil (66–35 Pa·s^n), while the minimum values (20–35 Pa·s^n) were determined in the mayonnaises elaborated containing 15% oil (Table 2). Uribe-Wandurraga et al. [38] prepared low-fat mayonnaises (30% oil) using *Chlorella*, *Spirulina*, and *Dunaliella* microalgae, demonstrating higher K (74–94.5 Pa·s^n) than those found in this work, probably because these products also contained 4% starch, which increased the viscosity of the product. However, the K values of our samples agree with previous reports of low-fat mayonnaises (K = 31–49 Pa·s^n) elaborated with different gums such as xanthan gum, guar gum, and corn-dextran [25,48]. Like the behavior observed herein, Park et al. [8] also determined that the consistency index of low-fat mayonnaise can reduce from 90–51 and 87–23 Pa·s^n when the oil content is reduced from 54 to 38%. This could be explained by the role of oil particle-particle interactions in developing the viscosity of mayonnaise [49]. Hence, if the oil content is low, the number of oil particle-particle interactions will be reduced, and the system's viscosity will be lower.

It was found that the use of vegetable proteins, the WHC, and the OHC showed a negative correlation (CI = −0.936, −0.732, and −0.571, respectively) with the ΔK (loss of consistency index), indicating that proteins helped to mitigate the loss of viscosity. For instance, the reduction of oil content (from 30 to 15%) in samples with SPI (D, E, and F) and SPC (D′, E,′ and F′) results in minor losses of consistency index (ΔK = 3–15 Pa·s^n) in comparison to samples containing only sodium alginate (ΔK = 31–45 Pa·s^n). This was probably because the techno-functional properties of proteins might increase the number and strength of inter and intramolecular interactions [50,51].

The viscosity showed to be positively affected by the oil content (p < 0.05); as can be seen in Figure 4, the viscosity of the systems showed to reduce at higher shear stress; moreover, the viscosity at zero shear rate tends to reduce with reduced oil concentrations. The highest viscosities at zero shear rate were obtained in samples A (175 Pa·s) and D′ (125 Pa·s), while the lowest values were found in the D (~100 Pa·s). In summary, the samples A, D′ and E′ were demonstrated to have similar rheological characteristics (consistency index, viscosity, and power law index) to the commercial product.

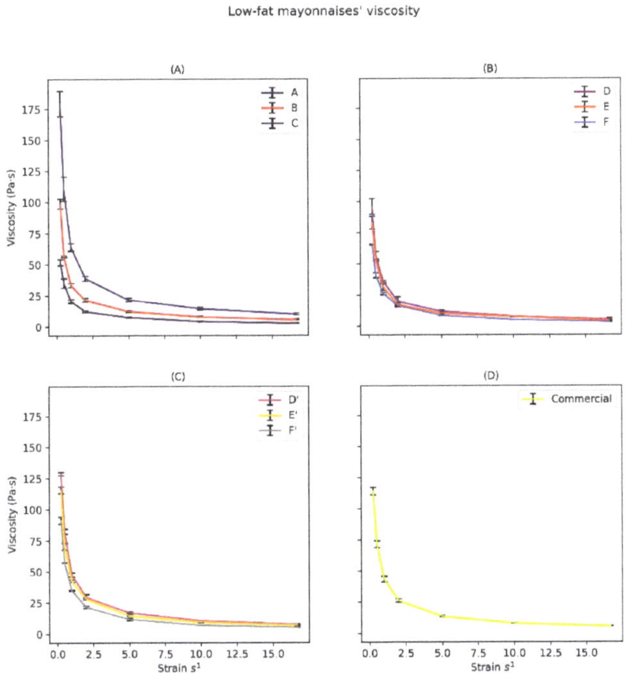

Figure 4. The viscosity of different low-fat mayonnaises at different strain stress. (**A**) Sodium alginate low-fat mayonnaises (A, B, C = 30, 22.5, 15% oil, respectively). (**B**) Low-fat mayonnaises with soy protein isolate (D, E, F = 30, 22.5, 15% oil, respectively). (**C**) Low-fat mayonnaises with spirulina protein concentrate (D′, E′, F′ = 30, 22.5, 15% oil, respectively). (**D**) Commercial low-fat mayonnaise.

3.2.4. Sensory Analysis

The results obtained from the sensory evaluation of the low-fat mayonnaises are shown in Figure 5; in addition, the relationship between the sensory scores and the products' characteristics (oil concentration, presence of protein, and viscosity) was determined using a correlation matrix (Figure 6). It is noteworthy that sample A was not included in the correlation matrix because it was detected as an outlier for the correlation between *K* with the taste score and the acceptability of the product (the full correlation matrix can be consulted in Figure S1).

The appearance of all mayonnaises containing SPC (D′, E,′ and F′) was significantly lower ($p < 0.05$) in comparison to the rest of the samples, probably due to the green color of the SPC (Figure 7); therefore, proteins with creamy-like color such as from amaranth, oat, or rice could be considered for further reformulations. The rest of the products demonstrated identical evaluations ($p > 0.05$) compared to the commercial product. This sensory attribute was demonstrated to positively correlate with the oil concentration, the WHC, and the consistency index (+0.29, +0.49, and +0.47, respectively). This result probably because these factors enhance the thickness of the system and mimic the appearance of the mayonnaise. In contrast, the presence of proteins had a negative correlation (CI = −0.46) with the appearance, and the same trend was found for the OHC (CI = −0.84). In this regard, the negative effect of proteins and the OHC could result from the low scoring in the mayonnaises containing SPC.

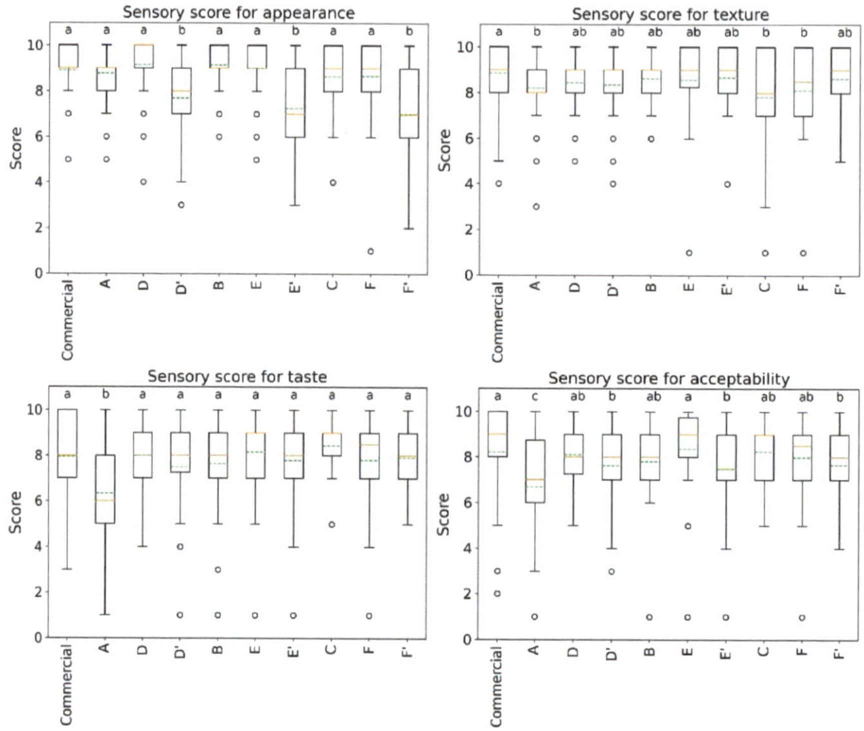

Figure 5. Structured scale results for sensory evaluation of low-fat mayonnaises. Dotted lines represent the arithmetic mean, and the solid yellow line indicates the median. Different letter in the same panel indicates a significant difference ($p < 0.05$).

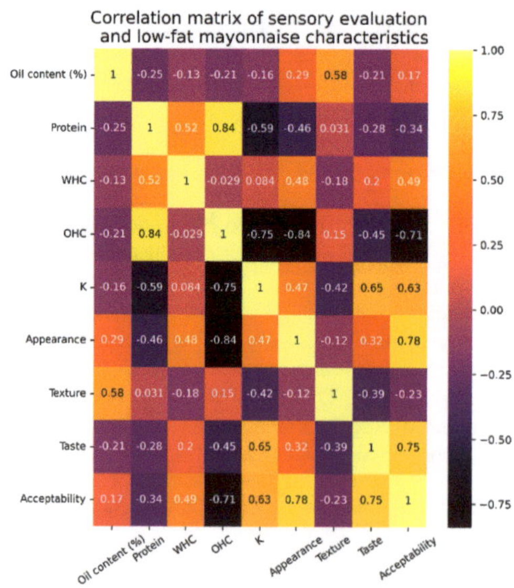

Figure 6. Correlation matrix of low-fat mayonnaise characteristics and sensory attributes.

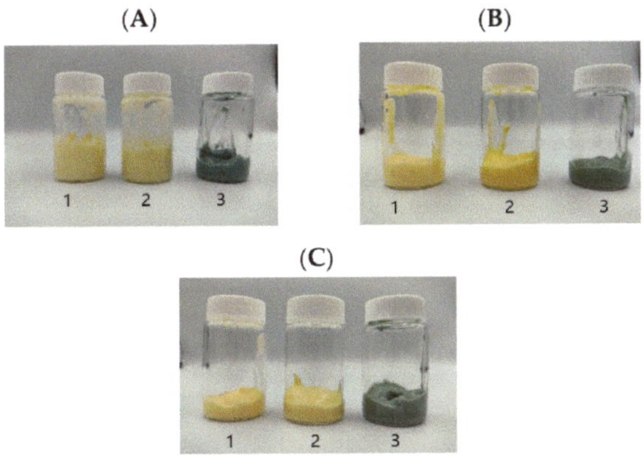

Figure 7. Low-fat mayonnaises obtained with 30%, 22.5%, and 15% oil (panels (**A**), (**B**), and (**C**), respectively). 1 = A, B, and C samples. 2 = D, E, and F samples. 3 = D′, E′, and F′ samples.

The texture of the samples B, D, D′, E, E′, and F′ demonstrated similar scoring to the commercial product ($p > 0.05$). Conversely, the A, C, and F scored significantly lower than the commercial mayonnaise ($p < 0.05$). The main contributor to desired texture was the oil concentration (CI = +0.58), followed by the OHC of proteins (CI = +0.15). The K was negatively correlated with this sensory attribute (CI = −0.42), suggesting that even if the hydrogel formed provided similar viscosity to a commercial low-fat mayonnaise, using only sodium alginate did not mimic the lubricity and the smooth mouthfeel expected for this product [52]. However, the techno-functionality of the proteins could develop these textural properties [53].

All samples containing SPC (D′, E′, and F′) generally showed an acceptability score higher than 7. However, these values were significantly lower than the commercial mayonnaise ($p < 0.05$).

One of the decisive factors was the product's appearance; nevertheless, the panelists commented that these samples with SPC could be directed toward producing flavored mayonnaises, such as jalapeño, coriander, and avocado taste. The other samples showed similar acceptability with the commercial mayonnaise ($p > 0.05$), except for the A sample. The most correlated characteristic with the acceptability was the appearance and the taste scores (CI = +0.78 and +0.75, respectively). Additionally, WHC and the consistency index (K) (CI = +0.49–+0.63, respectively) contributed to the acceptability mainly because they enhanced the products' appearance and taste. In addition, the texture score showed an inverse relation with the acceptability; this could suggest that the consumers are more interested in the product's appearance and taste than the texture.

The sensory evaluation of our products demonstrated better results compared to similar formulations. For example, Yang et al. [42] developed mayonnaises with an emulsion-filled gel approach using sodium alginate and 30% oil; these products showed acceptability around 4.9–6.9. On the other hand, comparable acceptability values (7–8.5) in microparticulate low-fat mayonnaises (20–40% oil reduction) using whey protein isolate are reported by [46].

Finally, the E, C, D, and the D′ showed similar or higher purchase (>65%) intention in comparison to the commercial low-fat mayonnaise (68%), indicating that these selected products have promising opportunities in the market (Figure 8). In contrast, the rest of the samples showed a lower probability of being purchased by the consumers.

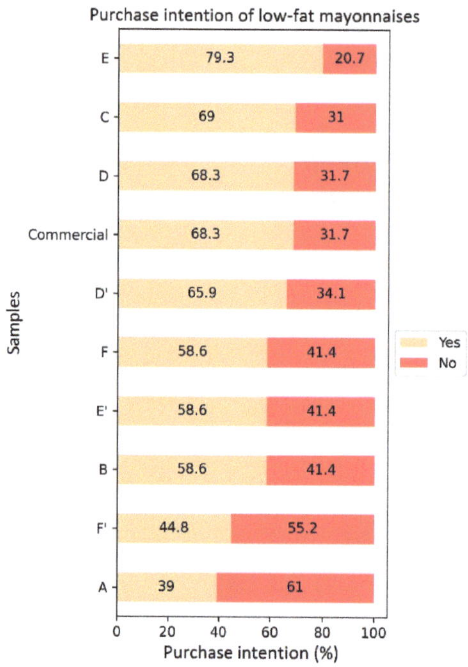

Figure 8. Purchase intention of different low-fat mayonnaise formulations.

3.2.5. Amino Acid Bioavailability

For evaluating the essential amino acids (EAA) provided by the proteins used, it was decided to study the bioavailability of EAA in the sample with higher purchase intention containing soy protein (sample E) and SPC (sample D′). Furthermore, these were compared to the EAA bioavailability in the commercial product (Table 3). Control samples (not subjected to digestion) type E (soy) contained all essential amino acids; in the case of sample D′ (spirulina), only His was not detected. The bioavailability profile showed a nutritional improvement compared to commercial samples, which lacked four essential amino acids (His, Phe, Ile, and Lys). The total EAA ($\sum$ EAA) in low-fat mayonnaises showed around 21 mg/25 g, which overcomes ($p < 0.05$) to the $\sum$ EAA of commercial products (14.645 $\pm$ 0.465 mg/25 g).

After the mouth digestion (time 1), there was an increase in the concentration of Val, Trp, Phe, Ile, and Lys in the D′ sample. Similarly, sample E slightly increased Trp, Phe, Ile, and Leu, but the His was not detected in this sample. Probably the amylase and environment conditions (pH) promoted hydrogel hydrolysis (artificial saliva) formed by sodium alginate gum, releasing the protein and thus increasing its availability [27,54]. In this step, the highest $\sum$ EAA found was in D′ (25.132 $\pm$ 2.181 mg/25 g), followed by E (19.684 $\pm$ 1.929 mg/25 g), remaining the lowest in the commercial low-fat mayonnaise (15.352 $\pm$ 0.601 mg/25 g).

Table 3. Essential amino acid profile for E, D′ and Commercial low-fat mayonnaises.

Amino Acid (mg/25 g Product)	D′	E	Commercial
		Time 0	
His	ND	1.740 ± 0.068 a	ND
Thr	2.613 ± 0.096 a	2.266 ± 0.059 b	2.191 ± 0.157 b
Val	1.566 ± 0.098 a	1.102 ± 0.043 b	0.656 ± 0.076 c
Met	0.774 ± 0.155 a	0.621 ± 0.131 a	0.511 ± 0.054 a
Trp	8.149 ± 1.478 ab	7.533 ± 0.535 b	9.278 ± 0.106 a
Phe	2.120 ± 0.177 b	2.537 ± 0.083 a	ND
Ile	1.142 ± 0.300 a	1.059 ± 0.048 a	ND
Leu	3.400 ± 0.091 a	2.699 ± 0.187 b	2.007 ± 0.069 c
Lys	1.729 ± 0.108 a	1.506 ± 0.170 a	ND
∑ EAA	21.496 ± 1.203 a	21.066 ± 1.328 a	14.645 ± 0.464 b
		Time 1	
His	ND	ND	ND
Thr	2.451 ± 0.162 a	2.217 ± 0.212 a	1.520 ± 0.033 b
Val	1.302 ± 0.071 a	1.032 ± 0.077 b	0.709 ± 0.004 c
Met	0.914 ± 0.149 a	0.753 ± 0.024 a	0.628 ± 0.005 b
Trp	10.624 ± 0.095 a	7.731 ± 0.997 b	10.966 ± 0.492 a
Phe	3.029 ± 0.190 a	2.435 ± 0.021 b	ND
Ile	1.331 ± 0.107 a	1.186 ± 0.038 a	ND
Leu	3.176 ± 0.478 a	2.825 ± 0.463 a	1.527 ± 0.064 b
Lys	2.303 ± 0.930 a	1.501 ± 0.093 a	ND
∑ EAA	25.132 ± 2.181 a	19.684 ± 1.929 b	15.352 ± 0.601 c
		Time 2	
His	ND	ND	ND
Thr	3.219 ± 0.164 a	3.038 ± 0.131 a	2.481 ± 0.499 a
Val	1.326 ± 0.141 a	1.274 ± 0.115 a	0.610 ± 0.166 b
Met	0.850 ± 0.146 a	0.779 ± 0.024 a	ND
Trp	13.222 ± 0.048 a	13.354 ± 1.291 a	2.471 ± 0.301 b
Phe	3.587 ± 0.630 a	3.113 ± 0.352 a	ND
Ile	1.355 ± 0.140 a	1.094 ± 0.187 a	ND
Leu	3.714 ± 0.235 a	3.324 ± 0.065 a	ND
Lys	1.992 ± 0.191 a	2.076 ± 0.105 a	ND
∑ EAA	29.265 ± 1.695 a	28.052 ± 2.270 a	5.562 ± 0.966 b

ND = Not Detected. E = mayonnaise with soy protein isolate and 22.5% oil. D′ = mayonnaise with spirulina protein concentrate and 30% oil. ∑ EAA = sum of Essential Amino Acids. Different letters in the same row indicates significant difference among samples ($p < 0.05$).

After gastric digestion (time 2), all EAA increased, standing out samples E and D′ (made with plant proteins). Only three EAA (Thr, Val, Trp) remained in commercial samples, while E and D′ samples were complete (except for His). The ∑ EAA of the commercial product (5.562 ± 0.966 mg/25 g) was significantly lower ($p < 0.05$) in comparison to our low-fat mayonnaises (∑ EAA = 28–29 mg/25 g). The increase of EAAs after gastric digestion could be explained by the disruption of the peptide bonds induced by the pepsin activity (proteolytic), provoking the release of a higher amount of EAAs from the low-fat mayonnaise [55,56].

This digestion phase is essential because es followed by the ileal digestion and absorption of amino acids; hence, our products could provide a higher amount of available EAAs. Even though the concentrations of EAAs in the mayonnaises did not fulfill the FAO requirements [57], the formulations proposed herein could slightly contribute to EAA intake in the diet with approximately one tablespoon.

4. Conclusions

The present study demonstrated that the emulsion-filled gels could be used to formulate low-fat mayonnaises. These products showed good physical stability, similar rheological characteristics to commercial products, and an increase in bioavailability of essential amino acid content (after buccal and gastric digestion). Moreover, it was possible to obtain well-accepted products with vegetable proteins and low oil concentration. Specifically, the purchase intention demonstrated that new products could be developed with the following formulations: soy protein isolate with 22.5–30% oil or *A. platensis* protein concentrate with 30% oil. Further research could be conducted on using different proteins with high WHC and creamy-like color. These could enhance the taste and appearance, which were the most relevant attributes for the final acceptability of the product.

Supplementary Materials: The following supporting information can be downloaded at: https://www.mdpi.com/article/10.3390/app12157456/s1. Figure S1. Correlation matrix of all low-fat mayonnaise (including sample A) characteristics and sensory attributes. Table S1. Amino acid profile of low-fat mayonnaises.

Author Contributions: Conceptualization, D.B.-A. and J.M.-O.; methodology, D.B.-A., J.M.-O. and L.R.-O.; validation, D.B.-A. and M.R.-R.; formal analysis, D.B.-A. and M.R.-R.; investigation, J.M.-O. and L.R.-O.; resources, D.B.-A. and M.R.-R.; data curation, D.B.-A. and J.M.-O.; writing—original draft preparation, J.M.-O. and D.B.-A.; writing—review and editing, D.B.-A. and M.R.-R.; visualization, J.M.-O.; supervision, D.B.-A. and M.R.-R.; project administration, D.B.-A.; funding acquisition, D.B.-A. All authors have read and agreed to the published version of the manuscript.

Funding: This research and the APC was funded by "Consejo Nacional de Ciencia y Tecnología" CONACyT, grant number FOINS 4950.

Institutional Review Board Statement: The Research and Ethics Committee from Health Sciences Department in the "Universidad de Las Americas Puebla" approved the protocol for the sensory evaluation conducted in this work on 4 May 2021 (document number P-001).

Informed Consent Statement: Informed consent was obtained from all participants at the beginning of the sensory evaluation.

Data Availability Statement: Not applicable.

Acknowledgments: The authors acknowledge to "Universidad De Las Americas Puebla (UDLAP)" and "Consejo Nacional de Ciencia y Tecnología (CONACyT)" for the scholarships granted to J. Metri Ojeda and the access to laboratories, as well as the funds granted to the project FOINS 4950.

Conflicts of Interest: The authors declare no conflict of interest.

References

1. Maruyama, S.; Streletskaya, N.A.; Lim, J. Clean label: Why this ingredient but not that one? *Food Qual. Prefer.* **2020**, *87*, 104062. [CrossRef]
2. Taslikh, M.; Mollakhalili-Meybodi, N.; Alizadeh, A.M.; Mousavi, M.-M.; Nayebzadeh, K.; Mortazavian, A.M. Mayonnaise main ingredients influence on its structure as an emulsion. *J. Food Sci. Technol.* **2021**, *59*, 2108–2116. [CrossRef]
3. Mirzanajafi-Zanjani, M.; Yousefi, M.; Ehsani, A. Challenges and approaches for production of a healthy and functional mayonnaise sauce. *Food Sci. Nutr.* **2019**, *7*, 2471–2484. [CrossRef]
4. Teklehaimanot, W.H.; Duodu, K.G.; Emmambux, M.N. Maize and teff starches modified with stearic acid as potential fat replacer in low calorie mayonnaise-type emulsions. *Starch-Stärke* **2013**, *65*, 773–781. [CrossRef]
5. Agyei-Amponsah, J.; Macakova, L.; DeKock, H.L.; Emmambux, M.N. Effect of Substituting Sunflower Oil with Starch-Based Fat Replacers on Sensory Profile, Tribology, and Rheology of Reduced-Fat Mayonnaise-Type Emulsions. *Starch-Stärke* **2021**, *73*, 2000092. [CrossRef]
6. Su, H.; Lien, C.; Lee, T.; Ho, J. Development of Low-fat Mayonnaise Containing Polysaccharide Gums as Functional Ingredi-ents. *J. Sci. Food Agric.* **2010**, *90*, 806–812. [CrossRef]
7. Bajaj, R.; Singh, N.; Kaur, A. Properties of octenyl succinic anhydride (OSA) modified starches and their application in low fat mayonnaise. *Int. J. Biol. Macromol.* **2019**, *131*, 147–157. [CrossRef]
8. Park, J.J.; Olawuyi, I.; Lee, W.Y. Characteristics of low-fat mayonnaise using different modified arrowroot starches as fat replacer. *Int. J. Biol. Macromol.* **2020**, *153*, 215–223. [CrossRef]

9. Varela, P.; Fiszman, S. Exploring consumers' knowledge and perceptions of hydrocolloids used as food additives and ingredients. *Food Hydrocoll.* **2013**, *30*, 477–484. [CrossRef]
10. Farjami, T.; Madadlou, A. An overview on preparation of emulsion-filled gels and emulsion particulate gels. *Trends Food Sci. Technol.* **2019**, *86*, 85–94. [CrossRef]
11. Geremias-Andrade, I.M.; Souki, N.P.; Moraes, I.C.; Pinho, S.C. Rheology of Emulsion-Filled Gels Applied to the Development of Food Materials. *Gels* **2016**, *2*, 22. [CrossRef]
12. Li, A.; Gong, T.; Hou, Y.; Yang, X.; Guo, Y. Alginate-stabilized thixotropic emulsion gels and their applications in fabrication of low-fat mayonnaise alternatives. *Int. J. Biol. Macromol.* **2019**, *146*, 821–831. [CrossRef]
13. Yang, X.; Gong, T.; Lu, Y.-H.; Li, A.; Sun, L.; Guo, Y. Compatibility of sodium alginate and konjac glucomannan and their applications in fabricating low-fat mayonnaise-like emulsion gels. *Carbohydr. Polym.* **2020**, *229*, 115468. [CrossRef]
14. Benslima, A.; Sellimi, S.; Hamdi, M.; Nasri, R.; Jridi, M.; Cot, D.; Li, S.; Nasri, M.; Zouari, N. The brown seaweed Cystoseira schiffneri as a source of sodium alginate: Chemical and structural characterization, and antioxidant activities. *Food Biosci.* **2020**, *40*, 100873. [CrossRef]
15. Ramírez-Rodrigues, M.; Estrada-Beristain, C.; Metri-Ojeda, J.; Pérez-Alva, A.; Baigts-Allende, D. *Spirulina platensis* Protein as Sustainable Ingredient for Nutritional Food Products Development. *Sustainability* **2021**, *13*, 6849. [CrossRef]
16. Torres-Tiji, Y.; Fields, F.J.; Mayfield, S.P. Microalgae as a future food source. *Biotechnol. Adv.* **2020**, *41*, 107536. [CrossRef] [PubMed]
17. Benelhadj, S.; Gharsallaoui, A.; Degraeve, P.; Attia, H.; Ghorbel, D. Effect of pH on the functional properties of Arthrospira (Spirulina) platensis protein isolate. *Food Chem.* **2016**, *194*, 1056–1063. [CrossRef]
18. Böcker, L.; Bertsch, P.; Wenner, D.; Teixeira, S.; Bergfreund, J.; Eder, S.; Fischer, P.; Mathys, A. Effect of Arthrospira platensis microalgae protein purification on emulsification mechanism and efficiency. *J. Colloid Interface Sci.* **2020**, *584*, 344–353. [CrossRef]
19. Menegotto, A.L.L.; de Souza, L.E.S.; Colla, L.M.; Costa, J.A.V.; Sehn, E.; Bittencourt, P.R.S.; Flores, L.D.M.; Canan, C.; Colla, E. Investigation of techno-functional and physicochemical properties of Spirulina platensis protein concentrate for food enrichment. *LWT* **2019**, *114*, 108267. [CrossRef]
20. Yücetepe, A.; Saroğlu, Ö.; Özçelik, B. Response surface optimization of ultrasound-assisted protein extraction from Spirulina platensis: Investigation of the effect of extraction conditions on techno-functional properties of protein concentrates. *J. Food Sci. Technol.* **2019**, *56*, 3282–3292. [CrossRef] [PubMed]
21. Ojeda, J.M.; Rodrigues, M.R.; Allende, D.B. Study of the perception and the acceptability of mayonnaise ingredients among Mexican consumers and its global preference. *Rev. Española De Nutr. Hum. Y Dietética* **2022**, *26*, 1620. [CrossRef]
22. Stone, A.K.; Avarmenko, N.A.; Warkentin, T.D.; Nickerson, M.T. Functional properties of protein isolates from different pea cultivars. *Food Sci. Biotechnol.* **2015**, *24*, 827–833. [CrossRef]
23. Nourbehesht, N.; Shekarchizadeh, H.; Soltanizadeh, N. Investigation of stability, consistency, and oil oxidation of emulsion filled gel prepared by inulin and rice bran oil using ultrasonic radiation. *Ultrason. Sonochemistry* **2018**, *42*, 585–593. [CrossRef]
24. Xu, W.; Xiong, Y.; Li, Z.; Luo, D.; Wang, Z.; Sun, Y.; Shah, B.R. Stability, microstructural and rheological properties of complex prebiotic emulsion stabilized by sodium caseinate with inulin and konjac glucomannan. *Food Hydrocoll.* **2020**, *105*, 105772. [CrossRef]
25. Kumar, Y.; Roy, S.; Devra, A.; Dhiman, A.; Prabhakar, P.K. Ultrasonication of mayonnaise formulated with xanthan and guar gums: Rheological modeling, effects on optical properties and emulsion stability. *LWT* **2021**, *149*, 111632. [CrossRef]
26. Sukkwai, S.; Chonpracha, P.; Kijroongrojana, K.; Prinyawiwatkul, W. Influences of a natural colourant on colour and salty taste perception, liking, emotion and purchase intent: A case of mayonnaise-based dipping sauces. *Int. J. Food Sci. Technol.* **2017**, *52*, 2256–2264. [CrossRef]
27. Rui, X.; Zhang, Q.; Huang, J.; Li, W.; Chen, X.; Jiang, M.; Dong, M. Does lactic fermentation influence soy yogurt protein digestibility: A comparative study between soymilk and soy yogurt at different pH. *J. Sci. Food Agric.* **2018**, *99*, 861–867. [CrossRef]
28. Shim, S.-M.; Choi, M.-H.; Park, S.-H.; Gu, Y.-U.; Oh, J.-M.; Kim, S.; Kim, H.-Y.; Kim, G.-H.; Lee, Y. Assessing the digestibility of genetically modified soybean: Physiologically based in vitro digestion and fermentation model. *Food Res. Int.* **2010**, *43*, 40–45. [CrossRef]
29. Matsushita, K. Automatic Precolumn Derivatization of Amino Acids and Analysis by Fast LC Using the Agilent 1290 Infinity LC System. *Agil. Tech. Note* **2010**, *5990*, 1–4.
30. Zaiontz, C. Real Statistics Software. Available online: https://www.real-statistics.com/appendix/citation-real-statistics-software-website/ (accessed on 5 February 2022).
31. Safi, C.; Ursu, A.V.; Laroche, C.; Zebib, B.; Merah, O.; Pontalier, P.-Y.; Vaca-Garcia, C. Aqueous extraction of proteins from microalgae: Effect of different cell disruption methods. *Algal Res.* **2014**, *3*, 61–65. [CrossRef]
32. Bleakley, S.; Hayes, M. Functional and Bioactive Properties of Protein Extracts Generated from *Spirulina platensis* and *Isochrysis galbana T-Iso*. *Appl. Sci.* **2021**, *11*, 3964. [CrossRef]
33. Lozober, H.S.; Okun, Z.; Shpigelman, A. The impact of high-pressure homogenization on thermal gelation of Arthrospira platensis (Spirulina) protein concentrate. *Innov. Food Sci. Emerg. Technol.* **2021**, *74*, 102857. [CrossRef]
34. Stone, A.K.; Karalash, A.; Tyler, R.T.; Warkentin, T.D.; Nickerson, M.T. Functional attributes of pea protein isolates prepared using different extraction methods and cultivars. *Food Res. Int.* **2015**, *76*, 31–38. [CrossRef]

35. Alfaro-Diaz, A.; Urías-Silvas, J.; Loarca-Piña, G.; Gaytan-Martínez, M.; Prado-Ramirez, R.; Mojica, L. Techno-functional properties of thermally treated black bean protein concentrate generated through ultrafiltration process. *LWT* **2020**, *136*, 110296. [CrossRef]
36. Yu, D.; Zhao, Y.; Li, T.; Li, D.; Chen, S.; Wu, N.; Jiang, L.; Wang, L. Effect of electrochemical modification on the structural characteristics and emulsion storage stability of soy protein isolate. *Process. Biochem.* **2018**, *75*, 166–172. [CrossRef]
37. Bühler, J.M.; Dekkers, B.L.; Bruins, M.E.; van der Goot, A.J. Modifying Faba Bean Protein Concentrate Using Dry Heat to Increase Water Holding Capacity. *Foods* **2020**, *9*, 1077. [CrossRef]
38. Uribe-Wandurraga, Z.N.; Martínez-Sánchez, I.; Savall, C.; García-Segovia, P.; Martínez-Monzó, J. Microalgae fortification of low-fat oil-in-water food emulsions: An evaluation of the physicochemical and rheological properties. *J. Food Sci. Technol.* **2020**, *58*, 3701–3711. [CrossRef]
39. Shen, Y.; Babu, K.S.; Amamcharla, J.; Li, Y. Emulsifying properties of pea protein/guar gum conjugates and mayonnaise application. *Int. J. Food Sci. Technol.* **2022**, *57*, 3955–3966. [CrossRef]
40. Carcelli, A.; Crisafulli, G.; Carini, E.; Vittadini, E. Can a Physically Modified Corn Flour Be Used as Fat Replacer in a May-onnaise? *Eur. Food Res. Technol.* **2020**, *246*, 2493–2503. [CrossRef]
41. Sato, A.C.; Moraes, K.; Cunha, R. Development of gelled emulsions with improved oxidative and pH stability. *Food Hydrocoll.* **2012**, *34*, 184–192. [CrossRef]
42. Yang, X.; Li, A.; Yu, W.; Li, X.; Sun, L.; Xue, J.; Guo, Y. Structuring Oil-in-Water Emulsion by Forming Egg Yolk/Alginate Complexes: Their Potential Application in Fabricating Low-Fat Mayonnaise-like Emulsion Gels and Redispersible Solid Emulsions. *Int. J. Biol. Macromol.* **2020**, *147*, 595–606. [CrossRef] [PubMed]
43. Li, J.; Wang, Y.; Jin, W.; Zhou, B.; Li, B. Application of micronized konjac gel for fat analogue in mayonnaise. *Food Hydrocoll.* **2014**, *35*, 375–382. [CrossRef]
44. Chen, H.; Mao, L.; Hou, Z.; Yuan, F.; Gao, Y. Roles of additional emulsifiers in the structures of emulsion gels and stability of vitamin E. *Food Hydrocoll.* **2020**, *99*, 105372. [CrossRef]
45. Zhong, Y.; Xiang, X.; Wang, X.; Zhang, Y.; Hu, M.; Chen, T.; Liu, C. Fabrication and characterization of oil-in-water emulsions stabilized by macadamia protein isolate/chitosan hydrochloride composite polymers. *Food Hydrocoll.* **2020**, *103*, 105655. [CrossRef]
46. Sun, C.; Liu, R.; Liang, B.; Wu, T.; Sui, W.; Zhang, M. Microparticulated whey protein-pectin complex: A texture-controllable gel for low-fat mayonnaise. *Food Res. Int.* **2018**, *108*, 151–160. [CrossRef]
47. Drozłowska, E.; Bartkowiak, A.; Łopusiewicz, Ł. Characterization of Flaxseed Oil Bimodal Emulsions Prepared with Flaxseed Oil Cake Extract Applied as a Natural Emulsifying Agent. *Polymers* **2020**, *12*, 2207. [CrossRef]
48. Schädle, C.N.; Bader-Mittermaier, S.; Sanahuja, S. Characterization of Reduced-Fat Mayonnaise and Comparison of Sensory Perception, Rheological, Tribological, and Textural Analyses. *Foods* **2022**, *11*, 806. [CrossRef]
49. Katsaros, G.; Tsoukala, M.; Giannoglou, M.; Taoukis, P. Effect of storage on the rheological and viscoelastic properties of mayonnaise emulsions of different oil droplet size. *Heliyon* **2020**, *6*, e05788. [CrossRef]
50. León, O.; Soto, D.; López, D.; Muñoz-Bonilla, A.; Fernández-García, M. Fat-Replacer Properties of Oxidized Cassava Starch Using Hydrogen Peroxide/Sodium Bicarbonate Redox System in Mayonnaise Formulation and Its Stability. *Starch-Stärke* **2019**, *71*, 1900112. [CrossRef]
51. Chen, B.; Cai, Y.; Liu, T.; Huang, L.; Deng, X.; Zhao, Q.; Zhao, M. Improvements in physicochemical and emulsifying properties of insoluble soybean fiber by physical-chemical treatments. *Food Hydrocoll.* **2019**, *93*, 167–175. [CrossRef]
52. Du, M.; Lu, W.; Zhang, Y.; Mata, A.; Fang, Y. Natural polymer-sourced interpenetrating network hydrogels: Fabrication, properties, mechanism and food applications. *Trends Food Sci. Technol.* **2021**, *116*, 342–356. [CrossRef]
53. Wang, Y.; Jiao, A.; Qiu, C.; Liu, Q.; Yang, Y.; Bian, S.; Zeng, F.; Jin, Z. A combined enzymatic and ionic cross-linking strategy for pea protein/sodium alginate double-network hydrogel with excellent mechanical properties and freeze-thaw stability. *Food Hydrocoll.* **2022**, *131*, 107737. [CrossRef]
54. Zhu, Y.; Marin, L.; Xiao, Y.; Gillies, E.; Siqueira, W. pH-Sensitive Chitosan Nanoparticles for Salivary Protein Delivery. *Nanomaterials* **2021**, *11*, 1028. [CrossRef] [PubMed]
55. Fang, M.; Xiong, S.; Hu, Y.; Yin, T.; You, J. In vitro pepsin digestion of silver carp (Hypophthalmichthys molitrix) surimi gels after cross-linking by Microbial Transglutaminase (MTGase). *Food Hydrocoll.* **2019**, *95*, 152–160. [CrossRef]
56. Ketnawa, S.; Ogawa, Y. Evaluation of protein digestibility of fermented soybeans and changes in biochemical characteristics of digested fractions. *J. Funct. Foods* **2018**, *52*, 640–647. [CrossRef]
57. World Health Organization. *Protein and Amino Acids Requirement in Human Nutrition*; WHO Technical Report Series; World Health Organization: Geneva, Switzerland, 2002; p. 935.

applied sciences

MDPI

Article

Advances in the Characterization of *Usnea barbata* (L.) Weber ex F.H. Wigg from Călimani Mountains, Romania

Violeta Popovici [1], Laura Bucur [2,*], Cerasela Elena Gîrd [3,*], Suzana Ioana Calcan [4], Elena Iulia Cucolea [4], Teodor Costache [4], Dan Rambu [4], Mădălina Ungureanu-Iuga [5,6,*], Mircea Oroian [5,*], Silvia Mironeasa [5,†], Verginica Schröder [7,*], Emma Adriana Ozon [8,†], Aureliana Caraiane [9,†] and Victoria Badea [1,†]

1 Department of Microbiology and Immunology, Faculty of Dental Medicine, Ovidius University of Constanta, 7 Ilarie Voronca Street, 900684 Constanta, Romania; violeta.popovici@365.univ-ovidius.ro (V.P.); victoria.badea@365.univ-ovidius.ro (V.B.)
2 Department of Pharmacognosy, Faculty of Pharmacy, Ovidius University of Constanta, 6 Capitan Al. Serbanescu Street, 900001 Constanta, Romania
3 Department of Pharmacognosy, Phytochemistry, and Phytotherapy, Faculty of Pharmacy, Carol Davila University of Medicine and Pharmacy, 6 Traian Vuia Street, 020956 Bucharest, Romania
4 Research Center for Instrumental Analysis SCIENT, 1E Petre Ispirescu Street, 077167 Ilfov, Romania; suzana.calcan@scient.ro (S.I.C.); iulia.cucolea@scient.ro (E.I.C.); teodor.costache@scient.ro (T.C.); dan.rambu@scient.ro (D.R.)
5 Faculty of Food Engineering, Stefan cel Mare University of Suceava, 13th University Street, 720229 Suceava, Romania; silviam@fia.usv.ro
6 Integrated Center for Research, Development, and Innovation in Advanced Materials, Nanotechnologies and Distributed Systems for Fabrication and Control (MANSiD), Stefan cel Mare University of Suceava, 13th University Street, 720229 Suceava, Romania
7 Department of Cellular and Molecular Biology, Faculty of Pharmacy, Ovidius University of Constanta, 6 Capitan Al. Serbanescu Street, 900001 Constanta, Romania
8 Department of Pharmaceutical Technology and Biopharmacy, Faculty of Pharmacy, Carol Davila University of Medicine and Pharmacy, 6 Traian Vuia Street, 020956 Bucharest, Romania; emma.budura@umfcd.ro
9 Department of Oral Rehabilitation, Faculty of Dental Medicine, Ovidius University of Constanta, 7 Ilarie Voronca Street, 900684 Constanta, Romania; aureliana.caraiane@365.univ-ovidius.ro
* Correspondence: laurabucur@univ-ovidius.ro (L.B.); cerasela.gird@umfcd.ro (C.E.G.); madalina.iuga@usm.ro (M.U.-I.); m.oroian@fia.usv.ro (M.O.); verginica.schroder@univ-ovidius.ro (V.S.)
† These authors contributed equally to this work.

Abstract: *Usnea barbata* (L.) Weber ex F.H. Wigg (*U. barbata*) is a medicinal representative of the lichens from the *Usnea* genus (*Parmeliaceae*, lichenized *Ascomycetes*), containing bioactive secondary metabolites. The aim of this study is a comparative analysis between two separated parts of the thallus layers: medulla–cortex (*mc*UB) and central cord (*cc*UB) and the whole dried *U. barbata* thallus (*d*UB). These three samples were examined regarding color differences. The *U. barbata* thallus morphology was examined through fluorescent microscopy (FM) and scanning electron microscopy (SEM). The mineral content was measured using inductively coupled plasma mass spectrometry (ICP-MS), and Fourier transform infrared spectroscopy (FT-IR) preliminarily established the differences in the metabolite content. Finally, extracts in different solvents (ethanol and acetone) were obtained from all studied samples, and their total phenolic content (TPC) and free radical scavenging activity (antiradical activity, AA) were evaluated by spectrophotometry. The ICP-MS results showed that from 23 elements analyzed, 18 minerals were quantified in *mc*UB, 13 in *d*UB, and only 12 in *cc*UB. The *cc*UB fraction recorded the lowest mineral content, color intensity (chroma), luminosity (L^*), and TPC value, followed in increasing order by *d*UB and *mc*UB. FT-IR spectra displayed different peaks in *cc*UB and *d*UB samples compared to *mc*UB. The *mc*UB fraction also showed the highest TPC, significantly correlated with AA. However, *d*UB had the highest antiradical activity, followed by *mc*UB and *cc*UB, with noticeable differences in the acetone extract. The final correlation between all variable data obtained indicates that 99.31% of the total variance was associated with all minerals, total phenolics, and color parameters and was also related to the antiradical activity. These obtained results complete our previous studies on autochthonous *U. barbata*. Moreover, being a source of bioactive metabolites, extracting them from the *mc*UB fraction could increase this process's yield and selectivity.

Keywords: *Usnea barbata* (L.) Weber ex F.H. Wigg; morphology; color; minerals; phenolic secondary metabolites; free radical scavenging activity

1. Introduction

Lichens (also known as lichenized fungi [1]) are symbiont organisms between fungi and algae/cyanobacteria [2]. They could be considered a potential source of new drugs [3] due to their bioactive secondary metabolites with various pharmacological effects [4]. One of the most significant representatives in the lichens world is the *Usnea* genus (*Parmeliaceae*, lichenized *Ascomycetes*), a potent phytomedicine with various pharmacological activities [5]. It has more than 600 species, endemic to Asia, Africa, Europe, and America and is used worldwide for traditional medicines (except in Australia) and food preparation [6]. *Usnea* sp. can be transplanted for air monitoring [7], and, at the same time, their culture conditions could be optimized, aiming for biomass increasing and antioxidant metabolite production [8]. Due to their significant diversity, the *Usnea* lichens are mainly studied in various Earth regions: Europe [9], India [10], Russia [11], Taiwan [12], Turkey [13], and New Zeeland [14]. *U. barbata* is a valuable representative of the *Usnea* genus, used in ethnomedicine, homeopathy, and the cosmetic industry [15]. Moreover, Redzic et al. [16] proved that *U. barbata* was used as human food (mush and lichen bread) for survival in four-year-long (1992–1995) war conditions in Bosnia.

All *U. barbata* benefits are due to its chemical composition, which is strongly dependent on growth conditions [17]. Bazarnova et al. [18] quantified in *U. barbata* various primary and secondary metabolites. The most well-known phenolic secondary metabolites reported are oligosaccharides, monosaccharides (glucose and galactose), citric, succinic, alpha-linolenic, and stearic acids. According to Salgado et al. [19], various polyhydroxylated lipids, fatty acids, and other undetected compounds can be added to the *U. barbata* phytochemical profile.

The fungal partner (mycobiont) synthesizes these specific compounds mainly due to symbiosis. As a response to biological and abiotic stress, the secondary metabolites are extracellular and are found within the lichen thalli as crystals on the fungal hyphae surface [20]. The primary metabolites are required for lichen growth and nutrition. They are intracellular in fungal or algal (photobiont) cells [20]. By weight, the mycobiont predominates in *U. barbata*, compared to photobionts, assuring the form of the thallus built due to both symbiotic partners' activity. *U. barbata* is a fruticose lichen; its thallus has a heteromerous structure with three layers: cortex, medulla, and central axis (central cord). As a cover, *U. barbata* has a thin cortex (external layer) with a subcortical, low represented algal zone. These structures are followed by a medulla and a cartilaginous central axis (central cord) [9]. The upper cortex has a compact structure formed by a dense pack of fungal hyphae, with a protective role. Algal cells interweaved with fungal hyphae compose the green algal zone [9]. The medulla is thick and loose, with rare fungal hyphae. Finally, the central axis (central cord) is formed by longitudinal fungal hyphae that are densely packed, forming a compact cartilaginous structure [21]. Around this central cord, the other layers (medulla and cortex) and algal zone have a radial disposition. Both associated organisms function in the symbiotic process. The algal partner provides carbohydrates resulting from photosynthesis; they are directly transferred to the lichen-forming fungus in a specific form, thus increasing the resistance to desiccation [22]. The mycobiont assures the nutrients and metals required for algal metabolic processes. Moreover, the fungal partner synthesizes secondary metabolites to protect the photobiont against extreme temperatures, UV exposure, desiccation, and herbivores [23]. When a high heavy metal accumulation in the lichen thallus occurs, these lichen metabolites form complexes, increasing lichen metal tolerance [24]. When heavy metals pass intracellularly, they are expected to influence the lichen metabolic processes.

Numerous authors analyzed the lichen thallus components in their studies for various reasons. Hájek et al. [25] showed the influence of low temperature on the *Usnea sp.* thalli structure. Zverina et al. [26] and Carreras et al. [27] described heavy metal accumulation in *Usnea sp.* and the lichen thallus destruction by heavy metal stress. Bubach et al. [21] performed a correlation between the matrix of the biological elements (Na, Mg, K, Ca, Fe) and geographical parameters in both fractions (cortex–medulla and central axis) of *Usnea* sp. affected by a volcanic eruption.

Our study's novelty consists of separating and analyzing these two different parts of the *U. barbata* thallus (medulla–cortex and central cord). We aimed to explore the morphology of *U. barbata* lichen native to the volcanic Călimani Mountains using fluorescent microscopy (FM) and electronic scanning microscopy (SEM) images. The obtained micrographs provided the *U. barbata* integral thallus details compared to ground samples with different particle sizes. The mineral content, phenolic secondary metabolites, and free radical scavenging activity evaluation was completed with an overview of the correlation between the data obtained for all lichen samples. By establishing the fraction with the highest content of phenolic compounds, our study could increase the extraction yield of pharmacologically active secondary metabolites.

2. Materials and Methods

2.1. Lichen Sample Preparation

U. barbata is native to Calimani Mountains, Romania (900 m altitude, 47°29′ N, 25°12′ E). The lichen harvesting was performed in March 2020 from a conifer forest belonging to a natural peat bog region [28]. The freshly collected thalli were cleaned and naturally dried in an airy room sheltered from direct sunlight at 18–25 °C [29].

The dried lichen preservation for an extended period was performed in large paper bags under similar conditions. At the Ovidius University of Constanta, Department of Pharmaceutical Botany, Faculty of Pharmacy, *U. barbata* identification was realized using standard methods [9]. Samples are preserved at Ovidius University of Constanta, in the Herbarium of Pharmacognosy Department, Faculty of Pharmacy (Popovici 2/2020, Ph-UOC) [29]. The dried lichen thalli were ground in a laboratory mill, LM 120 (Perkin Elmer, Waltham, MA, USA) for 5 min [30]. After grinding, the dried *U. barbata* lichen was passed through a sieve (no. 5 [31]) and separated into two fractions. The first fraction of the dried lichen (*d*UB)—medulla–cortex (*mc*UB)—was passed again through the meshes of the same sieve and separated as a moderately fine (315 μm) powder. The central cord represents the second fraction (*cc*UB).

All lichen parts were ground again in the laboratory mill and passed through a sieve (no. 7), thus obtaining a fine powder with particle size < 180 μm. The samples were kept in small paper bags until analysis (Supplementary Material, Figure S1).

2.2. Lichen Morphology

2.2.1. Fluorescent Microscopy

Fluorescent microscopy images were obtained using an OPTIKA B-350 microscope (Ponteranica, BG, Italy) blue filter (λex = 450–490 nm; λem = 515–520 nm) and green filter (λex = 510–550 nm; λem = 590 nm). The dried *U. barbata* lichen thalli (Figure 1a) were washed with deionized water (DIW, Merck Millipore, Burlington, MA, USA). The hand-cut cross-sections were hydrated with phosphate saline buffer (Thermo Fisher Scientific, Waltham, MA, USA), pH = 7.4, and stained with 3% acridine orange (Merck Millipore, Burlington, MA, USA) for 5 min. The samples were rinsed with DIW and placed on the microscope slides. The FM images were obtained at 100× and 400× magnification and processed with Optikam Pro 3 Software (OPTIKA S.R.L., Ponteranica, BG, Italy). All observations were performed in triplicate.

(**a**) (**b**) (**c**) (**d**)

Figure 1. *U. barbata* dried lichen—entire thallus (**a**) and ground—*d*UB (**b**); both separated fractions: medulla-cortex—*mc*UB (**c**) and central cord—*cc*UB (**d**).

2.2.2. Scanning Electron Microscopy

We aimed to recognize the specific *U. barbata* morphology by examining the ground samples, *d*UB, *mc*UB, and *cc*UB, with particles of different sizes: approximately 315 μm (Figure 1b–d) and 180 μm (Figure S1, Supplementary Material) using scanning electron microscopy.

Scanning electron microscopy (SEM) images were obtained using a VEGA II LSH (Tescan, Czech Republic) device. The samples were fixed on double-sided carbon adhesive bands, and the acceleration tension was 30 kV. The SEM images were obtained at different magnifications (200×, 300×, 1 k×, 2 k×, 5 k×) and scale bars (200 μm–5 μm). All observations were performed in triplicate.

2.3. Color Evaluation

The color properties of *U. barbata* dried lichen and both parts (*mc*UB and *cc*UB) as a fine powder (particle size < 180 μm) were determined in triplicate in the CIELab system by using a Konica Minolta CR-400 (Konica Minolta, Tokyo, Japan) colorimeter. The color properties in terms of L^* (lightness, 0 for absolute black, 100 for absolute white), a^* (red-green intensity), and b^* (yellow-blue intensity) were recorded in triplicate. The hue angle (0°—red, 90°—yellow, 180°—green, and 270°—blue) and chroma value (0—gray, 100—pure color) were calculated using Equations (1) and (2):

$$h_{ab} = arctan\left(\frac{b^*}{a^*}\right) \quad (1)$$

$$C^* = \sqrt{(a^*)^2 + (b^*)^2} \quad (2)$$

where h_{ab}—hue angle, C^*—Chroma, L^*—lightness, a^*—positive values describe red and negative, green nuance, b^*—positive values represent yellow and negative, blue nuance.

2.4. Elemental Analysis

ICP-MS analyzed twenty-three elements in lichen samples: calcium (Ca), iron (Fe), (magnesium (Mg), manganese (Mn), zinc (Zn), aluminum (Al), silver (Ag), barium (Ba), cobalt (Co), chromium (Cr), copper (Cu), lithium (Li), nickel (Ni), thallium (Tl), vanadium (V), molybdenum (Mo), palladium (Pa), platinum (Pt), antimony (Sb), arsenic (As), lead (Pb), cadmium (Cd), and mercury (Hg). Our previous study described this method [28], and detailed data can be found in Supplementary Material, Tables S1–S4.

The NexION™ 300S inductively coupled plasma mass spectrometer (PerkinElmer, Inc., Hopkinton, MA, USA) was the platform for elemental analysis; dried lichen sample digestion was performed with 65% HNO_3 and 30% H_2O_2. [28].

The obtained results from the ICP-MS analysis were processed with Syngistix Software (PerkinElmer, Inc, Hopkinton, MA, USA) Version 2.3. This determination was done in triplicate, and the mineral concentrations were expressed as the mean (n = 3) ± SD [28].

2.5. FT-IR Analysis

Ground lichen samples (particle size ≤180 μm) were analyzed in triplicate by Fourier-transform infrared spectroscopy (FT-IR), by attenuated total reflection (ATR), on a Thermo Scientific Nicolet iS20 (Waltham, MA, USA) spectrometer. The spectra were recorded at 4 cm^{-1} intervals, in the range of 650 cm^{-1} to 4000 cm^{-1}. ATR correction was applied, and the average spectra were extracted using Omnic software.

2.6. Total Phenolic Content

2.6.1. Dried Lichen Extracts in Ethanol and Acetone

Two series of approximately 1 g *d*UB, *mc*UB, and *cc*UB were refluxed for 1 h with 100 mL solvent (96% ethanol and acetone). The resulting extractive solutions for each lichen part (*d*UB, *mc*UB, and *cc*UB) were filtered and then made up to 100 mL in a volumetric flask with each corresponding solvent (Figure S3, Supplementary Material).

2.6.2. Folin–Ciocâlteu Method

According to a previously described method, the total phenolic content was determined using the Folin–Ciocâlteu reagent (Merck, Darmstadt, Germany) [32]. Pyrogallol (Merck, Darmstadt, Germany) was used as a standard, and the TPC values were calculated as μg of pyrogallol equivalents (PyE) per g dried sample. For this analysis, in three volumetric flasks of 25 mL, 2 mL of each ethanol extract (from *d*UB, *mc*UB, and *cc*UB) was added, and then 1 mL of Folin–Ciocâlteu reagent, 10 mL water, and 12 mL 290 g/L of Na_2CO_3 solution up to the mark. In each volumetric flask, a blue coloration appeared. After 30 min of reaction in a dark place at room temperature, the absorbance values (each value being A1 in the calculation formula) were read at 760 nm using a Jasco V630 UV-Vis Spectrophotometer (JASCO Corporation, Tokyo, Japan) with Spectra Manager™ Software. A similar determination of phenolic contents was performed for *d*UB, *mc*UB, and *cc*UB acetone extracts. The total polyphenol content (TPC) determination was done in triplicate, and the obtained data were expressed as the mean (n = 3) ± SD.

2.7. Free Radical Scavenging Activity

The free radical scavenging activity of the dUB, mcUB, and ccUB ethanol and acetone extracts was determined on a Jasco V630 UV-Vis Spectrophotometer (JASCO Corporation, Tokyo, Japan) using the 2,2-diphenyl-1-picrylhydrazyl (DPPH) free radical scavenging assay [28]. The DPPH solution was obtained by dissolution of DPPH (Sigma Aldrich, St. Louis, MO, USA) in methanol to assess an absorbance value of 0.80 ± 0.02; then, 3.90 mL of DPPH solution with 0.1 mL of each dried lichen extract was vortexed for 30 s. The reaction time in a dark place at room temperature was 30 min; finally, the absorbances values at 515 nm were measured. The DPPH solution in methanol with no added dried lichen extract was used as a standard, while methanol was the blank. Two dilutions in ethanol and acetone were obtained (1:2 and 1:4) for both lichen extracts, and the DPPH radical scavenging activity was calculated according to Equation (3).

$$\text{Scavenging of DPPH free radicals } (\%) = 100 \times \frac{A\ control - A\ sample}{A\ control} \quad (3)$$

A control and *A sample* are the absorbance values at 515 nm for the DPPH and sample solutions. This determination was performed in triplicate; the obtained data are expressed as the mean (n = 3) ± SD.

2.8. Statistical Analysis

Mean values of three determinations were compared using analysis of variance (ANOVA) with the Tukey test and Student's t-test. The differences were considered significant when $p < 0.05$. Principal component analysis (PCA) was performed using XLSTAT for Excel software, 2022 version (Addinsoft, New York, NY, USA).

3. Results

3.1. Lichen Samples

Dried *U. barbata* lichen (Figure 1a) had a green-gray color; after grinding for 5 min, the ground *U. barbata* thallus (*d*UB) obtained is shown in Figure 1b. After passing through the meshes of sieve no. 5, two fractions were separated: *mc*UB (Figure 1c) as a moderate fine green-gray powder and *cc*UB (Figure 1d) as filiform, white-yellow fragments, about 5 mm long.

3.2. Lichen Morphology

The *U. barbata* lichen thallus morphology, which allows the separation of both fractions (*mc*UB and *cc*UB), was visualized using two modern microscopy techniques. We performed fluorescent microscopy (FM) for the entire dried lichen and scanning electron microscopy (SEM) for ground lichen and fractions with different particle sizes.

3.2.1. Fluorescence Microscopy

Fluorescence microscopy is an efficient method for viewing the lichen internal anatomy [33]. The FM images (Figure 2) were obtained using inherent autofluorescence and acridine-orange-induced fluorescence. Thus, the *U. barbata* layers (Figure 2a,b) and their cellular and subcellular structures (Figure 2c) can be highlighted.

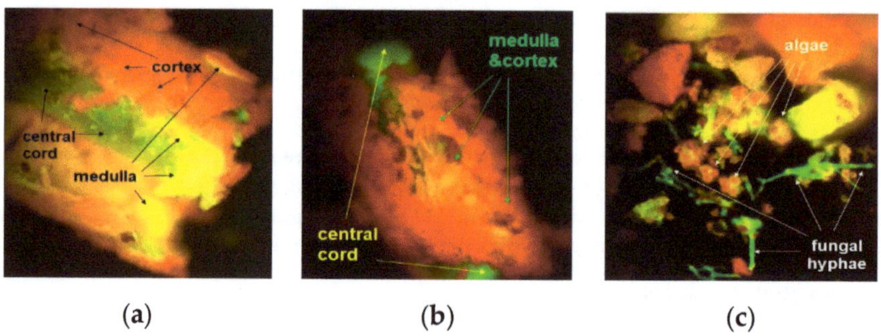

Figure 2. Microstructure of unground *U. barbata* lichen thallus—FM images with different magnifications: 100× (**a**) and 400× (**b**,**c**).

Figure 2a indicates the three layers of the lichen thallus: cortex, medulla, and central cord. Due to the thick and loose medulla, the three layers separating into two distinct parts—medulla–cortex (*mc*UB) and central cord (*cc*UB)—is clear, as shown in Figure 2a,b. With rare fungal hyphae, the loose medulla is the morphological property of *U. barbata* and several *Usnea* sp. (*U. cornuta*, *U. fragilescens*). *Usnea* lichens can have a compact medulla (*U. longissima*, *U. rubicunda*) or a dense one (*U. ceratina*, *U. florida*) [9].

The algal zone between the cortex and medulla in *U. barbata* is visible in Figure 2c. It is a subcortical structure formed by algal cells intertwined with fungal hyphae (Figure 2c).

3.2.2. Scanning Electron Microscopy

Scanning electron microscopy (SEM) is an optimal hybrid method for examining lichen thalli at high resolution [34], showing the micro-morphology of *U. barbata* thallus layers [35]. The obtained SEM images combined the optical microscopy visualization performance and facility with high-resolution electron microscopy [36].

Thus, the morphological characteristics were easily identified in all ground *U. barbata* samples with particle sizes around 315 μm. The SEM micrographs are displayed in Figures 3 and 4.

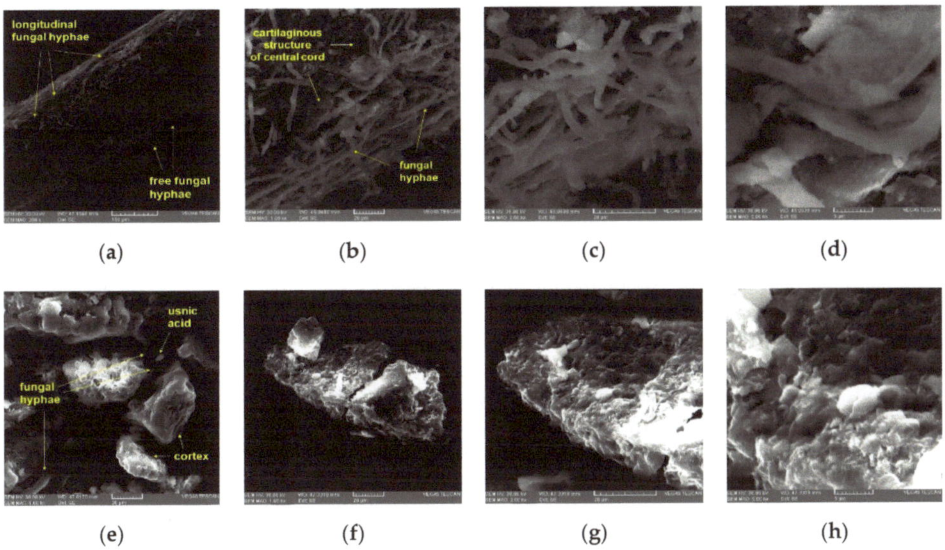

Figure 3. (**a**–**d**) Microstructure of the central cord (*cc*UB) with particle size > 315 µm, in SEM images with different magnifications: (**a**) 300× (scale bar = 100 µm), (**b**) 1 k× (scale bar = 20 µm), (**c**) 2 k× (scale bar = 20 µm), (**d**) 5 k× (scale bar = 5 µm); (**e**–**h**) SEM micrographs of *mc*UB (medulla–cortex) with particle size ≤ 315 µm: (**e**) cortex parts with rare fungal hyphae on the surface, 1 k× (scale bar = 20 µm), (**f**–**h**) cortex fragment at different magnifications: (**f**) 1 k× (scale bar = 20 µm), (**g**) 2 k× (scale bar = 20 µm), (**h**) 5 k× (scale bar = 5 µm).

Figure 4. SEM micrographs in *d*UB (particle size > 315 µm): (**a**–**c**) cortex fragments, central cord, and fungal hyphae, at different magnifications: (**a**) 1 k× (scale bar = 20 µm), (**b**) 2 k× (scale bar = 20 µm), (**c**) 5 k× (scale bar = 5 µm); (**d**–**f**) central cord with fungal hyphae between cortex fragments, at different magnifications: (**d**) 200 k× (scale bar = 200 µm), (**e**) 1 k× (scale bar = 20 µm), (**f**) 5 k× (scale bar = 5 µm).

Figure 3 shows SEM images of the central cord, *cc*UB (particle size > 315 μm), and medulla–cortex, *mc*UB (particle size ≤ 315 μm). The SEM micrographs of *cc*UB (Figure 3a–d) were obtained at the following magnifications and scale bars: 300× (scale bar = 100 μm), 1 k× (scale bar = 20 μm), 2 k× (scale bar = 20 μm), 5 k× (5 μm). The cartilaginous structure of the central cord can be observed and, on its surface, free fungal hyphae (Figure 3a,b). This particular structure of the central axis and the aspect of free fungal hyphae are detailed in Figure 3c,d at 2 k× and 5 k× magnification (scale bar = 20 μm and 5 μm).

In *mc*UB (particle size ≤ 315 μm), cortex fragments with a compact structure and rare fungal hyphae can be seen (Figure 3e–h). On the surface of fungal hyphae, secondary metabolite crystals can be observed (Figure 3e). In *Usnea* lichens, usnic acid is the most common secondary metabolite in high concentrations; for this reason, usnic acid is commonly indicated in the corresponding figures [37]. Figure 3f–h displays a cortex fragment at different magnifications: 1 k× (scale bar = 20 μm), 2 k× (scale bar = 20 μm), 5 k× (scale bar = 5 μm). At high magnifications, the structural differences between the central axis and cortex (both constituted by fungal hyphae)—which are the basis of their different properties—can be observed.

In *d*UB (particle size > 315 μm), we can identify the morphology of *U. barbata* thallus: central cord parts, cortex fragments, and rare fungal hyphae (Figure 4a–c). An isolated fragment of the central axis can be observed at different magnifications (Figure 4d–f).

The structures of both layers (central cord and cortex) formed by mycelial hyphae are essential, supporting dried lichen separation in *cc*UB and *mc*UB fractions. Their longitudinal arrangement and interweaving in the central axis give it a particular strength and elasticity. This property explains that it separates after a short grinding time and cannot be finely ground unless the grinding time is long.

The SEM micrographs of all ground lichen samples (*cc*UB, *mc*UB, and *d*UB) as a fine powder (particle size ≤ 180 μm) are shown in Figure 5. SEM images displayed in Figure 5 highlighted the microstructure of *cc*UB (Figure 5a–c) and *mc*UB (Figure 5d,e) achieved at different magnifications: 1 k× (scale bar = 20 μm), 2 k× (scale bar = 20 μm), and 5 k× (scale bar = 5 μm). A noticeable difference can be observed between the central cord fragments (Figure 5a–c) and the cortex ones (Figure 5d–f). Due to the compact cartilaginous structure of the intertwined longitudinal mycelial hyphae, the fragments of the central axis appear as pieces of a broken vessel, with a glossy, deep surface, irregular sharpness, or rounded edges. The structure from Figure 3d can be recognized in Figure 5c. The cortex fragments are porous, and the compact structure's rupture during grinding occurs along with a series of pores on its thickness. The images of *d*UB powder (Figure 5g–i) show differentiated fragments belonging to both layers of the lichen thalli (central axis and cortex) and fungal hyphae.

3.3. Color Evaluation

The data obtained for the color parameters of *U. barbata* fractions are displayed in Table 1. Significant differences ($p < 0.05$) between all samples were observed for all color parameters considered, supported by the images from Figure 1. The highest luminosity was observed for *cc*UB (70.29), while the *mc*UB fraction presented the lowest value (59.08). The samples containing the medulla–cortex (*d*UB and *mc*UB) exhibited a green nuance indicated by the negative values of the a^* parameter. In contrast, the positive values observed for *cc*UB suggested a red nuance. Both fractions and integral lichen showed a yellow nuance, as the positive values of the b^* parameter showed, the nuance being more pronounced in *cc*UB (13.92). Compared to the *cc*UB fraction, the *mc*UB sample showed a lower hue angle and chroma value.

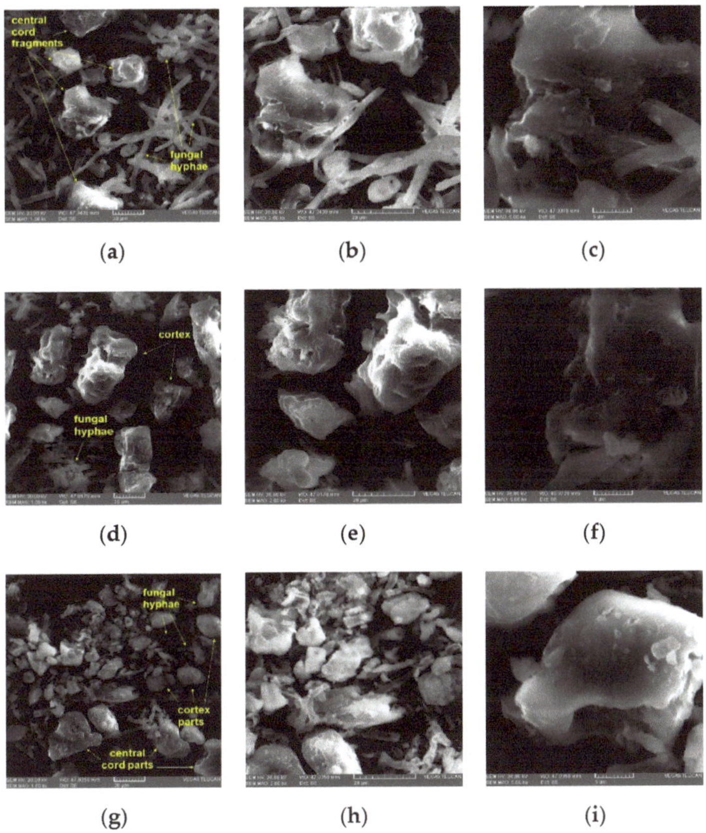

Figure 5. SEM micrographs of *cc*UB, *mc*UB, and *d*UB as a fine powder (particle size ≤ 180 μm): *cc*UB (**a–c**); *mc*UB (**d–f**); *d*UB (**g–i**) with different magnifications: 1 k× and scale bar = 20 μm (**a**,**d**,**g**), 2 k× and scale bar = 20 μm (**b**,**e**,**h**), and 5 k× and scale bar = 5 μm (**c**,**f**,**i**).

Table 1. Color evaluation of *U. barbata* samples.

Property	*d*UB	*mc*UB	*cc*UB	*F*-Value
	Color Properties			
L^* (adim.)	63.02 ± 0.11 ᵇ	59.08 ± 0.15 ᶜ	70.29 ± 0.13 ᵃ	5703.14 ***
a^* (adim.)	-1.29 ± 0.02 ᵇ	-1.83 ± 0.03 ᶜ	2.12 ± 0.17 ᵃ	1414.01 ***
b^* (adim.)	13.60 ± 0.02 ᵇ	12.52 ± 0.04 ᶜ	13.92 ± 0.05 ᵃ	1086.01 ***
h_{ab} (°)	178.52 ± 0.00 ᶜ	178.57 ± 0.00 ᵇ	181.42 ± 0.01 ᵃ	165,195.71 ***
C^* (adim.)	13.66 ± 0.02 ᵇ	12.65 ± 0.05 ᶜ	14.08 ± 0.04 ᵃ	1125.62 ***

*d*UB—dried *U. barbata* integral lichen, *mc*UB—medulla–cortex, *cc*UB—central cord, L^*—lightness, a^*—positive values describe red and negative, green nuance, b^*—positive values represent yellow and negative, blue nuance; h_{ab}—hue angle, C^*—Chroma, adim.—adimensional. Mean values followed by different superscript letters are significantly different, *** significant at $p < 0.001$.

The green color is due to chlorophyll from the algal layer [22], and the yellow one to usnic acid, the most well-known phenolic secondary metabolite [38].

Chlorophyll degradation significantly increases in prolonged sunlight and heavy metal exposure of lichens [39], and their color can be modified. Due to usnic acid's photoprotective effect, *Usnea* species show a moderate level of photoinhibition [40]. Zulaini et al. [41]

reported that *U. diffracta thalli* change color chane from greyish-green to brownish-green through massive heavy metal accumulation due to chlorophyll breakdown.

3.4. Elemental Analysis

The elemental content of dried *U. barbata* lichen and both fractions is presented in Table 2.

Table 2. Mineral composition of *Usnea barbata* (L.) Weber ex F.H. Wigg dried lichen (*d*UB) and both fractions: medulla–cortex (*mc*UB) and central cord (*cc*UB).

Element Content (μg/g)	Sample			F-Value
	*d*UB	*mc*UB	*cc*UB	
Al	87.879 ± 1.152 [b]	285.828 ± 2.748 [a]	21.111 ± 0.608 [c]	18,444.13 ***
As	ND	0.219 ± 0.005	ND	-
Ba	3.782 ± 0.052 [b]	12.113 ± 0.604 [a]	0.596 ± 0.004 [c]	866.03 ***
Ca	979.766 ± 12.285 [b]	1549.600 ± 18.406 [a]	434.269 ± 22.222 [c]	2846.40 ***
Cd	ND	0.164 ± 0.001	ND	-
Co	ND	0.225 ± 0.001	ND	-
Cr	1.002 ± 0.008 [b]	2.494 ± 0.010 [a]	0.864 ± 0.007 [c]	34,521.18 ***
Cu	1.523 ± 0.013 [b]	2.246 ± 0.005 [a]	1.286 ± 0.014 [c]	5771.15 ***
Fe	52.561 ± 2.582 [b]	282.468 ± 1.149 [a]	33.107 ± 0.322 [c]	21,398.25 ***
Li	ND	0.185 ± 0.007	ND	-
Mg	172.721 ± 0.647 [b]	524.239 ± 2.419 [a]	53.482 ± 0.597 [c]	81,353.06 ***
Mn	101.425 ± 1.423 [b]	354.041 ± 1.083 [a]	29.098 ± 0.347 [c]	78,943.06 ***
Ni	0.449 ± 0.011 [b]	1.033 ± 0.005 [a]	0.355 ± 0.007 [c]	5658.89 ***
Pb	1.296 ± 0.007 [b]	2.177 ± 0.014 [a]	0.584 ± 0.004 [c]	21,958.36 ***
V	0.241 ± 0.004 [b]	1.237 ± 0.005 [a]	ND	92,152.36 ***
Zn	20.536 ± 0.125 [b]	33.223 ± 0.164 [a]	13.588 ± 0.097 [c]	17,179.89 ***
Hg	0.671 ± 0.020 [b]	0.708 ± 0.005 [a]	0.539 ± 0.017 [c]	99.48 ***
Sb	ND	0.152 ± 0.003	ND	-

The analysis was performed in triplicate. The results are shown as the mean ± standard deviation (SD). ND—non-detected, Al—aluminum, As—arsenic, Ba—barium, Ca—calcium, Cd—cadmium, Co—cobalt, Cr—chromium, Cu—copper, Fe—iron, Li—lithium, Mg—magnesium, Mn—manganese, Ni—nickel, Pb—lead, V—vanadium, Zn—zinc, Hg—mercury, Sb—antimony. Mean values followed by different letters are significantly different, *** significant at $p < 0.001$.

It can be noted that Ca, Fe, Mg, Mn, and Zn had the highest LOQ values (5 μg/g), followed by Al (1 μg/g) and the other metals (most numerous), which had LOQ = 0.1 μg/g. Generally, the LOQ value was directly proportional to their content (Table S5, Supplementary Material).

Comparing both fractions (*mc*UB and *cc*UB) with dried lichen (*d*UB), the data registered in Table 2 highlight other notable aspects. Thus, *mc*UB reported the highest metal contents and *cc*UB the lowest ones; these differences were significant for macroelements (*mc*UB > *d*UB > *cc*UB): Al (285.828 > 87.879 > 21.111 μg/g), Ca (1549.600 > 979.766 > 434.269 μg/g), Fe (282.468 > 52.561 > 33.107 μg/g), Mg (524.239 > 172.721 > 53.482 μg/g), and Mn (354.041 > 101.425 > 29.098 μg/g).

Generally, trace elements showed a similar distribution in *d*UB and both fractions, but the differences were significantly lower: Ba (12.113 > 3.782 > 0.596 μg/g), Cr (2.494 > 1.002 > 0.864 μg/g), Cu (2.246 > 1.523 > 1.286 μg/g), Ni (1.033 > 0.449 > 0.355 μg/g), Pb (2.177 > 1.296 > 0.584 μg/g), Zn (33.223 > 20.536 > 13.588 μg/g), Hg (0.708 > 0.671 > 0.539 μg/g). Another metal (V) was quantified only in *mc*UB and *d*UB (1.237 > 0.241 μg/g), being undetected in *cc*UB. Moreover, five trace elements undetected in *d*UB (As, Cd, Co, Li, Sb) were quantified in *mc*UB: As (0.219 μg/g), Cd (0.164 μg/g), Co (0.225 μg/g), Li (0.185 μg/g), and Sb (0.152 μg/g)

3.5. FT-IR Analysis

The molecular characteristics of *U. barbata* fractions are presented in Figure 6. Noticeable differences in absorbance values between fractions were observed for almost all peaks, with the highest intensities obtained for *mc*UB, followed by *d*UB and *cc*UB. Furthermore, in the regions 1380–1700 cm^{-1} and 2310–2380 cm^{-1}, different peaks appeared in *cc*UB and *d*UB samples compared to *mc*UB.

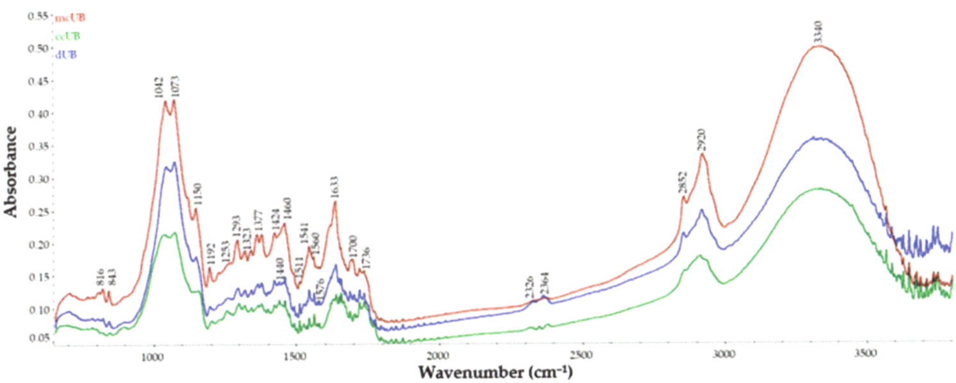

Figure 6. Overlay of FT-IR spectra of *U. barbata* dried lichen (*d*UB), medulla–cortex (*mc*UB), and central cord (*cc*UB).

3.6. Total Phenolic Content

The values of the total phenolic compounds extracted from dUB, mcUB, and ccUB in ethanol and acetone are displayed in Table 3.

Table 3. Total phenolic content (TPC, expressed as mg PyE/g dried sample) and free radical scavenging activity (AA, expressed as % DPPH radical scavenging) of *U. barbata* dried lichen and both fractions (*mc*UB and *cc*UB) extracted in ethanol and acetone.

Parameter	*d*UB	*mc*UB	*cc*UB	*F*-Value
$TPC_{ethanol}$	$23.481 \pm 0.039^{b,x}$	$25.212 \pm 0.084^{a,y}$	$18.383 \pm 0.004^{c,x}$	13200.49 ***
$TPC_{acetone}$	$22.675 \pm 0.108^{b,y}$	$36.243 \pm 0.093^{a,x}$	$15.170 \pm 0.129^{c,y}$	27784.01 ***
t-value	12.16 ***	152.42 ***	43.12 **	
$AA_{acetone}$ (% DPPH scavenging)	$16.878 \pm 0.204^{a,x}$	$12.747 \pm 0.221^{b,y}$	$11.336 \pm 0.174^{c,y}$	10.96 **
$AA_{ethanol}$ (% DPPH scavenging)	$15.985 \pm 0.197^{a,y}$	$15.735 \pm 0.185^{a,x}$	$15.080 \pm 0.326^{b,x}$	618.35 **
t-value	5.45 **	17.96 ***	17.55 ***	

*d*UB—dried *U. barbata* lichen, *mc*UB—medulla–cortex, *cc*UB—central cord, TPC—total phenolic content (mg PyE/g dried sample), mgPyE/g—milligram equivalent pyrogallol per 1 g lichen sample, AA—antioxidant activity. The mean values followed by superscript letters are significantly different: a–c in the same row for sample comparison, x–y in the same column for solvent comparison, *** significant at $p < 0.001$, ** significant at $p < 0.01$.

Phenolic values of both extracts decreased in the order *mc*UB, *d*UB, and *cc*UB. TPC values for *d*UB and *cc*UB were higher in ethanol extracts (23.481 and 18.383 mg PyE/g) than in acetone ones (22.675 and 15.170 mg PyE/g). The TPC value in *mcUB* acetone extract was 36.243 mg PyE/g, significantly higher than that in ethanol (25.212 mg PyE/g). These differences are due to the nature and amount of phenolic compounds in each sample extracted by each solvent. The coloration of the ethanol and acetone extracts from each of the three lichen samples (Figure S2, Supplementary Material) also supports this observation.

3.7. Free Radical Scavenging Activity

The obtained results are displayed in Table 3. Both extracts' free radical scavenging activity decreased in the order *d*UB, *mc*UB, and *cc*UB (Table 3). In ethanol extracts, % DPPH radical scavenging values registered were not significantly different between *d*UB (15.985%) and *mc*UB (15.735%), while *cc*UB showed a significantly different value (15.080%).

Moreover, only *mc*UB had AA highly correlated with TPC (R^2 = 0.976); the others showed a moderate correlation: R^2 = 0.776 (*d*UB) and R^2 = 0.738 (*cc*UB). Acetone extracts displayed considerable differences: 16.878% (*d*UB), 12.747% (*mc*UB), and 11.336% (*cc*UB). In this case, *cc*UB (R^2 = 0.997) showed the highest correlation between antiradical activity and the TPC value, followed by *mc*UB (R^2 = 0.969) and *d*UB (R^2 = 0.883). Only *d*UB acetone extract had AA higher than the ethanol one; *mc*UB and *cc*UB acetone extracts had lower antiradical activity compared to ethanol ones (Table 3). Finally, comparing the AA and TPC for all lichen samples in each solvent, we can see that only the antiradical activity of ethanol extracts had a moderate correlation with TPC (R^2 *ethanol* = 0.755).

3.8. Relationships between Variables

The similarities and dissimilarities between the characteristics of *U. barbata* samples are displayed in Figure 7.

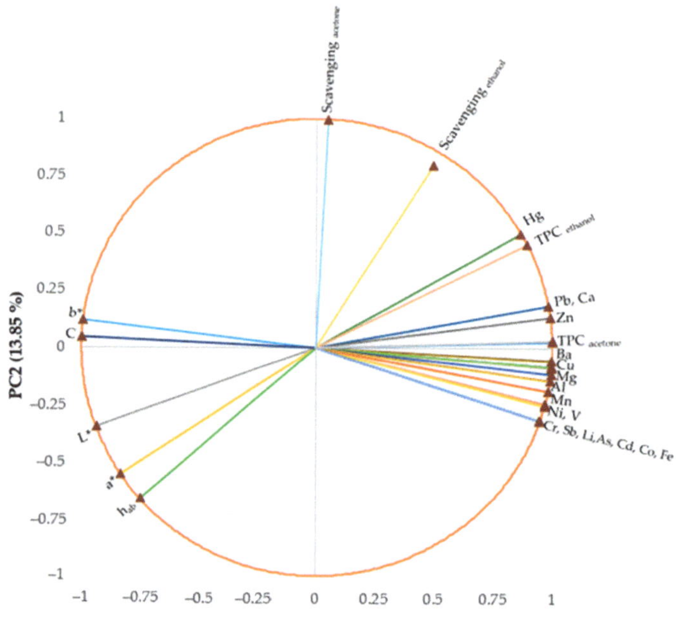

Figure 7. The Principal Components Analysis (PCA) plot.

The two principal components explained 99.31% of the total data variance, with 85.46% attributed to the first (PC1) and 13.85% to the second (PC2). The PC1 was associated with all minerals, total phenolics, and color parameters, while PC2 was related to scavenging activity. Luminosity (L^*), b^* parameter, and chroma (C^*) were negatively correlated ($r > -0.768$, $p < 0.05$) with all metals and the total phenolic content (TPC) in both solvents; the b^* parameter was negatively correlated ($r > 0.706$, $p < 0.05$) only with Al, Ba, Ca, Cu, Mg, Mn, Ni, Pb, V, Zn, Hg, Sb, and TPC. TPC values in both solvents were positively correlated ($r > 0.696$, $p < 0.05$) with all mineral contents.

An opposition between TPC in ethanol extracts and the a^* parameter and between b^* and C^* with some minerals (Cu, Mg, Al) was observed. The scavenging activity in ethanol extracts was positively correlated with the TPC content ($r > 0.775$, $p < 0.05$).

4. Discussion

Using advanced microscopical analysis, the present study provides an accurate morphological characterization of *U. barbata*, offering a complete structural investigation. The microstructure of *U. barbata* integral and fractions displayed in our study agreed with those reported by Ivanovic et al. [37] and Bubach et al. [21]. The whole lichen consists of 90% mycobiont and only 10% photobiont [24]. Of the separated fractions, one is formed exclusively of mycelial hyphae (*cc*UB), and the other (*mc*UB) has both symbionts. It is known that, in symbiotic partnerships, only the fungus synthesizes the lichen's secondary metabolites. Thus, we wanted to know which lichen fraction had phenolic metabolites and a mineral content higher than the entire lichen. Being a source of bioactive metabolites, extracting them from the fraction with the highest content could increase the process yield and selectivity.

Bubach et al. [21] revealed an entire micrograph of a cross-section of *Usnea* sp. thallus obtained by SEM-EDS analysis. These authors separated only two *U. barbata* thallus fractions (cortex–medulla and axis), similar to those described in our study (*mc*UB and *cc*UB). The cortex consisted of densely packed thick-walled fungal hyphae in their image, and the central cord had a compact, cartilaginous structure [21].

Most numerous organic compounds found in lichens are secondary metabolites of fungal origin [20]. Their amount is usually 0.1–10% in dried lichen and can reach 30% in some cases. These substances are mostly phenolic compounds, poorly soluble in water, and can be extracted using organic solvents [42]. They are deposited in the cortex and medulla as crystals on the hyphae surface. The lichen metabolites are differentially distributed in the thallus layers, correlated with their bioactivities [20]. Thus, the compounds from the upper cortex act as a light filter, having a photoprotective effect [40]. The most common cortical metabolites are usnic acid, atranorin, xanthones, pulvinic acid derivates, and anthraquinones. With few exceptions (atranorin), they are colored compounds (yellow, orange), giving the lichen thalli specific coloration [40]. These cortical constituents can be distinguished from numerous medullary ones [20].

Bioactive metabolites localization in the lichen thallus was explored using various methods. Some researchers used spot tests (directly applying suitable reagents to the lichen thallus) or after extraction and analyses of specific lichen parts [43]. Other studies described classical fluorescent techniques based on the molecule chromophore, where metabolites are only differentiated by their emission wavelength; then, the metabolite localization in living plant cells was obtained using multiphoton microscopy [44]. In 2016, based on the auto-ionisability of all main classes of lichenic bioactive compounds [45], Le Pogam et al. [46] achieved histolocalisation of the lichen *Ophioparma ventosa* by laser desorption ionization coupled to mass spectrometry imaging (LDI-MSI). They applied different slicing procedures to obtain lichen thalli transverse sections (cryosectioning and hand-cutting), fixed them on a carbon-conductive adhesive tape, and examined the metabolites' spatial distribution. MSI displayed the molecular images of usnic acid, thamnolic acid, divaricatic acid, miriquidic acid, haemoventosin, and all overlaid ions in cryosectioned pieces of the *O. ventosa* thallus [46]. Using the same method, Gadea et al. [43] obtained spatial mapping that revealed phenolic metabolites in the lichen *Pseudocyphellaria crocata*.

Moreover, for optimal extraction of localized lichen metabolites, different methods of sample preparation were elaborated. Komaty et al. [47] proposed two types of preparations for atranorin extraction from *Pseudevernia furfuracea*. In the first case, the lichen sample was milled (with a planetary ball mill), resulting in a homogeneous powder. The lichen thallus was ground in a blender in the second case, obtaining a mixture of medulla pieces and fine cortex powder. The higher extraction yield with acetone under microwave irradiation was obtained using cortex powder because atranorin is mainly a cortical metabolite. Ivanovic et al. [37] optimized usnic acid supercritical fluid extraction with CO_2 from *U. barbata*, using various grinding methods and conditions (temperature, CO_2 pressure).

In the case of pharmaceutical applications, either supplements or drugs, the regulatory agencies approve only one or a maximum of two unpolluted sources for the contained ingredients. The results are reproducible only for raw materials developed under the same conditions; therefore, restricting the harvesting area leads to accurate results, essential for human use formulations. However, only a few data on metal concentrations in *U. barbata* from unpolluted zones were found in previous studies. Culicov et al. [48] investigated *U. barbata* from the Mountain Natural Park in Bulgaria, Jayasekera et al. [49] examined *U. barbata* from the Sri Lanka rain forest, and Conti et al. [50] analyzed *U. barbata* from central and southern Tierra del Fuego, Patagonia, Argentina. Arsenic (As) was detected only in *mc*UB (0.219 μg/g), not in the whole dried lichen sample. However, this element was quantified in lichens belonging to the three mentioned regions: 0.134 μg/g in Sri Lanka, 0.823 μg/g in Tierra del Fuego, and 1.100 μg/g in Bulgaria Natural Park. Cadmium (Cd) was quantified in increased values in *U. barbata* from all three zones: 0.096 μg/g in Sri Lanka, 0.174 μg/g in Tierra del Fuego, and 0.590 μg/g in Bulgaria; in our study, Cd was detected only in *mc*UB (0.164 μg/g). The cobalt (Co) content in *mc*UB was 0.225 μg/g; however, *U. barbata* from all other zones displayed different Co contents: 0.522 μg/g in Tierra del Fuego, 0.260 μg/g in Sri Lanka, and 0.130 μg/g in Bulgarian Mountain Park. We found the highest lead content in dried *U. barbata* from Călimani Mountains (1.296 μg/g). In contrast, the lead amounts in the lichen from other zones were significantly lower (0.426 and 0.221 μg/g Pb in *U. barbata* from Sri Lanka and Tierra del Fuego). The zinc content in autochthonous lichen (20.536 μg/g) was lower than 42 μg/g (Tierra del Fuego) and 51 μg (Bulgaria) and higher than 15 μg/g (Sri Lanka). The *U. barbata* from Tierra del Fuego had a 0.827 μg/g nickel content, higher than the Călimani Mountains one (0.449 μg/g). Finally, an amount of 0.056 μg/g Sb was quantified by Conti et al. in *U. barbata* from Tierra del Fuego; only in the *mc*UB was 0.152 μg/g Sb detected.

From their native zones, *U. barbata* and other lichen species can be transplanted to various polluted zones for biomonitoring reasons [51]. Conti et al. [52] confirmed the considerable ability of *U. barbata* to reflect the levels of environmental elements on a global scale, signaling, at distant places, volcanic emissions. Previous studies reported that *Usnea* sp. was used in this scope, and their action was compared with other lichen species. Thus, *U. hirta* transplanted to a city in northern Italy had a similar capacity to accumulate various metals as *H. physodes* and *P. furfuracea* [53]. Meli et al. [54] showed that lichens, obtaining numerous nutrients from the air, significantly accumulate different air pollutants through the thallus surface. Lichens take in metals in different ways. The common ones are the physical passing of metal particles in the medulla intercellular spaces and their binding to extracellular sites of the mycobiont and photobiont [55]. Hence, the highest mineral content in the *mc*UB fraction, followed, in decreasing order, by *d*UB and *cc*UB, can be justified.

Accumulation of heavy/trace metals in lichens is essential for their use for nutrition and/or therapeutical purposes [56]. Moreover, usnic acid, the most bioactive metabolite from *Usnea* sp., has high hepatotoxicity [57]; that aspect represents another reason that further restricts their possible use in these scopes.

Heavy metal accumulation by food intake [58] disturbs various biochemical processes [59] in the human body. Thereby, permissible limits of heavy metals were established for edible plants by the Food and Agriculture Organization of the United Nations (FAO) and the World Health Organization (WHO, 1996) [60,61]. These values are displayed in Table 4.

As a culinary plant for bread and mush ingredients, *U. barbata* was used in the Balkan region [64]. In *d*UB and *cc*UB, the metal content values (Table 4) show that Cr, Pb, Cu, and Zn were lower than the allowable limit. Other heavy metals, Hg, Ni, and V, were found in higher contents compared to permissible ones. The other toxic elements (As, Cd, and Co) were not detected. On the other hand, *U. barbata* arsenic contents of 0.823 μg/g in Tierra del Fuego and 1.100 μg/g in Bulgaria Natural Park were higher than the acceptable limit. The cadmium and cobalt contents were higher than the permissible limit in *U. barbata* from all three zones compared with the Călimani mountains, while zinc and nickel contents were only measured in the Natural Park of Bulgaria and in Tierra del Fuego. It can be seen that the heavy metal content in *U. barbata* from other unpolluted earth zones was significantly higher than in our one. In *mc*UB, Cd and Co were higher than the permissible limits. The same fraction had other heavy metals higher than the allowable limits: Cr, Cu, Ni, Pb, Hg, and V; only Zn and As contents were lower than the permissible limits (Table 4).

Table 4. Permissible limits for heavy metals in edible and medicinal plants and their content in *U. barbata* whole lichen (*d*UB) and fractions (*mc*UB and *cc*UB).

Legislative Authority	**WHO/EU [60,61]**	**WHO [62]**	**Eu.Ph. [63]**	**Ch.Ph. [59]**	**Heavy Metals Content in *U. barbata* (µg/g)**		
Heavy Metal	**Permissible Limits (µg/g)** **Edible Plants**	**Medicinal Plants**			***d*UB**	***mc*UB**	***cc*UB**
As	0.5	10		2	ND	0.219	ND
Cd	0.02	0.2	1	0.3	ND	0.164	ND
Cr	1.3		2		ND	0.225	ND
Co	0.01				1.002	2.494	0.864
Cu	10	20		20	1.523	2.246	1.286
Hg	0.03	1	0.1	0.2	0.671	0.708	0.539
Ni	0.1				0.449	1.033	0.355
Pb	2	10	5		1.296	2.177	0.584
V	0.03				0.241	1.237	ND
Zn	50	50			20.536	33.223	13.588

WHO = World and Health Organisation, EU = European Union, Eu.Ph.= European Pharmacopoeia, Ch.Ph. = Chinese Pharmacopoeia, *d*UB—dried *U. barbata* lichen, *mc*UB—medulla-cortex, *cc*UB—central cord, ND = not detected.

The approvable limits for heavy metals in medicinal plants are higher compared to edible ones (Table 4).

Dobrescu et al. [65] reported that *U. barbata* is commonly used as an antiseptic (in Spain and USA). According to the same authors, it is a good remedy for various symptoms (insomnia, whooping cough, bleeding, jaundice, and nausea; in Europe). Our results showed that most heavy metals in autochthonous *U. barbata* were present lower amounts than permissible limits values. However, the Cr content in *mc*UB exceeded the permissible limit according to European Pharmacopoeia; the mercury contents in all three lichen fractions were lower than the WHO's and the FDA's acceptable limits [62] and higher than the ones according to the Chinese Pharmacopoeia [59], and European Pharmacopoeia [63] (Table 4).

The metals can generate complexes with polysaccharides, lichen secondary metabolites, and organic acids [55]. Bačkor and Fahselt (2004) [24] proved that, in *C. pleurota*, usnic acid might be associated with Fe, Cu, Ni, and Al. Thus, the significant positive correlation ($p > 0.05$) between the total phenolic and mineral contents highlighted in the final PCA plot could be explained.

As a preview of lichen constituents from all analyzed samples, the FT-IR spectra of *mc*UB showed different peaks compared to the *cc*UB fraction, while *d*UB presented all peaks found in both previous fractions. The peaks observed at 3340 cm^{-1} given by the O-H and N-H stretching vibration and 2852 and 2920 cm^{-1} due to the C-H stretching vibration could depict the presence of phenols, water, carbohydrates, polysaccharides [66], and lipids [67] in *U. barbata* fractions. Similar results were reported by Zaini et al. [68] for usnic acid extracted from *Usnea* sp. The peak at 3320 cm^{-1}, along with the ones at 2920 and 2852 and 1700 cm^{-1}, may be related to the placodiolic and usnic acid characteristics of *U. barbata* lichen [69]. The bands observed at 1633, 1700, and 1736 cm^{-1} attributed to the N-H and C=O bending vibration could indicate amino acids and esters [67]. The peaks found at 1700 and 1736 cm^{-1} can be attributed to the C=O vibrations in lipids, their presence being suggested by the bands observed at 1323 and 1377 cm^{-1} given by the CH_3 bending [67]. The peaks observed at 1633 and 1700 cm^{-1} can be assigned to the C=O, C-N, and/or C-O-NH- vibrations which may be due to proteins [67]. The aromatic and N-H bending vibrations found at 1511, 1541, 1560, and 1576 cm^{-1} can depict the presence of amino acids [67]. The bands at 1511 and 1576 cm^{-1} were observed only for *cc*UB and *d*UB, indicating the differences in amino acids and proteins compared to the *mc*UB sample; this hypothesis is supported by the appearance of the peak at 1440 cm^{-1}. The peaks in the regions 1377–1460 cm^{-1} can also be assigned to the CH_3 lipids/proteins and COO- of amino acids, while the vibrations of N-H and C-N had peaks at 1460–1576 cm^{-1} [64] and reveal the presence of proteins in *U. barbata* fractions. The bands observed in the region 1323–1440 cm^{-1} could be attributed to the primary or secondary O–H bending (in-plane) and phenol or tertiary alcohol (O–H bend) [67]; the peak at 1440 cm^{-1} was observed only in *cc*UB and *d*UB samples. The secondary aromatic amine and CN stretching vibration could be responsible for the peaks obtained at 1293 and 1323 cm^{-1}, while the band observed at 1253 cm^{-1}, attributed to the C-O stretching vibrations, could be due to acids or esters [67]. The specific bands at 1293 and 1323 cm^{-1} could be associated with usenamine A, a detected compound in *U. barbata* lichen [70]. The peaks obtained at 1042 and 1073 cm^{-1} can be due to the C-O stretching vibrations of carbohydrates and glycoproteins.

The most important phenolic secondary metabolites (usnic acid, for *Usnea* sp.) synthesized by mycobionts are deposited on the hyphae surface in the cortex and medulla [37,71]. Therefore, the highest TPC values in *mc*UB in both solvents, followed by *d*UB and *cc*UB, could be justified. Usnic acid and other specific secondary metabolites have a higher solubility in acetone than ethanol; this property can explain that *mc*UB has a higher TPC value in acetone extract than ethanol extract [32]. The free radical scavenging activity [72] was evaluated, and, in both extracts, dUB recorded the most significant antiradical potential [73], followed by *mc*UB and *cc*UB. However, due to the various free radical scavenging ability of phenolic metabolites found in each lichen extract (ethanol and acetone), AA values considerably differed. Consequently, the correlation between antiradical activity and TPC recorded significant variations. The highest AA of *d*UB in both extracts could be induced by the synergic interaction between the primary metabolites with a non-phenolic structure, the secondary ones, and the metals found in the whole lichen.

5. Conclusions

Our study could complete the scientific database, which must be constantly updated with the characteristics of *U. barbata* separated fractions (medulla–cortex and central cord) compared to whole lichen. Proving that the medulla–cortex fraction has the highest phenolic metabolite content, the obtained results could help increase their extraction yield. Further studies could quantify usnic acid and other bioactive metabolites in each separated lichen fraction. Moreover, this separation process could be investigated as a potential pre-treatment method for *U. barbata* dried lichen thalli, aiming for their extraction process optimization.

Supplementary Materials: The following supporting information can be downloaded at https://www.mdpi.com/article/10.3390/app12094234/s1, Figure S1. Ground *U. barbata* samples, as a fine powder with particle size < 180 μm (Dub—dried *U. barbata* lichen, *mc*UB—medulla–cortex, *cc*UB—central cord); Figure S2. The calibration curves of 23 metals analyzed in *U. barbata* dried samples using the ICP-MS method; Figure S3. Extracts in ethanol 96% (A) and acetone (B) from dried *U. barbata* lichen—*d*UB (a), medulla–cortex—*mc*UB (b), and central cord—*cc*UB (c). Table S1. Dried *U. barbata* sample digestion conditions for ICP-MS mineral analysis; Table S2. ICP-MS Standard solutions; Table S3. Preparation of calibration standard solutions (E1–E5); Table S4. Concentrations of calibration standard solutions (E1–E5) for different elements; Table S5. Calibration Curve Range, R^2, LOD, and LOQ (μg/L and μg/g) for each element.

Author Contributions: Conceptualization, V.P. and M.U.-I.; methodology, L.B., C.E.G., D.R., S.I.C., E.I.C., T.C., M.U.-I., M.O. and V.S.; software, V.P., L.B., D.R., S.I.C., E.I.C., T.C., M.U.-I., M.O., S.M. and V.S.; validation, L.B., D.R., S.I.C., E.I.C., T.C., M.U.-I. and M.O.; formal analysis, V.P. and V.S. investigation, V.P., E.A.O. and A.C.; resources, V.P., L.B., M.U.-I. and M.O.; data curation, C.E.G., S.M., E.A.O. and A.C.; writing—original draft preparation, V.P., D.R., S.I.C. and M.U.-I.; writing—review and editing, V.P., C.E.G., L.B., M.U.-I., M.O. and S.M.; visualization, L.B., C.E.G., M.O., S.M., V.S., E.A.O., A.C. and V.B.; supervision, E.I.C., T.C., L.B., C.E.G., M.O., S.M., V.S., E.A.O., A.C. and V.B.; project administration, V.B.; funding acquisition, V.P. All authors have read and agreed to the published version of the manuscript.

Funding: This work is supported by the project ANTREPRENORDOC, in the framework of the Human Resources Development Operational Programme 2014–2020, financed by the European Social Fund under the contract number 36355/23.05.2019 HRD OP/380/6/13—SMIS Code: 123847.

Institutional Review Board Statement: Not applicable.

Informed Consent Statement: Not applicable.

Data Availability Statement: Not applicable.

Acknowledgments: This study was performed in collaboration with the Research Centre of Instrumental Analysis SCIENT, Ilfov, Romania, Faculty of Food Engineering and Integrated Center for Research, Development, and Innovation in Advanced Materials, Nanotechnologies and Distributed Systems for Fabrication and Control (MANSiD), Stefan cel Mare University of Suceava, Romania, and Faculty of Pharmacy, Carol Davila University of Medicine and Pharmacy, Bucharest, Romania.

Conflicts of Interest: The authors declare no conflict of interest.

References

1. Grube, M.; Wedin, M. Lichenized Fungi and the Evolution of Symbiotic Organization. *Microbiol. Spectr.* **2016**, *4*, funk-0011-2016. [CrossRef] [PubMed]
2. Garrido-Benavent, I.; Pérez-Ortega, S. Past, present, and future research in bipolar lichen-forming fungi and their photobionts. *Am. J. Bot.* **2017**, *104*, 1660–1674. [CrossRef]
3. Shukla, V.; Joshi, G.P.; Rawat, M.S.M. Lichens as a potential natural source of bioactive compounds: A review. *Phytochem. Rev.* **2010**, *9*, 303–314. [CrossRef]
4. Zambare, V.P.; Christopher, L.P. Biopharmaceutical potential of lichens. *Pharm. Biol.* **2012**, *50*, 778–798. [CrossRef]
5. Prateeksha; Paliya, B.S.; Bajpai, R.; Jadaun, V.; Kumar, J.; Kumar, S.; Upreti, D.K.; Singh, B.N.R.; Nayaka, S.; Joshi, Y.; et al. The genus *Usnea:* A potent phytomedicine with multifarious ethnobotany, phytochemistry and pharmacology. *RSC Adv.* **2016**, *6*, 21672–21696. [CrossRef]
6. Yang, M.X.; Devkota, S.; Wang, L.S.; Scheidegger, C. Ethnolichenology—the use of lichens in the himalayas and southwestern parts of china. *Diversity* **2021**, *13*, 330. [CrossRef]
7. Huang, Y.P.; Xiang, J.T.; Wang, C.H.; Ren, D.; Johnson, D.; Xu, T. Lichen as a Biomonitor for Vehicular Emission of Metals: A Risk Assessment of Lichen Consumption by the Sichuan Snub-Nosed Monkey (*Rhinopithecus roxellana*). *Ecotoxicol. Environ. Saf.* **2019**, *180*, 679–685. [CrossRef]
8. Behera, B.C.; Verma, N.; Sonone, A.; Makhija, U. Optimization of culture conditions for lichen *Usnea ghattensis* G. Awasthi to increase biomass and antioxidant metabolite production. *Food Technol. Biotechnol.* **2009**, *47*, 7–12.
9. Randlane, T.; Tõrra, T.; Saag, A.; Saag, L. Key to European *Usnea* species. *Bibl. Lichenol.* **2009**, *100*, 419–462.
10. Mesta, A.R.; Rajeswari, N.; Kanivebagilu, V.S. Distribution of bioactive compounds in usneoid lichens from Western Ghats. *Plant Arch.* **2019**, *19*, 2163–2168.
11. Shcherbokova, A.; Koptina, A.; Samiento-Diaz, M.; Ulrich-Merzenich, G. Pharmacological activities of lichens from Russia. *Planta Med.* **2015**, *81*, PM_134. [CrossRef]
12. Ohmura, Y. A synopsis of the lichen genus *Usnea* (*Parmeliaceae, Ascomycota*) in Taiwan. *Mem. Natl. Museum Nat. Sci. Tokyo* **2012**, *48*, 91–137.
13. Cansaran, D.; Kahya, D.; Yurdakulol, E.; Atakol, O. Identification and quantitation of usnic acid from the lichen *Usnea* species of Anatolia and antimicrobial activity. *Z. Naturforsch. Sect. C J. Biosci.* **2006**, *61*, 773–776. [CrossRef] [PubMed]
14. Rafat, A. Ecological and Biological Studies of New Zealand Lichens in the Genus *Usnea*. Ph.D. Thesis, Lincoln University, Chester County, PA, USA, 2014; pp. 18–153.
15. Đorđević, S.; Ivanović, J.; Kukić-Marković, J.; Petrović, S.; Žižović, I.; Tadić, V.; Marković, G. HPLC determination of usnic acid content in different extracts of *Usnea barbata*. *Planta Med.* **2010**, *76*, LS1. [CrossRef]
16. Redzic, S.; Barudanovic, S.; Pilipovic, S. Wild Mushrooms and Lichens used as Human Food for Survival in War Conditions; Podrinje. *Hum. Ecol. Rev.* **2010**, *17*, 175–181.
17. Gómez-Serranillos, M.P.; Fernández-Moriano, C.; González-Burgos, E.; Divakar, P.K.; Crespo, A. *Parmeliaceae* family: Phytochemistry, pharmacological potential and phylogenetic features. *RSC Adv.* **2014**, *4*, 59017–59047. [CrossRef]
18. Bazarnova, Y.; Politaeva, N.; Lyskova, N. Research for the lichen *Usnea barbata* metabolites. *Z. Naturforsch. Sect. C J. Biosci.* **2018**, *73*, 291–296. [CrossRef]
19. Salgado, F.; Albornoz, L.; Cortéz, C.; Stashenko, E.; Urrea-Vallejo, K.; Nagles, E.; Galicia-Virviescas, C.; Cornejo, A.; Ardiles, A.; Simirgiotis, M.; et al. Secondary metabolite profiling of species of the genus *Usnea* by UHPLC-ESI-OT-MS-MS. *Molecules* **2018**, *23*, 54. [CrossRef]
20. Ranković, B. *Lichen Secondary Metabolites: Bioactive Properties and Pharmaceutical Potential*; Springer: Cham, Switzerland, 2015; pp. 1–29.
21. Bubach, D.; Dufou, L.; Catán, S.P. Evaluation of dispersal volcanic products of recent events in lichens in environmental gradient, Nahuel Huapi National Park, Argentina. *Environ. Monit. Assess.* **2014**, *186*, 4997–5007. [CrossRef]
22. Sanders, W.B.; Masumoto, H. Lichen algae: The photosynthetic partners in lichen symbioses. *Lichenologist* **2021**, *53*, 347–393. [CrossRef]

23. Lyskova, N.S.; Bazarnova, I.G.; Kruchina-Bogdanov, I.V. Study of the composition and properties of secondary metabolites of the Lichen *Usnea barbata*. *Khimiya Rastit. Syr'ya* **2018**, *1*, 121–127.
24. Bačkor, M.; Fahselt, D. Lichen photobionts and metal toxicity. *Symbiosis* **2008**, *46*, 1–10.
25. Hájek, J.; Hojdová, A.; Trnková, K.; Váczi, P.; Bednaříková, M.; Barták, M. Responses of thallus anatomy and chlorophyll fluorescence-based photosynthetic characteristics of two antarctic species of genus *Usnea* to low temperature. *Photosynthetica* **2021**, *59*, 95–105. [CrossRef]
26. Coufalík, P.; Uher, A.; Zvěřina, O.; Komárek, J. Determination of cadmium in lichens by solid sampling graphite furnace atomic absorption spectrometry (SS-GF-AAS). *Environ. Monit. Assess.* **2020**, *192*, 222. [CrossRef]
27. Carreras, H.A.; Wannaz, E.D.; Perez, C.A.; Pignata, M.L. The role of urban air pollutants on the performance of heavy metal accumulation in *Usnea amblyoclada*. *Environ. Res.* **2005**, *97*, 50–57. [CrossRef]
28. Popovici, V.; Bucur, L.; Calcan, S.I.; Cucolea, E.I.; Costache, T.; Rambu, D.; Schröder, V.; Gîrd, C.E.; Gherghel, D.; Vochita, G.; et al. Elemental Analysis and In Vitro Evaluation of Antibacterial and Antifungal Activities of *Usnea barbata* (L.) Weber ex F.H. Wigg from Călimani Mountains, Romania. *Plants* **2022**, *11*, 32. [CrossRef]
29. Popovici, V.; Bucur, L.; Gîrd, C.E.; Rambu, D.; Calcan, S.I.; Cucolea, E.I.; Costache, T.; Ungureanu-Iuga, M.; Oroian, M.; Mironeasa, S.; et al. Antioxidant, Cytotoxic, and Rheological Properties of Canola Oil Extract of *Usnea barbata* (L.) Weber ex F. H. Wigg from Călimani Mountains, Romania. *Plants* **2022**, *11*, 854. [CrossRef]
30. Pop, A.; Fizeșan, I.; Vlase, L.; Rusu, M.E.; Cherfan, J.; Babota, M.; Gheldiu, A.M.; Tomuta, I.; Popa, D.S. Enhanced recovery of phenolic and tocopherolic compounds from walnut (*Juglans regia* L.) male flowers based on process optimization of ultrasonic assisted-extraction: Phytochemical profile and biological activities. *Antioxidants* **2021**, *10*, 607. [CrossRef]
31. Farmacopeea Rom. Ed. X, 1993. Available online: https://ro.scribd.com/doc/215542717/Farmacopeea-Romana-X (accessed on 2 February 2022).
32. Popovici, V.; Bucur, L.; Popescu, A.; Schröder, V.; Costache, T.; Rambu, D.; Cucolea, I.E.; Gîrd, C.E.; Caraiane, A.; Gherghel, D.; et al. Antioxidant and cytotoxic activities of *Usnea barbata* (L.) F.H. Wigg. dry extracts in different solvents. *Plants* **2021**, *10*, 909. [CrossRef]
33. Pegg, T.J.; Gladish, D.K.; Baker, R.L. Algae to angiosperms: Autofluorescence for rapid visualization of plant anatomy among diverse taxa. *Appl. Plant Sci.* **2021**, *9*, 1–8. [CrossRef]
34. Pathan, A.K.; Bond, J.; Gaskin, R.E. Sample preparation for SEM of plant surfaces. *Mater. Today* **2010**, *12*, 32–43. [CrossRef]
35. Yuan, J.; Wang, X.; Zhou, H.; Li, Y.; Zhang, J.; Yu, S.; Wang, M.; Hao, M.; Zhao, Q.; Liu, L.; et al. Comparison of Sample Preparation Techniques for Inspection of Leaf Epidermises Using Light Microscopy and Scanning Electronic Microscopy. *Front. Plant Sci.* **2020**, *11*, 133. [CrossRef]
36. Thiberge, S.; Nechushtan, A.; Sprinzak, D.; Gileadi, O.; Behar, V.; Zik, O.; Chowers, Y.; Michaeli, S.; Schlessinger, J.; Moses, E. Scanning electron microscopy of cells and tissues under fully hydrated conditions. *Proc. Natl. Acad. Sci. USA* **2004**, *101*, 3346–3351. [CrossRef] [PubMed]
37. Ivanovic, J.; Meyer, F.; Misic, D.; Asanin, J.; Jaeger, P.; Zizovic, I.; Eggers, R. Influence of different pre-treatment methods on isolation of extracts with strong antibacterial activity from lichen Usnea barbata using carbon dioxide as a solvent. *J. Supercrit. Fluids* **2013**, *76*, 1–9. [CrossRef]
38. Ingoífsdotir, K. Molecules of Interest Usnic acid. *Phytochemistry* **2002**, *61*, 729–736. [CrossRef]
39. Gauslaa, Y.; Solhaug, K.A. Differences in the Susceptibility to Light Stress Between Epiphytic Lichens of Ancient and Young Boreal Forest Stands. *Funct. Ecol.* **1996**, *10*, 344. [CrossRef]
40. Varol, M.; Tay, T.; Candan, M.; Türk, A.; Koparal, A.T. Evaluation of the sunscreen lichen substances usnic acid and atranorin. *Biocell* **2015**, *39*, 25–31.
41. Zulaini, A.A.M.; Muhammad, N.; Asman, S.; Hashim, N.H.; Jusoh, S.; Abas, A.; Yusof, H.; Din, L. Evaluation of transplanted lichens, *Parmotrema tinctorum* and *Usnea diffracta* as bioindicator on heavy metals accumulation in southern Peninsular Malaysia. *J. Sustain. Sci. Manag.* **2019**, *14*, 1–13.
42. Bačkorová, M.; Jendželovský, R.; Kello, M.; Bačkor, M.; Mikeš, J.; Fedoročko, P. Lichen secondary metabolites are responsible for induction of apoptosis in HT-29 and A2780 human cancer cell lines. *Toxicol. Vitr.* **2012**, *26*, 462–468. [CrossRef]
43. Gadea, A.; Fanuel, M.; Le Lamer, A.C.; Boustie, J.; Rogniaux, H.; Charrier, M.; Devehat, F.L. Le Mass spectrometry imaging of specialized metabolites for predicting lichen fitness and snail foraging. *Plants* **2020**, *9*, 70. [CrossRef]
44. Talamond, P.; Verdeil, J.L.; Conéjéro, G. Secondary metabolite localization by autofluorescence in living plant cells. *Molecules* **2015**, *20*, 5024–5037. [CrossRef] [PubMed]
45. Le Pogam, P.; Legouin, B.; Le Lamer, A.C.; Boustie, J.; Rondeau, D. Analysis of the cyanolichen *Lichina pygmaea* metabolites using in situ DART-MS: From detection to thermochemistry of mycosporine serinol. *J. Mass Spectrom.* **2015**, *50*, 445–462. [CrossRef]
46. Le Pogam, P.; Legouin, B.; Geairon, A.; Rogniaux, H.; Lohézic-Le Dévéhat, F.; Obermayer, W.; Boustie, J.; Le Lamer, A.C. Spatial mapping of lichen specialized metabolites using LDI-MSI: Chemical ecology issues for *Ophioparma ventosa*. *Sci. Rep.* **2016**, *6*, 37807. [CrossRef] [PubMed]
47. Komaty, S.; Letertre, M.; Dang, H.D.; Jungnickel, H.; Laux, P.; Luch, A.; Carrié, D.; Merdrignac-Conanec, O.; Bazureau, J.P.; Gauffre, F.; et al. Sample preparation for an optimized extraction of localized metabolites in lichens: Application to *Pseudevernia furfuracea*. *Talanta* **2016**, *150*, 525–530. [CrossRef] [PubMed]

48. Culicov, O.A.; Yurukova, L. Comparison of element accumulation of different moss and lichen-bags, exposed in the city of Sofia (Bulgaria). *J. Atmos. Chem.* **2006**, *55*, 1–12. [CrossRef]
49. Jayasekera, R.; Rossbach, M. Background levels of heavy metals in plants of different taxonomic groups from a montane rain forest in Sri Lanka. *Environ. Geochem. Health* **1996**, *18*, 55–62. [CrossRef] [PubMed]
50. Conti, M.E.; Finoia, M.G.; Bocca, B.; Mele, G.; Alimonti, A.; Pino, A. Atmospheric background trace elements deposition in Tierra del Fuego region (Patagonia, Argentina), using transplanted *Usnea barbata* lichens. *Environ. Monit. Assess.* **2012**, *184*, 527–538. [CrossRef]
51. Zambrano, A.; Nash, T.H. Lichen responses to short-term transplantation in Desierto de los Leones, Mexico City. *Environ. Pollut.* **2000**, *107*, 407–412. [CrossRef]
52. Conti, M.E.; Jasan, R.; Finoia, M.G.; Iavicoli, I.; Plá, R. Trace elements deposition in the Tierra del Fuego region (south Patagonia) by using lichen transplants after the Puyehue-Cordón Caulle (north Patagonia) volcanic eruption in 2011. *Environ. Sci. Pollut. Res.* **2016**, *23*, 6574–6583. [CrossRef]
53. Bergamaschi, L.; Rizzio, E.; Giaveri, G.; Loppi, S.; Gallorini, M. Comparison between the accumulation capacity of four lichen species transplanted to a urban site. *Environ. Pollut.* **2007**, *148*, 468–476. [CrossRef]
54. Meli, M.A.; Desideri, D.; Cantaluppi, C.; Ceccotto, F.; Feduzi, L.; Roselli, C. Elemental and radiological characterization of commercial *Cetraria islandica* (L.) Acharius pharmaceutical and food supplementation products. *Sci. Total Environ.* **2018**, *613–614*, 1566–1572. [CrossRef] [PubMed]
55. Rola, K. Insight into the pattern of heavy-metal accumulation in lichen *thalli*. *J. Trace Elem. Med. Biol.* **2020**, *61*, 126512. [CrossRef] [PubMed]
56. Zhao, Y.; Wang, M.; Xu, B. A comprehensive review on secondary metabolites and health-promoting effects of edible lichen. *J. Funct. Foods* **2021**, *80*, 104283. [CrossRef]
57. Kwong, S.P.; Wang, C. Review: Usnic acid-induced hepatotoxicity and cell death. *Environ. Toxicol. Pharmacol.* **2020**, *80*, 103493. [CrossRef] [PubMed]
58. Marinescu, E.; Elisei, A.M.; Aprotosoaie, A.C.; Cioancă, O.; Trifan, A.; Miron, A.; Robu, S.; Ifrim, C.; Hăncianu, M. Assessment of heavy metals content in some medicinal plants and spices commonly used in Romania. *Farmacia* **2020**, *68*, 1099–1105. [CrossRef]
59. Luo, L.; Wang, B.; Jiang, J.; Fitzgerald, M.; Huang, Q.; Yu, Z.; Li, H.; Zhang, J.; Wei, J.; Yang, C.; et al. Heavy Metal Contaminations in Herbal Medicines: Determination, Comprehensive Risk Assessments, and Solutions. *Front. Pharmacol.* **2021**, *11*, 1–14. [CrossRef]
60. FAO/WHO, General Standard for Contaminants and Toxins in Food and Feed. Codex Alimentarius Commission, Adopted in 1995. Revised in 1997, 2006, 2008, 2009 Amended in 2010, 2012, 2013, 2014, 2015. Available online: http://www.fao.org/fao-who-codexalimentarius/shproxy/en/?lnk=1&url=https%253A%252F%252Fworkspace.fao.org%252Fsites%252Fcodex%252FStandards%252FCXS%2B193-1995%252FCXS_193e.pdf (accessed on 2 April 2022).
61. Union, E. Health and Food Safety; (EC) No 118/2006-Maximum Levels for Contaminants in Food Incl Mycotoxins. Off. J. Eur. Union 2006, 5–24. Available online: https://eur-lex.europa.eu/eli/reg/2021/1323/oj (accessed on 2 April 2022).
62. FAO/WHO Guidelines for Assessing Quality of Herbal Medicines with Reference to Contaminants and Residues. WHO Libr. Cat. 2007, 118. Available online: https://apps.who.int/iris/bitstream/handle/10665/43510/9789241594448_eng.pdf?sequence=1&isAllowed=y (accessed on 2 April 2022).
63. European Pharmacopoeia (Ph. Eur.) 10th Edition | EDQM European Directorate for the Quality of Medicines. Available online: https://www.edqm.eu/en/european-pharmacopoeia-ph-eur-10th-edition (accessed on 17 November 2020).
64. Simkova, K.; Polesny, Z. Ethnobotanical review of wild edible plants used in the Czech Republic. *J. Appl. Bot. Food Qual.* **2015**, *88*, 49–67.
65. Dobrescu, D.; Tanasescu, M.; Mezdrea, A.; Ivan, C.; Ordosch, E.; Neagoe, F.; Rizeanu, A.; Trifu, L.; Enescu, V. Contributions to the complex study of some lichens-*Usnea* genus. Pharmacological studies on *Usnea barbata* and *Usnea hirta* species. *Rom. J. Physiol.* **1993**, *30*, 101–107.
66. Yang, Z.; Hu, Y.; Yue, P.; Luo, H.; Li, Q.; Li, H.; Zhang, Z.; Peng, F. Physicochemical Properties and Skin Protection Activities of Polysaccharides from *Usnea longissima* by Graded Ethanol Precipitation. *ACS Omega* **2021**, *6*, 25010–25018. [CrossRef]
67. Thummajitsakul, S.; Samaikam, S.; Tacha, S.; Silprasit, K. Study on FTIR spectroscopy, total phenolic content, antioxidant activity and anti-amylase activity of extracts and different tea forms of *Garcinia schomburgkiana* leaves. *LWT* **2020**, *134*, 110005. [CrossRef]
68. Zaini, E.; Nisak, R.K.; Utami, R.D.; Fitriani, L.; Ismed, F. Effect of milling on physicochemical properties of usnic acid isolated from *Usnea* sp. *Orient. J. Chem.* **2017**, *33*, 3031–3036. [CrossRef]
69. Popovici, V.; Bucur, L.; Vochita, G.; Gherghel, D.; Mihai, C.T.; Rambu, D.; Calcan, S.I.; Costache, T.; Cucolea, I.E.; Matei, E.; et al. In vitro anticancer activity and oxidative stress biomarkers status determined by *Usnea barbata* (L.) F.H. Wigg. dry extracts. *Antioxidants* **2021**, *10*, 1141. [CrossRef] [PubMed]
70. Kondratiuk, A.S.; Savchuk, O.M.; Hur, J.S. Optimization of protein extraction for lichen thalli. *Mycobiology* **2015**, *43*, 157–162. [CrossRef] [PubMed]
71. He, Y.; Tang, H.; Zhang, Z. Ultramicromorphological observation of *Usnea longissima* Ach. *Afr. J. Biotechnol.* **2012**, *11*, 7102–7108.

72. Rahman, M.M.; Islam, M.B.; Biswas, M.; Khurshid Alam, A.H.M. In vitro antioxidant and free radical scavenging activity of different parts of *Tabebuia pallida* growing in Bangladesh. *BMC Res. Notes* **2015**, *8*, 621. [CrossRef]
73. Tirzitis, G.; Bartosz, G. Determination of antiradical and antioxidant activity: Basic principles and new insights. *Acta Biochim. Pol.* **2010**, *57*, 139–142. [CrossRef]

applied sciences

MDPI

Review

Jerusalem Artichoke (*Helianthus tuberosus* L.) as a Promising Dietary Feed Ingredient for Monogastric Farm Animals

Gabriela Maria Cornescu *, Tatiana Dumitra Panaite, Cristina Soica, Ana Cismileanu and Cristina Camelia Matache

The National Research-Development Institute for Animal Biology and Nutrition (INCDBNA-IBNA Balotești), 077015 Balotesti, Romania
* Correspondence: gabriela_cornescu@yahoo.com

Abstract: In recent years, there has been significant attention toward the incorporation of alternative functional feed ingredients in monogastric diets. The objective is to improve sustainability and optimize animal performance both under normal conditions and in heat stress situations. Among these alternatives, Jerusalem artichoke (*Helianthus tuberosus* L.) has emerged as a promising candidate due to its nutritional composition and potential health benefits. This review aims to investigate the potential utilization of Jerusalem artichoke in monogastric diets and the impact on productive performance parameters. Moreover, the potential prebiotic effects of Jerusalem artichoke on the composition and activity of monogastric gut microbiota are revealed, showing its implications for gut health and reduction in pathogenic bacteria. The incorporation of Jerusalem artichoke in monogastric diets poses several challenges, such as limitation of the dietary inclusion rate. However, there are also future perspectives to consider, such as optimizing processing techniques, evaluating the effects of different cultivars, and exploring potential synergies with other dietary feed ingredients. In summary, this study provides a comprehensive overview of the key findings and unique perspectives on the utilization of Jerusalem artichoke in monogastric diets, highlighting its potential as a valuable feed ingredient.

Keywords: Jerusalem artichoke; monogastric; feed ingredients; alternative; sustainable

1. Introduction

Jerusalem artichoke (*Helianthus tuberosus* L.), also known as sunroot, sunchoke, sunflower species, is a perennial plant (Asteraceae family, genus Helianthus) originating from North America, and it is used mostly for its edible tubers in human and animal nutrition. In a French translation dating from 1919, Lacaita [1] discovered extensive discussions about Jerusalem artichoke, referred to, in those times, as "Truffe du Canada". Prior research, conducted on various farm animals, regarded Jerusalem artichoke (J.A.) as a potential feed ingredient substitute (both fresh/silage stems and leaves can be incorporated into the diet of dairy cows and pigs, including fresh/powder tubers for pigs or poultry diets) because of its distinct composition and the potential health advantages it offers, including for pigs [2–7], laying hens [8–10], broilers [11–13], Japanese quail [14], sheep [15–17], fish [18–22], rabbits [23–25], cattle [26], horses [27–29], goats [30–32], bees, and wasps [33]. Lindberg [34] suggests that innovative and sustainable feed sources can be derived from traditional food sources, agro-industrial byproducts, aquaculture, biotechnology, and novel technologies to supply the necessary nutrients for animal food production. In the poultry and swine farming sector, it is crucial to consider ensuring the essential amino acid availability of the novel proposed feed ingredients to be tested and introduced into diet formulations while simultaneously maintaining productive performances. This will enable the adoption of low- or reduced-protein diets, while guaranteeing that there are no negative impacts on the overall productivity [35]. The proximal analyses of J.A. showed that the plant presents almost all indispensable amino acids, including threonine and tryptophan; however, it is not recommended to be used as complete feed, but as a supplementary one [36], as a strategic approach to complement the diet, substituting more costly energy sources. Sawicka [37] considered J.A. an ideal raw material that is affordable, with lots of natural bioactive ingredients (inulin, vitamins B_1, B_2, B_7, C, organic acids, polyphenols, flavonoids, anthocyanins, phenolic compounds, coumarins, terpenes, dietary fiber, enzymes, polysaccharides, minerals, and essential amino acids), and a highly versatile plant with minimal waste (e.g., tubers for ethanol production, human food and animal feed; leaves as soil fertilizer (nitrogen source); stalk mass as protein for animal feed). Shazoo [38] identified several primary utilizations of J.A.; among them, the most important ones include inulin, fructose syrup, yeast, and powder/capsules/tablets production for human pharmaceutical purposes. The potential applications of inulin, derived from both the tubers and aerial components of *Helianthus tuberosus* L., encompass various effects such as antidiabetic [39–41], antioxidant [42,43], anti-inflammatory [44,45], weight loss [46], anti-constipation [47], and increasing immunity benefits [48]. Manokhina [49] highlighted the importance of J.A. as a valuable bioenergetic crop due to its inulin, fructose, and protein content, global cultivation, and adaptability, emphasizing its utilization as a valuable feeding solution for farmers. Jerusalem artichoke is utilized for both its above-ground biomass and tubers, exhibiting high tolerance to water stress and remarkable adaptability to extreme temperatures (35–45 °C for the plants; −30 to −45 °C for the tubers). It has low soil, minimal fertilizer, water, and pesticide requirements, and can be cultivated on marginal lands [50]. Negosanu et al. [51], from the Vegetable Research Development Station (V.R.D.S.) in Buzău, Romania, conducted extensive research on Jerusalem artichoke, which is locally known as "nap-porcesc", "morcov-porcesc", "gulii", "pere iernatice" and "cartoful săracilor" since 1996. They registered three varieties, namely Rareș, Dacic, and Flavius, in the Official Catalogue of Species and Varieties of Cultivated Crops [52]. Moreover, in 2018–2020, a study conducted by Dima et al. [53] at the Research and Development Station for Plant Culture on Sandy Soils (S.C.D.C.P.N.) Dăbuleni found that four Jerusalem artichoke genotypes (Dacic, Olimp, Rustic, and a local population) had an average inulin content of 12.93%, emphasizing its functional potential as a valuable fiber source. Agapie et al. [52] highlighted that J.A. exhibits a remarkable ability to adapt to challenging environmental conditions, able to resistant water scarcity and extreme temperature fluctuations, and showing resilience against diseases and pests. Consequently, it is suggested as a practical option for addressing climate change-related issues in Romania.

This review aims to disclose the potential advantages of incorporating Jerusalem artichoke into the diet of monogastric animals, highlighting potential benefits in terms of productivity and gastrointestinal health.

2. Literature Review Methodology

An extensive search across Medline, PubMed Central (PMC), Web of Science, Scopus, and Google Scholar was conducted without time or language constraints. For our database bibliography exploration, the keywords included were as follows: Jerusalem artichoke, inulin, co-products, waste, biofuel, bioethanol, diabetes, performances, monogastric, gut microbiota, immune function, pathogenic bacteria, lipid metabolism, and nutrient absorption.

3. Aspects Regarding the Nutritional Dynamics and Harvest Variability of Jerusalem Artichoke Tubers and Aerial Parts (Steams, Branches, and Leaves)

In order to unlock the full nutritional value of novel and underutilized feed sources, and optimize their ideal formulation for diets, it is crucial to comprehensively assess and compare their chemical and physical properties, and nutritional composition to other varieties or raw materials commonly used, especially in monogastric rations.

Kays and Nottingham [54] affirmed that the proximate analyses of J.A. tubers show a significant variation in data composition compared to other vegetables due to differences in cultivars, harvest period, production conditions, postharvest treatment, and preparation methods.

According to Ma [36], J.A. tubers are traditionally harvested in late autumn, especially during the period when carbohydrates are transferred from the plant's aerial parts to underground tubers. Farzinmehr et al. [55] found that harvesting J.A. forage every 60 days optimally increased annual yields of dry matter, protein, and energy biomass in both tubers and forage, influenced by the maturity stage and harvesting frequency. According to Marien [56], the proximate analysis values for the average metabolizable energy (ME) content of J.A. tubers was 15.0 MJ/kg, with a CP content of 69 g/kg in dry matter. The crude protein content of J.A. leaves is about four times higher compared to tubers [57] and three times higher compared to stems [58], and is especially rich in lysine and methionine [59]. It is well known that ensuring an optimum dietary level of methionine and lysine is crucial for promoting growth, maximizing feed efficiency, and maintaining overall health in monogastric animals. These amino acids are often supplemented in animal feeds to meet their specific requirements and optimize their performance.

Table 1 presents the nutritional composition of Jerusalem artichoke's fresh aerial parts and tubers.

Table 1. Proximal composition of Jerusalem artichoke's aerial parts and tubers (per 100 g/fresh weight).

Nutrient	Determined	
	In Aerial Parts	**In Tubers**
DM, % as fed	32.3	16.8–28.2
CP, %DM	7.0–23.0	6.6–8.3
CF, %DM	11.2–17.8	3.8–6.6
NDF, %DM	34.1–51.0	8.5–9.5
ADF, %DM	23.3–41.2	5.3–6.0
Lignin, %DM	11.0–12.0	0.9–1.2
EE, %DM	1.6–3.4	0.5–1.4
Ash, %DM	8.0–18.5	5.8–6.1
GE, MJ/kg DM	16.8	17.1
Calcium, g/kg DM	16.8–20.2	1.4–2.1
Phosphorus, g/kg DM	2.8–3.9	2.9–3.5
Potassium, g/kg DM	-	22.7
Magnesium, g/kg DM	4.0–6.2	1.8
Zinc, g/kg DM	40.0–50.0	0.013
Iron, g/kg DM	110–150	0.039

DM, dry matter; CP, crude protein; CF, crude fiber; NDF, nitrogen detergent fiber; ADF, acid nitrogen fiber; EE, ether extract; GE, gross energy. Data on proximal composition of Jerusalem artichoke adapted from Heuzé et al. [60] and Feedipedia, a program by INRAE, CIRAD, AFZ, and FAO, https://www.feedipedia.org/node/544 (accessed on 21 July 2023).

The proximal composition between the two parts is highly significant when taking into consideration their utilization for different purposes or in farm animal feed, such as forage or raw/powder tuber consumption. Authors such as Keys and Nottingham [54] stated that information regarding the proximal composition of J.A. tubers is notably limited in comparison to other vegetables. They found that the CP content exhibits variability, ranging from 1.6 to 2.4 g per 100 g of fresh weight, but being constant during all developing stages [61]. Other authors, such as Okrouhlá et al. [62], reported the following composition for J.A. tubers: 90.00%DM, 8.57% CP, 0.33% EE, 3.51% CF, 7.47% ash, 70.12% N-free extract, 55.76% fructans, and 13.79 MJ/kg of metabolizable energy for pigs. In their study on the chemical composition of J.A. tubers concerning the effects of variety and harvest date, Florkiewicz et al. [63] recorded the following information regarding composition: 7.1–7.2 g protein/100 g dry matter, 21.5–25.6 g dry matter/100 g, 16.3–16.4 g dietary fiber/100 g dry matter, 7.1–8.1 mg vitamin C/100 g, 218–225.9 mg total polyphenols/100 g, 45.8–47.4 g fructans/100 g dry matter, 0.47–0.65 g glucose/100 g dry matter, and 13.4–14.5 g sucrose/100 g dry matter. The research unveiled significant variations in the content of the analyzed components, which were dependent on both the variety and harvest time of the tubers. Tubers harvested in spring exhibited higher amounts of glucose, fructose, sucrose, dietary fiber, phenolic compounds, and nitrate (V). Conversely, tubers harvested in autumn showed higher levels of fructans, vitamin C, and nitrate (III). The fiber content of J.A. tubers ranges from 11.4 to 20.8 g per 100 g of dry matter. This fiber mainly consists of cellulose, hemicellulose, pectin, and lignins, which are linked to cellulose but are composed of phenylpropan alcohol instead of monosaccharides [64]. Farzinmehr et al. [55] reported that the best harvesting interval for J.A. forage to achieve tubers with the highest yearly yield, water-soluble carbohydrates, and digestibility was every 120 days, while the highest nutritive value and yield of the forages were observed when harvesting J.A. every 60 days, with the DM biomass of 27.16 t/ha, CP of 98.6–145 g/kg DM, and OM digestibility between 0.607 and 0.691. The average value of the ME (MJ/kg DM) of J.A. tubers ranged from 12.1 to 12.6, depending on whether the tubers were harvested from plots with four, three, or two forage cuts per year. According to Kays and Nottingham [54], J.A. aerial parts exhibited higher total digestible nutrient levels while having lower protein concentrations compared to alfalfa (*Medicago sativa* L.). Thus far, the studies presented provide evidence that the properties of Jerusalem artichoke depend on various influencing factors, making these factors important to take into consideration (e.g., climate, cultivar [53], harvesting [55], technology, and growing conditions [65]) when assessing its nutritional value and potential applications. Furthermore, as demonstrated, the tubers and leaves of Jerusalem artichoke exhibit diverse nutrient compositions, and their utilization in farm animal diets can be optimized by taking into account both their nutritional content and growth development stage. Figure 1 presents the amino acid content of both the aerial parts and tubers of J.A. The literature regarding the amino acid content in J.A. is limited.

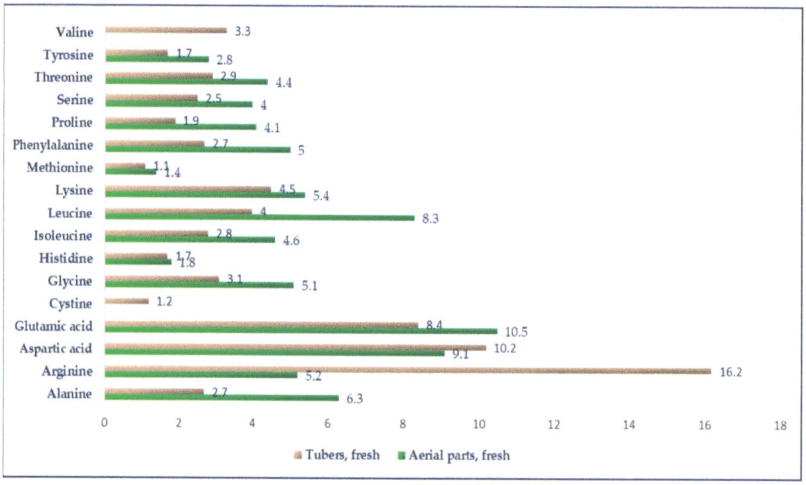

Figure 1. The amino acid composition of fresh tubers and aerial parts of Jerusalem artichoke (% protein). Heuzé et al. [60] and Feedipedia, a program by INRAE, CIRAD, AFZ, and FAO, https://www.feedipedia.org/node/544 (accessed on 21 July 2023).

Cieslik et al. [64] analyzed red variety J.A. tubers, and observed variations in amino acid composition, influenced by cultivar and storage time. The significant presence of essential amino acids, such as methionine (72.6 mg/100 g), lysine (286.6 mg/100 g), leucine

(255.2 mg/100 g), isoleucine (180.8 mg/100 g), threonine (182.6 mg/100 g), phenylalanine (173.9 mg/100 g), and valine (209.7 mg/100 g), suggests that these tubers could be a valuable source of high-quality protein. However, the absence of detectable tryptophan indicates a potential limitation regarding in this specific amino acid. These findings underscore the importance of considering the complete amino acid profile for a comprehensive understanding of the nutritional value of these tubers. Additionally, the tubers demonstrate a favorable EAA:TAA ratio of 47.5% and an EAA:NEAA ratio of 90.7%. This aligns with the recommended ideal amino acid pattern (EAA:TAA ratio of 40%; and EAA:NEAA ratio > 60%) by the Food and Agriculture Organization (FAO) and the World Health Organization (WHO) [66], indicating that J.A. tubers can be considered a valuable source of high-quality protein. Danilcenko et al. [67] stated that 55% of the total amino acids in J.A. tubers are represented by essential amino acids. Regardless of the cultivar and storage period, arginine was the predominant essential amino acid, while asparagine, glutamine, and alanine were dominant nonessential amino acids. Tyrosine and methionine made up the lowest content in tubers. When harvested, glutamine (9.18 mg kg^{-1} dry weight), glycine (2.31 mg kg^{-1} dry weight), and leucine (2.77 mg kg^{-1} dry weight) levels were the highest. Regarding the mineral content of J.A., some authors [68] registered that J.A. tubers had a high content of potassium (23.35 g/kg DM), phosphorus (3.89 g/kg DM), magnesium (1.89 g/kg DM), and calcium (1.86 g/kg DM), as well as a representative content of iron (39.22 mg/kg DM), zinc (13.56 g/kg DM), and copper (7.18 mg/kg DM).

4. Jerusalem Artichoke as an Important Source of Inulin for Monogastric Farm Animals

De Leenheer and Hoebregs [69] stated that the earliest-documented records of J.A. as a source of inulin dates back to approximately 1870. However, it was not until the 1950s that the complete structure of inulin was fully clarified. Van Loo et al. [70] mentioned that the inulin content of tubers varies from 7 to 30% by fresh weight and approximative 50% by dry weight. Its versatile compounds offer a range of positive effects and applications in both human and animal nutrition [37]. According to Long et al. [71], inulin constituted 80% of the carbohydrates found in tubers. Additionally, as stated by Somda et al. [72], inulin serves as the primary storage carbohydrate in J.A., with the majority of carbon in tubers being present in the form of inulin [73].

Jerusalem artichoke represents a significant source of inulin and oligofructose, providing high concentrations, at 16.0–22.0 g/100 g and 12.0–15.0 g/100 g, respectively. This makes it prominent among various food sources, surpassing many, including bananas (0.3–0.7 g/100 g), asparagus (2.0–3.0 g/100 g), garlic (9.0–16.0 g/100 g), and leeks (3.0–10.0 g/100 g). Moreover, J.A. competes favorably or exceeds others in this aspect, such as chicory root (35.7–47.6 g/100 g) and dandelion greens (12.0–15.0 g/100 g), showcasing it as a notable prebiotic fiber contributor [73]. These inulin-type fructans, functioning as both functional fructans and soluble dietary fibers, comprise a combination of inulin, oligofructose, and fructooligosaccharide with β-configuration, classified based on their chain length or degree of polymerization as follows: small (3–5), medium (6–10), and long chains (11–60). The length or degree of polymerization is influenced by various factors such as genotype, environmental conditions, harvest time, and storage processes [74]. Bhanja et al. [75] stated that inulin serves as a functional bioingredient in both food and feed applications, acting as a prebiotic agent that can promote the growth of beneficial gut microflora, contributing to the overall health benefits. The inulin content in the tubers was estimated based on the analysis of the sugar content, which was approximately 803.4 g/kg in dry matter, indicating a corresponding inulin content of 650 g/kg in dry matter [57]. The proven benefits linked to inulin were confirmed through a range of conducted experiments in monogastric and ruminant animals as well.

Figure 2 presents the advantages of incorporating J.A. in farm diets with positive results in alterations of intestinal microflora composition, which subsequently influences their metabolism and physiology. Extracts derived from J.A. tubers have the potential to enhance feed efficiency, improve digestion, and lower the occurrence of diarrhea in monogastric animals.

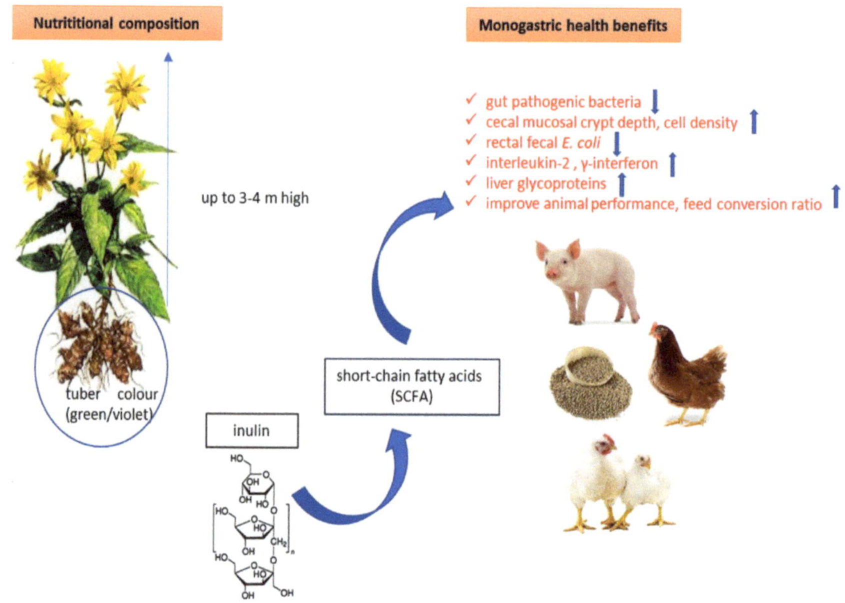

Figure 2. The effects of incorporating J.A. into the diet of monogastric animals, and its benefits linked to inulin.

Flickinger [76] suggests that extracts derived from J.A. tubers hold potential as supplements for animal feed, primarily due to their potential to promote bifidogenic effects. Although inulin and fructooligosaccharides, typically sourced from chicory, are frequently incorporated into the diets of farm animals, their precise health advantages in animals are not as comprehensively studied as they are in humans. It is crucial to acknowledge that different animal species possess distinct digestive tract structures, meaning that a single fructooligosaccharide supplement may not consistently produce optimal results across all species. Depending on the species, such supplements might have suboptimal effects or even lead to digestive issues like loose stools or excessive flatulence. According to Vhile et al. [77], fructooligosacharides are resistant to enzyme hydrolysis in the small intestine, but they stimulate the growth of bifidobacteria in the hindgut, which facilitates inulin fermentation.

5. Jerusalem Artichoke Supplementation Effects in Monogastric Animals on Productive Performances

5.1. Effects of Jerusalem Artichoke Inclusion in Pig Diets

According to Kosaric [61], J.A. tubers are acknowledged to possess a high nutritional value for domestic animals. Historically, it was common practice to let pigs forage in fields where the aerial parts of J.A. plants had been cut, and the pigs would dig up and consume the tubers directly on the field [78]. Jerusalem artichoke tubers were considered a viable alternative feed for pigs during periods of potato scarcity [79]. In a study conducted by Kongsted et al. [7], the impact of free-range foraging on J.A. and the effects of feeding concentrates either restrictively (at 30% of energy recommendations) or ad libitum on pigs was investigated. The results revealed that pigs fed restrictively exhibited a significantly lower daily gain (560 vs. 1224 g per pig) and an improved feed conversion ratio (17.6 vs. 42.8 MJ ME concentrate per kg live weight gain). In their study, Farworth et al. [3] studied the impact of supplementing Jerusalem artichoke tuber flour (1.5%), which is rich in fructooligosaccharides, to 28-day-old weanling pigs and registered a decrease in daily feed intake, daily weight gain, and feed conversion ratio as a result of the supplementation. Again, Farnworth [5] studied raw tuber supplementation in 28-day-old weaned pigs and noticed significantly increased body weight gain, but a reduction in feed efficiency. In a second trial, the pigs were fed a diet containing dried J.A. (10 to 60 g/kg) and a significant increase in weight gain was observed, and although feed efficiency showed improvement, the increase was not statistically significant. Okrouhlá et al. [62] observed that the daily weight gain of the group fed with 12.2% J.A. showed significantly lower results ($p = 0.042$).

The digestion of intact J.A. tubers in pigs is clearly dependent on microbial activity in the digestive tract. Jerusalem artichoke tubers are very rich in fructan, and fructose liberated from fructans is 50% partially digested up to the ileum. A very low ileal digestibility of a diet containing fresh tubers of Jerusalem artichokes was obtained. By analyzing short fatty acids, it was observed that some of the resulting carbohydrates were subjected to microbial transformation in the fore-gut. Table 2 presents values of intestinal digestibility in pigs fed Jerusalem artichoke tubers in the study of Ly et al. [80].

Table 2. The ileal and fecal digestibility of J.A. tubers in pigs.

	Jerusalem Artichoke Tubers
	Ileal Digestibility, g/kg
Dry matter	465
Ash	385
Organic matter	485
Nitrogen	475
Crude fiber	69
Energy	489
	Faecal Digestibility, g/kg
Dry matter	859
Ash	706
Organic matter	860
Nitrogen	682
Crude fiber	665
Energy	864

A summary of the previous studies presented concludes that incorporating J.A. into pig diets presents opportunities as well as challenges. It served as a valuable alternative when potatoes were scarce; however, its impact on pig nutrition varies across studies. While some indicate enhanced feed efficiency and reduced undesirable compounds like skatole, others report fluctuations in weight gain and feed efficiency. Achieving the best results requires meticulous attention to diet composition and supplementation levels when using Jerusalem artichoke as a feed resource for pigs.

5.2. Effects of Jerusalem Artichoke in Reducing Skatole Levels

Feeding J.A. to male pigs resulted in a dose-dependent decrease in skatole levels in the hindgut and adipose tissue. This reduction in skatole levels may be attributed to the decrease in Clostridium perfringens and the increase in short-chain fatty acids (SCFAs), with a subsequent reduction in pH [77]. Adding 4% of dried Jerusalem artichoke tubers for 40 days to diet of piglets (28 days of age) decreased skatole levels in the colon and feces, as well as a decreased the dry matter content and pH in the colon [81]. Additionally, there was a reduction in enterobacteria levels in the colon with increasing levels of J.A. [79].

Okrouhlá et al. [62] supplemented different levels of J.A. (4.1%, 8.1%, or 12.2%) in pig diets for 13 days before slaughter to reduce skatole and indole levels responsible for boar taint, which can affect the taste and odor of pork. The group fed with 12.2% J.A. showed a decrease in skatole concentration in adipose tissue ($p = 0.003$) and reduced *E. coli* levels in feces ($p \leq 0.001$). A significant correlation was found between J.A. concentration, *E. coli* levels, and skatole levels in adipose tissue.

The inclusion of J.A. in pig diets presents opportunities for reducing skatole levels and modifying the intestinal microbiota, but careful attention to diet composition and supplementation levels is necessary for achieving optimal results.

5.3. *Effects of Jerusalem Artichoke Inclusion in Poultry's Diets*

Poultry studies also explored the effects of incorporating dietary Jerusalem artichoke on the productivity of poultry, as well as the quality characteristics of eggs and meat. Initially considered an alternative to synthetic antibiotics, Jerusalem artichoke was recognized for its natural, less toxic, and residue-free effects, making it an ideal feed additive for its utilization in poultry production [82]. Jawad and Al-Abboodi [83] indicated that J. A. supplementation in poultry diets can facilitate improved diet metabolism, which has a positive impact on the feed conversion ratio and other production parameters. In a 16-week feeding trial conducted by Yildiz et al. [84], the impact of dried J.A. (5%) in laying hen diets was examined, along with vetch (5%), and various combinations of J.A. (5% and 10%) and vetch (5% and 10%) to mitigate the diarrheal effects of inulin. The experimental group with a combination of 10% J.A. and 10% vetch showed a lower live weight by 4.36–10.09% compared to the other groups, but it exhibited improved feed efficiency. Notably, there were no adverse effects observed on egg quality of the hens across all groups. Sritiawthai et al. [10] studied the effects of supplementing dried J.A. (50 ppm and 100 ppm) as a dietary prebiotic in laying hen feed. The researchers examined the impact of J.A. on productive performance parameters, egg quality characteristics, and intestinal microflora. The results showed that there were no significant effects on feed intake, final body weight, egg production, egg weight, and egg mass with the inclusion of dried J.A. However, at the 100 ppm supplementation concentration, there was a significant increase in the albumen ratio, yolk–albumen ratio, and the albumen height of eggs. Additionally, the population of lactic acid bacteria in the caecum also significantly increased at this supplementation concentration.

Following the ban of antibiotics in broilers, researchers explored alternative methods for controlling gastrointestinal microorganisms. Kleessen et al. [11] conducted a study where they tested the use of 0.5% topinambur syrup in drinking water. The results showed a significant increase in the cecal counts of *B. bacteriovorus*, as well as significantly lower numbers of total aerobes, including *Enterobacteriaceae* and *C. perfringens*. Moreover, the study observed reduced levels of endotoxins in the blood and an increased body weight on day 35 of the trial period. Katiyanon et al. [85] researched the impact of supplementing J.A. (at 1%, 2%, 3%, or 4% concentrations) in broiler diet on production performances. The results revealed that supplementation at 1% and 4% concentrations led to an increased body weight gain and improved feed efficiency, along with a decrease in the mortality rate. Additionally, the serum cholesterol levels decreased significantly for all concentrations of J.A. supplementation. Notably, feeding J.A. at 3% in the broiler diet showed beneficial effects in terms of body weight gain, feed efficiency, and mortality. Al-Abboodi and Jawad [12] conducted a study to assess the efficiency of J.A. powder addition at different concentrations (0.5%, 1%, 1.5%, and 2%) in broiler diets (1 to 6 weeks of age). The supplementation of 2%, 0.5%, and 1.5% did not significantly influence the productive parameters. However, the experimental groups showed an overall improvement ranging from 2.5% to 4.2% in the final body weight, 5.1% to 18.5% in the production index, 0.1% to 0.2% in the growth rate, and 4.7% to 8.9% in the European production efficiency factor. The best results were obtained with 2% supplementation, which exhibited a higher improvement percentage for all production parameters of broiler chickens.

Therefore, including J.A. as a natural feed additive improved feed efficiency and overall production parameters. J.A. also has the potential to enhance egg quality and gut health. In broilers, it is connected to higher weight gain, better feed efficiency, and lower mortality rates, especially at certain supplementation concentrations.

6. Utilizing Jerusalem Artichoke as a Prebiotic to Enhance Gut Health in Monogastric Farm Animals

In modern intensive production systems, various feed solutions, such as prebiotics, have been employed to prevent intestinal disorders, enhance the gut ecosystem, and subsequently improve the health status and production performances of monogastric animals by modulating the intestinal microbiota [86]. The intestinal microbiota of animals is highly correlated with their physiological, developmental, nutritional, and immunological processes, directly impacting their health and productive performance [87].

Figure 3 presents various internal and external factors that influence animal microbiota, including the host genotype, diet, breeding environment, and exposure to antibiotics. According to Valdovska et al. [88], many environmental stressors in farm animals, including management practices and dietary changes, can disturb the intestinal ecosystem, increasing the risk of pathogen infections, and decreasing feed intake with negative impact on growth performances. As a result, the immune function and balance of intestinal microbiota may be negatively affected, causing gut disorders, infections, and diarrhea [89]. The weaning period for piglets highlights the vital role of gut microbiota as it involves separating them from sows and transitioning from milk to solid food [90]. Weaning is associated with reduced feed intake, growth, intestinal changes, diarrhea, and increased mortality. Piglets are particularly susceptible to *E. coli* overgrowth, including post-weaning colibacillosis [91]. According to Clavijo and Flórez [92], the dominant bacterial phyla found in chickens consist of Proteobacteria, Firmicutes, Actinobacteria, Bacteroidetes, and Cyanobacteria. The principal bacterial groups found in pigs' gastrointestinal tract mostly consist of Streptococcus, Lactobacillus, Eubacterium, Fusobacterium, Bacteroides, Bifidobacterium, Clostridium, Escherichia, Prevotella, Proteobacteria, and Ruminococcus [93]. The Firmicutes and Bacteroidetes bacterial phyla represent the intestinal microbiota in all monogastric animals [94]. According to Bamigbade et al. [95], prebiotics are considered a group of biological nutrients that promote the growth and activity of Lactobacilli and Bifidobacteria in the gastrointestinal tract of monogastric animals.

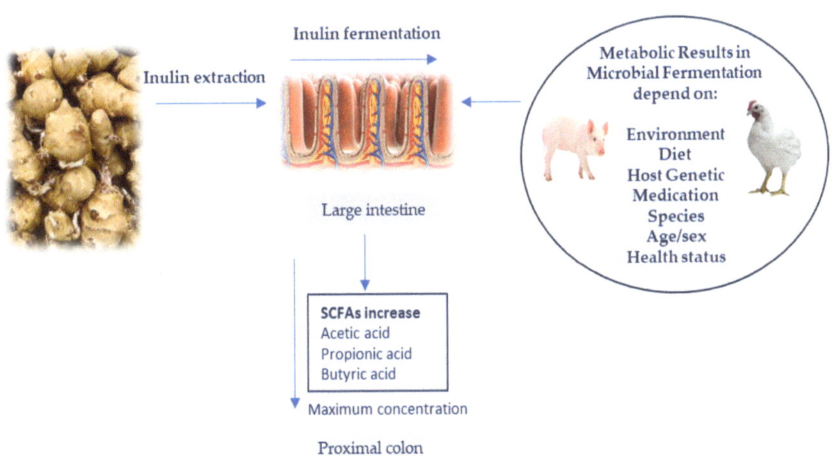

Figure 3. Several key factors that influence the gastrointestinal microbiota of monogastric animals.

When these prebiotics are broken down by microflora in the gut, they produce short-chain fatty acids (SCFAs) that are released into the colon and subsequently absorbed into the bloodstream. The two primary groups of prebiotics extensively studied in this regard are fructo-oligosaccharides (FOSs) and galactooligosaccharides (GOSs). In their study, Barszcz et al. [81] incorporated either 2% inulin from chicory root or 4% dried J.A. tubers into pig diets. The researchers observed an increase in cecal valeric acid levels and a decreasing in isoacids concentration in the colon. Additionally, they found a reduction in β-glucosidase and β-glucuronidase activity in the middle colon and an increase in *Bifidobacterium* spp. populations in both the proximal and distal colon. Valdovska et al. [88] found that although J.A. can impact the concentration of short-chain fatty acids in pig manure and alter the microbiota composition in the large intestine, its prebiotic potential remains incompletely understood.

Using J.A. and probiotics (*Lactobacillus reuteri* and *Pediococcus pentosaceus*) as alternatives to ameliorate gut microbiota in weaned piglets (7 weeks old), the same authors [88] obtained a significant improvement of the microbial contents, defense mechanism, and regeneration processes in pigs' intestines. In the study conducted by Farnworth et al. [3], the addition of J.A. tuber flour to the piglet diet did not result in any significant effect on the concentration of volatile fatty acids (VFAs). Despite observing an increase in propionic acid, isobutyric acid, butyric acid, iso-valeric acid, and valeric acid, along with a decrease in digesta acetic acid content, these experimental results were not statistically significant. Volatile fatty acids are the primary end products of carbohydrate digestion in ruminants, but they are also present in significant concentrations in pigs' large intestines. The cecum and colon in pigs transport volatile fatty acids at rates similar to, or even greater than, horses' large intestinal mucosa or cattle's rumen epithelium [96].

Kleessen et al. [11] conducted a study to examine the effects of 0.5% topinambur syrup added to drinking water. The findings showed a decrease in bacterial endotoxin and Enterobacteriaceae levels. In their experiment, Yildiz et al. [84] incorporated 0%, 5%, and 10% J.A. in laying hen diets. However, the dietary treatments did not lead to any changes in fecal and intestinal pH values.

Despite having a short large intestine, poultry respond well to prebiotic administration. Fructooligosaccharides, with their high digestibility, have been observed to enhance Lactobacillus production. The type and solubility of fiber influence morphological and physiological changes in the digestive tract [97,98]. While initially seen as an antibiotic alternative, the utilization of prebiotics in the poultry industry remains limited.

Sevane et al. [99], utilized chicory (*C. intybus* L.) roots as an inulin source in the broiler diet, incorporating it at a concentration of 746 g kg^{-1}. The results revealed that inulin-type fructans inclusion (5 g/kg diet) positively influenced the immune status, particularly by enhancing the production of long-chain fatty acids.

Other authors, such as Rebolé et al. [100], noticed that the addition of inulin-type fructans from dietary chicory roots (0, 10, or 20 g/kg) decreased serum cholesterol concentration and abdominal fat deposition in boilers. Both types of inulin showed beneficial prebiotic properties when experimented with on monogastric farm animals.

The wide variability in the results noted in studies involving the supplementation of monogastric animals with J.A. can be attributed to the differences in experimental variables such as the dietary level included, the period length of J.A. dietary supplementation, along with the specific animal species, age, and health status that were investigated.

7. Perspectives and Considerations regarding Jerusalem Artichoke Utilization

Despite the numerous favorable opinions in the literature regarding the utilization of J.A. in both human and animal nutrition, the cultivation of this crop currently holds a more important aspect from the standpoint of biofuel production. While this focus on biofuel is noteworthy, it is essential not to overlook the extensive benefits associated with J.A. incorporation proven by experimental studies. Its nutritional profile, including the presence of inulin, suggests potential improvements in gut health and overall animal well-being. Additionally, the crop's resistance to drought makes it particularly suitable for cultivation in Romania, offering a sustainable and resilient alternative for feed production.

It is imperative to reconsider and emphasize the multifaceted advantages of J.A. for animal nutrition. Exploring and highlighting these benefits can contribute to a more comprehensive understanding of the crop's potential and encourage its broader adoption in various agricultural applications.

8. Conclusions

In summary, considerable evidence affirms the prebiotic advantages of Jerusalem artichoke for monogastric animals, demonstrating favorable effects on gastrointestinal microbiota and varied outcomes on production parameters. However, it is noteworthy that the majority of studies are either dated or limited in scope, particularly with regard to each specific monogastric farm animal. This underscores the necessity of additional research, emphasizing the need for a nuanced approach in integrating Jerusalem artichoke into animal diets. Addressing the raised concerns and undertaking further studies will significantly contribute to a more thorough comprehension of the role of Jerusalem artichoke in enhancing gut health and overall animal performance.

Author Contributions: Conceptualization, G.M.C. and T.D.P.; methodology, G.M.C.; validation, G.M.C., T.D.P. and C.S.; investigation, A.C. and C.C.M.; resources, G.M.C. and A.C.; data curation T.D.P. and C.S.; writing—original draft preparation, G.M.C.; writing—review and editing, G.M.C., T.D.P. and C.S.; visualization, C.S.; supervision, T.D.P.; project administration, T.D.P. All authors have read and agreed to the published version of the manuscript.

Funding: This research was funded by the Ministry of Agriculture and Rural Development—ADER 8.1.6; Ministry of Research, Innovation and Digitization—Project PN 23 20 0101; National Research Development Project Projects to Finance Excellence: (PFE)-8/2021.

Institutional Review Board Statement: Not applicable.

Informed Consent Statement: Not applicable.

Data Availability Statement: Not applicable.

Conflicts of Interest: The authors declare no conflict of interest.

References

1. Lacaita, C.C. The "Jerusalem Artichoke"(*Helianthus tuberosus*). *Bull. Misc. Inform.* **1919**, 321–339. [CrossRef]
2. Ruszczyc, Z.; Glaps, J. Silage of Jerusalem artichoke tubers for fattening pigs. *Rocz. Nauk. Rol. Ser. B Zootech.* **1958**, *72*, 565–571.
3. Farnworth, E.R.; Modler, H.W.; Jones, J.D.; Cave, N.; Yamazaki, H.; Rao, A.V. Feeding Jerusalem artichoke flour rich in fructooligosaccharides to weanling pigs. *Can. J. Anim. Sci.* **1992**, *72*, 977–980. [CrossRef]
4. Farnworth, E.R.; Jones, J.D.; Modler, H.W.; Cave, N. The use of Jerusalem artichoke flour in pig and chicken diets. In *Studies in Plant Science*; Elsevier: Amsterdam, The Netherlands, 1993; Volume 3, pp. 385–389.
5. Farnworth, E.R.; Modler, H.W.; Mackie, D.A. Adding Jerusalem artichoke (*Helianthus tuberosus* L.) to weanling pig diets and the effect on manure composition and characteristics. *Anim. Feed Sci. Technol.* **1995**, *55*, 153–160. [CrossRef]
6. Ly, J. Nitrogen and energy balance in pigs fed Jerusalem artichokes (*Helianthus tuberosus* L.). *RCPP* **2000**, *7*, 32–37.
7. Kongsted, A.G.; Horsted, K.; Hermansen, J.E. Free-range pigs foraging on Jerusalem artichokes (*Helianthus tuberosus* L.) Effect of feeding strategy on growth, feed conversion and animal behaviour. *Acta Agric. Scand. A Anim. Sci.* **2013**, *63*, 76–83.
8. Yıldız, G.; Sacakli, P.; Gungor, T.; Uysal, H. The effect of jerusalem artichoke (*Helianthus tuberasus* L.) on blood parameters, liver enzymes and ıntestinal ph in laying hen. *J. Anim. Vet. Adv.* **2008**, *7*, 1297–1300.
9. Park, S.O.; Park, B.S. Effect of feeding inulin oligosaccharides on cecum bacteria, egg quality and egg production in laying hens. *Afr. J. Biotechnol.* **2012**, *11*, 9516–9521.
10. Sritiawthai, E.; Kaewtapee, C.; Bunchasak, C.; Poeikhampha, T. The effect of Jerusalem artichoke (*Helianthus tuberosus* L.) supplimentation on production performance, egg quality characteristics and intestinal microflora of laying hens. *J. Appl. Sci.* **2013**, *13*, 183–187. [CrossRef]
11. Kleessen, B.; Elsayed, N.A.A.E.; Loehren, U.; Schroedl, W.; Krueger, M. Jerusalem artichokes stimulate growth of broiler chickens and protect them against endotoxins and potential cecal pathogens. *J. Food Prot.* **2003**, *66*, 2171–2175. [CrossRef]
12. Al-abboodi, A.A.; Jawad, H.S. Effect of supplementing different levels of Jerusalem artichoke (*Helianthus tuberosus* L.) on broiler production performance. *Plant Arch.* **2018**, *18*, 1570–1574.
13. Saffah, S.F.; Jawad, H.S.; Maaeni, Y.M. Effect of supplementing two levels of jerusalem artichoke (*Helianthus tuberosus* L.) powder in broiler diets on histological parameters of small intestinal segments. *Plant Arch.* **2020**, *20*, 3543–3547.

14. Abdel-Wahab, A.A.; Elnesr, S.S.; Abdel-Kader, I.A. Effect of dietary supplementation of Jerusalem Artichoke extract on performance, blood biochemistry, antioxidant parameters, and immune response of growing Japanese quail. *J. Anim. Physiol. Anim. Nutr.* **2023**, *107*, 920–927. [CrossRef] [PubMed]
15. Razmkhah, M.; Rezaei, J.; Fazaeli, H. Use of Jerusalem artichoke tops silage to replace corn silage in sheep diet. *Anim. Feed Sci. Technol.* **2017**, *228*, 168–177. [CrossRef]
16. Abd El-Mola, A.M.; Aboul-Fotouh, G.E. Effect of feeding *Helianthus tuberosus* (Jerusalem artichoke) on rations nutritive value and some blood parameters of Ossimi rams. *Egypt. J. Nutr. Feed.* **2018**, *21*, 365–371.
17. Papi, N.; Kafilzadeh, F.; Fazaeli, H. Use of Jerusalem artichoke aerial parts as forage in fat-tailed sheep diet. *Small Rumin. Res.* **2019**, *174*, 1–6. [CrossRef]
18. Tiengtam, N.; Khempaka, S.; Paengkoum, P.; Boonanuntanasarn, S. Effects of inulin and Jerusalem artichoke (*Helianthus tuberosus*) as prebiotic ingredients in the diet of juvenile Nile tilapia (*Oreochromis niloticus*). *Anim. Feed Sci. Technol.* **2015**, *207*, 120–129. [CrossRef]
19. Goran, S.M.A.; Omar, S.S.; Anwer, A.Y. Water quality and physiological parameters of common carp fingerling fed on Jerusalem artichoke tubers. *Polytechnic* **2016**, *6*, 502–516.
20. Boonanuntanasarn, S.; Tiengtam, N.; Pitaksong, T.; Piromyou, P.; Teaumroong, N. Effects of dietary inulin and Jerusalem artichoke (*Helianthus tuberosus*) on intestinal microbiota community and morphology of Nile tilapia (*Oreochromis niloticus*) fingerlings. *Aquac. Nutr.* **2018**, *24*, 712–722. [CrossRef]
21. Sewaka, M.; Trullas, C.; Chotiko, A.; Rodkhum, C.; Chansue, N.; Boonanuntanasarn, S.; Pirarat, N. Efficacy of synbiotic Jerusalem artichoke and *Lactobacillus rhamnosus* GG-supplemented diets on growth performance, serum biochemical parameters, intestinal morphology, immune parameters and protection against *Aeromonas veronii* in juvenile red tilapia (*Oreochromis* spp.). *Fish Shellfish Immunol.* **2019**, *86*, 260–268.
22. Trullàs, C.; Sewaka, M.; Rodkhum, C.; Chansue, N.; Boonanuntanasarn, S.; Kamble, M.T.; Pirarat, N. Effects of Jerusalem artichoke (*Helianthus tuberosus*) as a prebiotic supplement in the diet of red tilapia (*Oreochromis* spp.). *Animal* **2022**, *12*, 2882. [CrossRef] [PubMed]
23. Dokoupilova, A.; Zita, L.; Kvaček, J.; Janda, K.; Hofmanova, B.; Masopustova, R. Jerusalem artichoke (*Helinathus tuberosus*) tops as a natural source of inulin in rabbit diet: Effect on growth performance and health status. *J. Central Eur. Agric.* **2019**, *20*, 796–801. [CrossRef]
24. Kurchaeva, E.E.; Vostroilov, A.V.; Vysotskaya, E.A.; Artemov, E.S.; Maksimov, I.V. Efficiency of application of the probiotic "Enzymsporin" and grass flour of Jerusalem artichoke to increase the productivity of rabbits. *IOP Conf. Ser. Earth Environ. Sci.* **2020**, *422*, 012059. [CrossRef]
25. Havlíček, Z.; Kvaček, J.; Dočkalová, H. Jerusalem artichoke (*Helianthus tuberosus*) in rabbit nutrition. *J. Cent. Eur. Agric.* **2023**, *24*, 95–103.
26. Ārne, A.; Ilgaža, A. Jerusalem artichoke flour feeding effects on calf development in the first months of life. *Res. Rural. Dev.* **2014**, *1*, 169–175.
27. Glatter, M.; Bochnia, M.; Goetz, F.; Gottschalk, J.; Koeller, G.; Mielenz, N.; Zeyner, A. Glycaemic and insulinaemic responses of adult healthy warm-blooded mares following feeding with Jerusalem artichoke meal. *J. Anim. Physiol. Anim. Nutr.* **2017**, *101*, 69–78. [CrossRef]
28. Ersahince, A.; Kara, K. Nutrient composition and in vitro digestion parameters of Jerusalem artichoke (*Helianthus tuberosus* L.) herbage at different maturity stages in horse and ruminant. *J. Anim. Feed Sci.* **2017**, *26*, 213–225. [CrossRef]
29. Glatter, M.; Borewicz, K.; van den Bogert, B.; Wensch-Dorendorf, M.; Bochnia, M.; Greef, J.M.; Zeyner, A. Modification of the equine gastrointestinal microbiota by Jerusalem artichoke meal supplementation. *PLoS ONE* **2019**, *14*, e0220553. [CrossRef]
30. Stefler, J. Effect of feeding artichoke leaves on the milk production of lactating goats. *Szaktanácsok* **1990**, *3–4*, 22–27.
31. Osman, A. Studies on the feeding value of green fodder from Jerusalem artichoke (*Helianthus tuberosus* L.). *J. Anim. Poult. Prod.* **2003**, *28*, 5277–5282. [CrossRef]
32. Kanokwan, K.; Paengkoum, S.; Kongmun, P.; Yu, Z.; Paengkoum, P. Effect of Jerusalem artichoke supplementation on methanogenic achaea in dairy goats using real time PCR technique. In Proceedings of the First Asia Dairy Goat Conference, Kuala Lumpur, Malaysia, 9–12 April 2012; Volume 9, p. 82.
33. Denisow, B.; Tymoszuk, K.; Dmitruk, M. Nectar and pollen production of *Helianthus tuberosus* L.—An exotic plant with invasiveness potential. *Acta Bot. Croat.* **2019**, *78*, 135–141. [CrossRef]
34. Lindberg, J.E. Nutrient and energy supply in monogastric food producing animals with reduced environmental and climatic footprint and improved gut health. *Animal* **2023**, *17*, 100832. [CrossRef] [PubMed]
35. Kidd, M.T.; Maynard, C.W.; Mullenix, G.J. Progress of amino acid nutrition for diet protein reduction in poultry. *J. Anim. Sci. Biotechnol.* **2021**, *12*, 45. [CrossRef] [PubMed]
36. Ma, X.Y.; Zhang, L.H.; Shao, H.B.; Xu, G.; Zhang, F.; Ni, F.T.; Brestic, M. Jerusalem artichoke (*Helianthus tuberosus*), a medicinal salt-resistant plan has high adaptability and multiple-use values. *J. Med. Plants Res.* **2011**, *5*, 1272–1279.
37. Sawicka, B.; Skiba, D.; Pszczółkowski, P.; Aslan, I.; Sharifi-Rad, J.; Krochmal-Marczak, B. Jerusalem artichoke (*Helianthus tuberosus* L.) as a medicinal plant and its natural products. *Cell Mol. Biol.* **2020**, *66*, 160–177. [CrossRef] [PubMed]
38. Shazzo, R.I.; Gish, R.A.; Ekutech, R.I.; Kornena, E.P.; Kajshev, V.G.; Krivorotov, S.B. *Jerusalem Artichoke: Biology, Agricultural Cultivation, Place in the Ecosystem, Processing Technologies (Yesterday, Today, Tomorrow)*; KSAU, RAAS, Krasnodar Scientific Research

Institute of Storage and Processing of Agricultural Products of the Russian Academy of Agricultural Sciences, Ministry of Agriculture of the Russian Federation: Krasnodar, Russia, 2013; 181p. (In Russian)
39. Shao, T.; Yu, Q.; Zhu, T.; Liu, A.; Gao, X.; Long, X.; Liu, Z. Inulin from Jerusalem artichoke tubers alleviates hyperglycaemia in high-fat-diet-induced diabetes mice through the intestinal microflora improvement. *Br. J. Nutr.* **2020**, *123*, 308–318. [CrossRef]
40. Ahn, H.Y.; Kim, M.; Seo, C.R.; Yoo, H.J.; Lee, S.H.; Lee, J.H. The effects of Jerusalem artichoke and fermented soybean powder mixture supplementation on blood glucose and oxidative stress in subjects with prediabetes or newly diagnosed type 2 diabetes. *Nutr. Diabetes* **2018**, *8*, 42. [CrossRef] [PubMed]
41. Takahashi, H.; Nakajima, A.; Matsumoto, Y.; Mori, H.; Inoue, K.; Yamanouchi, H.; Anzai, K. Administration of Jerusalem artichoke reduces the postprandial plasma glucose and glucose-dependent insulinotropic polypeptide (GIP) concentrations in humans. *Food Nutr. Res.* **2022**, *4*, 66. [CrossRef]
42. Nizioł-Łukaszewska, Z.; Furman-Toczek, D.; Zagórska-Dziok, M. Antioxidant activity and cytotoxicity of Jerusalem artichoke tubers and leaves extract on HaCaT and BJ fibroblast cells. *Lipids Health Dis.* **2018**, *17*, 280. [CrossRef]
43. Pan, L.; Sinden, M.R.; Kennedy, A.H.; Chai, H.; Watson, L.E.; Graham, T.L.; Kinghorn, A.D. Bioactive constituents of *Helianthus tuberosus* (Jerusalem artichoke). *Phytochem. Lett.* **2009**, *2*, 15–18. [CrossRef]
44. Saiki, P.; Yoshihara, M.; Kawano, Y.; Miyazaki, H.; Miyazaki, K. Anti-Inflammatory Effects of Heliangin from Jerusalem artichoke (*Helianthus tuberosus*) Leaves Might Prevent Atherosclerosis. *Biomolecule* **2022**, *12*, 91. [CrossRef]
45. Sasaki, H.; Lyu, Y.; Nakayama, Y.; Nakamura, F.; Watanabe, A.; Miyakawa, H.; Shibata, S. Combinatorial effects of soluble, insoluble, and organic extracts from Jerusalem artichokes on gut microbiota in mice. *Microorganisms* **2020**, *8*, 954. [CrossRef] [PubMed]
46. Guess, N.D.; Dornhorst, A.; Oliver, N.; Bell, J.D.; Thomas, E.L.; Frost, G.S. A randomized control trial: The effect of inulin on weight management and ectopic fat in subjects with prediabetics. *Nutr. Metab.* **2015**, *12*, 2–10. [CrossRef] [PubMed]
47. Konikoff, T.; Gophna, U. Oscillospira: A central, enigmatic component of the human gut microbiota. *Trends Microbiol.* **2016**, *24*, 523–524. [CrossRef]
48. Ali, S.S.R.; Ambasankar, K.; Musthafa, M.S.; Harikrishnan, R. Jerusalem artichoke enriched diet on growth performance, immuno-hematological changes and disease resistance against *Aeromonas hydrophila* in Asian seabass (*Lates calcarifer*). *Fish Shellfish Immunol.* **2007**, *70*, 335–342.
49. Manokhina, A.A.; Dorokhov, A.S.; Kobozeva, T.P.; Fomina, T.N.; Starovoitov, V.I. Jerusalem artichoke as a strategic crop for solving food problems. *Agronomy* **2022**, *12*, 465. [CrossRef]
50. Ciuciuc, E.; Drăghici, I.; Drăghici, R.; Croitoru, M.; Băjenaru, M.F. The behavior of varieties of Jerusalem artichoke on the sandy soils from south of Oltenia. *Ann. Univ. Craiova-Agric. Mont. Cadastre Ser.* **2020**, *49*, 52–57.
51. Negoşanu, G.; Vînătoru, C.; Barcanu, E.; Agapie, O.L.; Tănase, B.; Gherase, I. Evaluation of the main phenotypic and physico-chemical characteristics in the new genotypes of Jerusalem artichoke (*Helianthus tuberosus*) obtained at VRDS Buzău. *Sci. Pap.-Ser. B Hortic.* **2020**, *64*, 442–447.
52. Agapie, O.L.; Barcanu, E.; Gherase, I.; Tănase, B.; Dobre, G. *Helianthus tuberosus* L. as an Alternative to Climate Change in Romania. *Bull. Univ. Agric. Sci. Vet. Med. Cluj-Napoca. Hortic.* **2022**, *79*, 112–116.
53. Dima, M.; Croitoru, M.; Drăghici, R.; Drăghici, I.; Ciuciuc, E.; Băjenaru, M.F. Researches on the behavior of Jerusalem artichoke varieties grown on sandy soils in terms of nutritional quality of tubers. *Sci. Pap. Ser. Manag. Econom. Eng. Agric. Rural. Dev.* **2021**, *21*, 309–316.
54. Kays, S.J.; Nottingham, S.F. *Biology and Chemistry of Jerusalem artichoke: Helianthus tuberosus L.*; CRC Press: Boca Raton, FL, USA, 2007.
55. Farzinmehr, S.; Rezaei, J.; Fazaeli, H. Effect of harvesting frequency and maturity stage of Jerusalem artichoke forage on yield, chemical composition and in vitro fermentation of the tubers and forage. *SJAR* **2020**, *18*, e0602. [CrossRef]
56. Marien, C. Effects of tubers of the Jerusalem artichoke (*Helianthus tuberosus*) and Potatoes (*Solanum tuberosum*) on the Intestinal Microbiota of Pigs and Evaluation of a Procedure for Quantification of Microbial Mass in Pig Faeces. Ph.D. Thesis, Kassel University, Kassel, Germany, 2011.
57. Schweiger, P.; Stolzenburg, K. *Mineral Stoffgehalte und Mineralstoffentzüge Verschiedener Topinambursorten*; LAP: Forchheim, Germany, 2003.
58. Malmberg, A.; Theander, O. Differences in chemical composition of leaves and stem in Jerusalem artichoke and changes in low-molecular sugar and fructan content with time of harvest. *Wed. J. Agric. Res.* **1986**, *16*, 7–12.
59. Stauffer, M.D.; Chubey, B.B.; Dorrell, D.G. Growth, yield and compositional characteristics of Jerusalem artichoke as they relate to biomass production [*Helianthus tuberosus*]. *Am. Chem. Soc. Div. Fuel Chem. Prepr.* **1981**, *25*, CONF-800814-P3.
60. Heuzé, V.; Tran, G.; Chapoutot, P.; Bastianelli, D.; Lebas, F. Jerusalem Artichoke (*Helianthus tuberosus*). Feedipedia, a Programme by INRAE, CIRAD, AFZ and FAO. 2015. Available online: https://www.feedipedia.org/node/544 (accessed on 21 July 2023).
61. Kosaric, N.; Cosentino, G.P.; Wieczorek, A.; Duvnjak, Z. The Jerusalem artichoke as an agricultural crop. *Biomass* **1984**, *5*, 1–36. [CrossRef]
62. Okrouhlá, M.; Čítek, J.; Švejstil, R.; Zadinová, K.; Pokorná, K.; Urbanová, D.; Stupka, R. The effect of dietary *Helianthus tuberosus* L. on the populations of pig faecal bacteria and the prevalence of skatole. *Animals* **2020**, *10*, 693. [CrossRef] [PubMed]
63. Florkiewicz, A.; Cieslik, E.; Filipiak-Florkiewicz, A. Wpływ odmiany i terminu zbioru na skład chemiczny bulw topinamburu (*Helianthus tuberosus* L.). *ŻNTJ* **2007**, *3*, 71–81.

64. Cieślik, E.; Gębusia, A.; Florkiewicz, A.; Mickowska, B. The content of protein and of amino acids in Jerusalem artichoke tubers (*Helianthus tuberosus* L.) of red variety Rote Zonenkugel. *Acta Sci. Pol. Technol. Aliment.* **2011**, *10*, 433–441.
65. Sawicka, B.; Kalembasa, D. Fluctuations of selected microelements in *Helianthus tuberosus* L. tubers due to diverse nitrogen nutrition. *Adv. Food Sci.* **2011**, *33*, 166–173.
66. Joint WHO/FAO/UNU Expert Consultation. *Protein and Amino Acid Requirements in Human Nutrition*; Technical Report Series; World Health Organization: Geneva, Switzerland, 2007; 265p.
67. Danilcenko, H.; Jariene, E.; Gajewski, M.; Sawicka, B.; Kulaitiene, J.; Cerniauskiene, J. Changes in amino acids content in tubers of Jerusalem artichoke (*Helianthus tuberosus* L.) Cultivars during storage. *Acta Sci. Pol. Hortorum Cultus* **2013**, *12*, 97–105.
68. Harmankaya, M.; Juhaimi, F.A.; Özcan, M.M. Mineral contents of Jerusalem artichoke (*Helianthus tuberosus* L.) growing wild in Turkey. *Anal. Lett.* **2012**, *45*, 2269–2275. [CrossRef]
69. De Leenheer, L. Production and use of inulin: Industrial reality with a promising future. In *Carbohydrates as Organic Raw Materials III*; VCH Verlagsgesellschaft mbH: Weinheim, Germany; VCH Publishers Inc.: New York, NY, USA, 1996; pp. 67–92.
70. Van Loo, J.; Coussement, P.; De Leenheer, L.; Hoebregs, H.; Smits, G. On the presence of inulin and oligofructose as natural ingredients in the western diet. *Crit. Rev. Food Sci. Nutr.* **1995**, *35*, 525–552. [CrossRef] [PubMed]
71. Long, X.H.; Mehta, S.K.; Liu, Z.P. Effect of NO3–N enrichment onseawater stress tolerance of Jerusalem artichoke (*Helianthus tuberosus*). *Pedosphere* **2008**, *18*, 113–123. [CrossRef]
72. Somda, Z.C.; McLaurin, W.J.; Kays, S.J. Jerusalem artichoke growth, development, and field storage. II. Carbon and nutrient element allocation and redistribution. *J. Plant Nutr.* **1999**, *22*, 1315–1334. [CrossRef]
73. Moshfegh, A.J.; Friday, J.E.; Goldman, J.P.; Ahuja, J.K.C. Presence of inulin and oligofructose in the diets of Americans. *J. Nutr.* **1999**, *129*, 1407S–1411S. [CrossRef]
74. Du, G.; Sun, Z.; Bao, S.; Zhong, Q.; Yang, S. Diversity of bacterial community in Jerusalem artichoke (*Helianthus tuberosus* L.) during storage is associated with the genotype and carbohydrates. *Front. Microbiol.* **2022**, *13*, 986659. [CrossRef]
75. Bhanja, A.; Sutar, P.P.; Mishra, M. Inulin—A polysaccharide: Review on its functional and prebiotic efficacy. *J. Food Biochem.* **2022**, *46*, e14386. [CrossRef] [PubMed]
76. Flickinger, E.A.; Van Loo, J.; Fahey, G.C. Nutritional responses to the presence of inulin and oligofructose in the diets of domesticated animals: A review. *Crit. Rev. Food Sci. Nutr.* **2003**, *43*, 19–60. [CrossRef]
77. Vhile, S.G.; Kjos, N.P.; Sørum, H.; Øverland, M. Feeding Jerusalem artichoke reduced skatole level and changed intestinal microbiota in the gut of entire male pigs. *Animal* **2011**, *6*, 807–814. [CrossRef]
78. Shoemaker, D.N. *The Jerusalem Artichoke as a Crop Plant*; US Department of Agriculture: Washington, DC, USA, 1927; p. 33.
79. Scharrer, K.; Schreiber, R. The digestibility of fresh and silaged tubers of *Helianthus tuberosus* by pigs. *Landw. Forsch.* **1950**, *2*, 156–161.
80. Ly, J.; Macias, M.; Reyes, J.L.; Figueroa, V. Ileal and faecal digestibility of Jerusalem artichokes (*Helianthus tuberosus* L.) in pigs. *J. Anim. Feed Sci.* **1995**, *4*, 195–205. [CrossRef]
81. Barszcz, M.; Taciak, M.; Skomiał, J. The effects of inulin, dried Jerusalem artichoke tuber and a multispecies probiotic preparation on microbiota ecology and immune status of the large intestine in young pigs. *Arch. Anim. Nutr.* **2016**, *70*, 278–292. [CrossRef] [PubMed]
82. Wilkins, R.; Kays, S.F. Biology and chemistry of Jerusalem artichoke (*Helianthus tuberosus* L.). *Exper. Agric.* **2008**, *44*, 439. [CrossRef]
83. Jawad, H.S.; Al-Abboodi, A.A. Jerusalem artichoke (*Helianthus tuberosus* L.) use in the poultry farms. *J. Nutraceutical Food Sci.* **2017**, *2*, 12.
84. Yildiz, G.; Sacakli, P.; Gungor, T. The effect of dietary Jerusalem artichoke (*Helianthus tuberosus* L.) on performance, egg quality characteristics and egg cholesterol content in laying hens. *Czech J. Anim. Sci.* **2006**, *51*, 349. [CrossRef]
85. Katiyanon, P.; Khajarean, J.; Tengjaroenkul, B.; Pimpukdee, K. Effects of feeding Jerusalem artichoke (*Helianthus tuberosus* L.) on performance, carcass quality and health of broilers. *Kaen Kaset* **2006**, *34*, 199–204.
86. Azad, M.A.; Gao, J.; Ma, J.; Li, T.; Tan, B.; Huang, X.; Yin, J. Opportunities of prebiotics for the intestinal health of monogastric animals. *Anim. Nutr.* **2020**, *6*, 379–388. [CrossRef]
87. Gardiner, G.E.; Metzler-Zebeli, B.U.; Lawlor, P.G. Impact of intestinal microbiota on growth and feed efficiency in pigs: A review. *Microorganisms* **2020**, *8*, 1886. [CrossRef]
88. Valdovska, A.; Jemeljanovs, A.; Pilmane, M.; Zitare, I.; Konosonoka, I.H.; Lazdins, M. Alternative for improving gut microbiota: Use of Jerusalem artichoke and probiotics in diet of weaned piglets. *Pol. J. Vet. Sci.* **2014**, *17*, 61–69. [CrossRef]
89. Modesto, M.; D'Aimmo, M.R.; Stefanini, I.; Trevisi, P.; De Filippi, S.; Casini, L.; Mazzoni, M.; Bosi, P.; Biavati, B. A novel strategy to select Bifidobacterium strains and prebiotics as natural growth promoters in newly weaned pigs. *Livest. Sci.* **2009**, *122*, 248–258. [CrossRef]
90. Lallès, J.P.; Bosi, P.; Smidt, H.; Stokes, C.R. Nutritional management of gut health in pigs around weaning. *Proc. Nutr. Soc.* **2007**, *66*, 260–268. [CrossRef]
91. Hopwood, D.E.; Hampson, D.J. Interactions between the intestinal microflora, diet and diarrhoea, and their influences on piglet health in the immediate post-weaning period. In *Weaning the Pig: Concepts and Consequences*; Wageningen Academic Publisher: Wageningen, The Netherlands, 2003; pp. 199–217.
92. Clavijo, V.; Flórez, M.J.V. The gastrointestinal microbiome and its association with the control of pathogens in broiler chicken production: A review. *Poult. Sci.* **2018**, *97*, 1006–1021. [CrossRef] [PubMed]

93. Knecht, D.; Cholewińska, P.; Jankowska-Mąkosa, A.; Czyż, K. Development of swine's digestive tract microbiota and its relation to production indices—A review. *Animals* **2020**, *10*, 527. [CrossRef] [PubMed]
94. He, Y.; Maltecca, C.; Tiezzi, F. Potential use of gut microbiota composition as a biomarker of heat stress in monogastric species: A review. *Animals* **2021**, *11*, 1833. [CrossRef] [PubMed]
95. Bamigbade, G.B.; Subhash, A.J.; Kamal-Eldin, A.; Nyström, L.; Ayyash, M. An updated review on prebiotics: Insights on potentials of food seeds waste as source of potential prebiotics. *Molecules* **2022**, *27*, 5947. [CrossRef] [PubMed]
96. Imoto, S.; Namioka, S. VFA production in the pig large intestine. *J. Anim. Sci.* **1978**, *47*, 467–478. [CrossRef] [PubMed]
97. Tang, S.; Sieo, C.; Ramasamy, K.; Saad, W.; Wong, H.; Ho, Y. Performance, biochemical and haematological responses, and relative organ weights of laying hens fed diets supplemented with prebiotic, probiotic and synbiotic. *BMC Vet. Res.* **2017**, *13*, 248. [CrossRef]
98. Karimian, R.; Rezaeipour, V. Effects of dietary mannan-oligosacharides an phytase supplementation alone or in combination on growth performance, serum metabolites, cecal microbiota activity and intestinal morphology in broiler chickens. *Poult. Sci.* **2020**, *8*, 27–32. [CrossRef]
99. Sevane, N.; Bialade, F.; Velasco, S.; Rebolé, A.; Rodríguez, M.L.; Ortiz, L.T.; Dunner, S. Dietary inulin supplementation modifies significantly the liver transcriptomic profile of broiler chickens. *PLoS ONE* **2014**, *9*, e98942. [CrossRef]
100. Rebolé, A.; Ortiz, L.T.; Rodríguez, M.; Alzueta, C.; Treviño, J.; Velasco, S. Effects of inulin and enzyme complex, indi-vidually or in combination, on growth performance, intestinal microflora, cecal fermentation characteristics, and jejunal histo-morphology in broiler chickens fed a wheat-and barley-based diet. *Poult. Sci.* **2010**, *89*, 276–286. [CrossRef]